Introduction to Quantum Mechanics

Introduction to Quantum Mechanics

Kendrick Porter

NY RESEARCH
PRESS
New York

Published by NY Research Press
118-35 Queens Blvd., Suite 400,
Forest Hills, NY 11375, USA
www.nyresearchpress.com

Introduction to Quantum Mechanics
Kendrick Porter

International Standard Book Number: 978-1-63238-708-0 (Hardback)

Cataloging-in-Publication Data

Introduction to quantum mechanics / Kendrick Porter.
p. cm.
Includes bibliographical references and index.
ISBN 978-1-63238-708-0
1. Quantum theory. 2. Physics. 3. Mechanics. I. Porter, Kendrick.
QC174.12 .I58 2019
530.12--dc23

Printed in China.

Contents

Preface

Quantum mechanics is a field in physics that studies atomic and subatomic particles. A principal characteristic of this science is the quantization of the physical quantities of energy, angular momentum and momentum. The objects under study in quantum mechanics exhibit wave particle duality and there exists a fundamental limit to the precision in the determination of physical quantities. The study of quantum entanglement is an important and upcoming area in this field. The applications of quantum mechanics vary from electronics and cryptography to quantum computing. This book presents the complex subject of quantum mechanics in the most comprehensible and easy to understand language. It unfolds the innovative aspects of quantum physics which will be crucial for the holistic understanding of the subject matter. Coherent flow of topics, student-friendly language and extensive use of examples make this textbook an invaluable source of knowledge.

Given below is the chapter wise description of the book:

Chapter 1- Quantum mechanics is a branch of physics. It describes phenomena at the smallest scales of energy. This chapter has been carefully written to provide an easy understanding of the fundamentals of quantum mechanics, such as quantum state, two-state quantum system, quantum theory, relativistic quantum mechanics, etc.

Chapter 2- The Schrödinger equation is a mathematical equation used for the systems which have significant quantum effects. It describes the change in the system occurring over time. It can be represented in a time dependent and time independent form. This aim of this chapter is to explore the fundamentals of the time-dependent and time-independent Schrödinger equation and the non-linear Schrödinger equation.

Chapter 3- An understanding of quantum mechanics requires an examination of the principal quantum mechanical equations, such as Dirac equation, Klein–Gordon equation, Pauli equation and Rydberg equation. These have been thoroughly discussed in this chapter.

Chapter 4- In a quantum system, quantum numbers describe the values of conserved quantities. They can have integral or half-integral values. This chapter closely examines the different quantum numbers, such as principal, azimuthal, magnetic, spin and total angular momentum quantum numbers, for a complete understanding of quantum numbers.

Chapter 5- A wave is a disturbance in matter or space which is characterized by a transfer of energy. It is described by a wave equation which describes the change of the disturbance with time. This chapter has been structured in a way to provide a comprehensive understanding of wave mechanics.

Chapter 6- This chapter closely examines the key theories in quantum mechanics such as quantum field theory, perturbation theory and scattering theory for an extensive understanding of the subject. All the principles and results of these theories have been carefully analyzed in this chapter.

Chapter 7- Quantum effects are generally witnessed in small energy scales of the atom. This chapter discusses some of the significant quantum mechanical effects such as Casimir effect, Zeeman effect, Stark effect, quantum Hall effect, photoelectric effect, Aharonov-Bohm effect, etc. for a holistic understanding of quantum mechanics.

Chapter 8- The mathematical formulation of quantum mechanics allows a rigorous of quantum phenomena and quantum effects. Abstract mathematical structures like quantum operators and the infinite-dimensional Hilbert spaces are used. This chapter has been carefully designed to provide an extensive understanding of bra-ket notation, Hilbert space and quantum operators such as position operator, momentum operator, Hamiltonian, angular momentum operator, etc.

At the end, I would like to thank all those who dedicated their time and efforts for the successful completion of this book. I also wish to convey my gratitude towards my friends and family who supported me at every step.

Kendrick Porter

1

Quantum Mechanics

Quantum mechanics is a branch of physics. It describes phenomena at the smallest scales of energy This chapter has been carefully written to provide an easy understanding of the fundamentals of quantum mechanics, such as quantum state, two-state quantum system, quantum theory, relativistic quantum mechanics, etc.

Quantum mechanics is a physical theory developed in the 1920s to account for the behavior of matter on the atomic scale. It has subsequently been developed into arguably the most empirically successful theory in the history of physics. However, it is hard to understand quantum mechanics as a description of the physical world, or to understand it as a physical explanation of the experimental outcomes we observe. Attempts to understand quantum mechanics as descriptive and explanatory, to modify it such that it can be so understood, or to argue that no such understanding is necessary, can all be taken as versions of the project of interpreting quantum mechanics.

The problematic nature of quantum mechanics stems from the fact that the theory often represents the state of a system using a sum of several terms, where each term apparently represents a distinct physical state of the system. What's more, these terms interact with each other, and this interaction is crucial to the theory's predictions. If one takes this representation literally, it looks as if the system exists in several incompatible physical states at once. And yet when the physicist makes a measurement on the system, only one of these incompatible states is manifest in the result of the measurement. What makes this especially puzzling is that there is nothing in the physical nature of a measurement that could privilege one of the terms over the others.

According to the Copenhagen interpretation of quantum mechanics, the solution to this puzzle is that the quantum state should not be taken as a description of the physical system. Rather, the role of the quantum state is to summarize what we can expect if we make measurements on the system. According to the many-worlds interpretation, the quantum state *is* to be taken as a description of the system, and the solution to the puzzle is that each term in that description produces a corresponding measurement outcome. That is, for any quantum measurement there are generally multiple measurement results occurring on distinct "branches" of reality. According to hidden variable theories, the quantum state is a *partial* description of the system, where the rest of the description is given by the values of one or more "hidden" variables. The solution to the puzzle in this case is that the hidden variables pick out one of the physical states described by the quantum state as the actual *one*. According to spontaneous collapse theories, the quantum state is a complete description of the system, but the dynamical laws of quantum mechanics are incomplete, and need to be supplemented with a "collapse" process that eliminates all but one of the terms in the state during the measurement process.

These interpretations and others present us with very different pictures of the nature of the physical world (or in the Copenhagen case, no picture at all), and they have different strengths and weaknesses. The question of how to decide between them is an open one.

The Development of Quantum Mechanics

Quantum mechanics was developed in the early twentieth century in response to several puzzles concerning the predictions of classical (pre-20th century) physics. Classical electrodynamics, while successful at describing a large number of phenomena, yields the absurd conclusion that the electromagnetic energy in a hollow cavity is infinite. It also predicts that the energy of electrons emitted from a metal via the photoelectric effect should be proportional to the intensity of the incident light, whereas in fact the energy of the electrons depends only on the frequency of the incident light. Taken together with the prevailing account of atoms as clouds of positive charge containing tiny negatively charged particles (electrons), classical mechanics entails that alpha particles fired at a thin gold foil should all pass straight through, whereas in fact a small proportion of them are reflected back towards the source.

In response to the first puzzle, Max Planck suggested in 1900 that light can only be emitted or absorbed in integral units of hn, where n is the frequency of the light and h is a constant. This is the hypothesis that energy is *quantized*—that it is a discrete rather than continuous quantity—from which quantum mechanics takes its name. This hypothesis can be used to explain the finite quantity of electromagnetic energy in a hollow cavity. In 1905 Albert Einstein proposed that the quantization of energy can solve the second puzzle too; the minimum amount of energy that can be transferred to an electron from the incident light is hn, and hence the energy of the emitted electrons is proportional to the frequency of the light.

Ernest Rutherford's solution to the third puzzle in 1911 was to posit that the positive charge in the atom is concentrated in a small nucleus with enough mass to reflect an alpha particle that collides with it. According to Niels Bohr's 1913 elaboration of this model, the electrons orbit this nucleus, but only certain energies for these orbital electrons are allowed. Again, energy is quantized. The model has the additional benefit of explaining the spectrum of light emitted from excited atoms; since only certain energies are allowed, only certain wavelengths of light are possible when electrons jump between these levels, and this explains why the spectrum of the light consists of discrete wavelengths rather than a continuum of possible wavelengths.

But the quantization of energy raises as many questions as it answers. Among them: Why are only certain energies allowed? What prevents the electrons in an atom from losing energy continuously and spiraling in towards the nucleus, as classical physics predicts? In 1924 Louis de Broglie suggested that electrons are wave-like rather than particle-like, and that the reason only certain electron energies are allowed is that energy is a function of wavelength, and only certain wavelengths can fit without remainder in the electron orbit for a given energy. By 1926 Erwin Schrödinger had developed an equation governing the dynamical behavior of these matter waves, and quantum mechanics was born.

This theory has been astonishingly successful. Within a year of Schrödinger's formulation, Clinton Davisson and Lester Germer demonstrated that electrons exhibit interference effects just like light waves—that when electrons are bounced off the regularly-arranged atoms of a crystal, their waves reinforce each other in some directions and cancel out in others, leading to more electrons being detected in some directions than others. This success has continued. Quantum mechanics (in the form of quantum electrodynamics) correctly predicts the magnetic moment of the electron to an accuracy of about one part in a trillion, making it the most accurate theory in the history of science. And so far its predictive track record is perfect: no data contradicts it.

But on a descriptive and explanatory level, the theory of quantum mechanics is less than satisfactory. Typically when a new theory is introduced, its proponents are clear about the physical ontology presupposed—the kind of objects governed by the theory. Superficially, quantum mechanics is no different, since it governs the evolution of waves through space. But there are at least two reasons why taking these waves as genuine physical entities is problematic.

First, although in the case of electron interference the number of electrons arriving at a particular location can be explained in terms of the propagation of waves though the apparatus, each electron is detected as a *particle* with a precise location, not as a spread-out wave. As Max Born noticed in 1926, the intensity (squared amplitude) of the quantum wave at a location gives the probability that the particle is located there; this is the Born rule for assigning probabilities to measurement outcomes. The second reason to doubt the reality of quantum waves is that the quantum waves do not propagate through ordinary three-dimensional space, but though a space of $3n$ dimensions, where n is the number of particles in the system concerned. Hence it is not at all clear that the underlying ontology is genuinely of waves propagating through space. Indeed, the standard terminology is to call the quantum mechanical representation of the state of a system a wave function rather than a wave, perhaps indicating a lack of metaphysical commitment: the mathematical function that represents a system has the form of a wave, even if it does not actually represent a wave.

So quantum mechanics is a phenomenally successful theory, but it is not at all clear what, if anything, it tells us about the underlying nature of the physical world. Quantum mechanics, perhaps uniquely among physical theories, stands in need of an interpretation to tell us what it means. Four kinds of interpretation are the first two—the Copenhagen interpretation and the many-worlds interpretation—take standard quantum mechanics as their starting point. The third and fourth—hidden variable theories and spontaneous collapse theories—start by modifying the theory of quantum mechanics, and hence are perhaps better described as proposals for replacing quantum mechanics with a closely related theory.

Principles of Quantum Mechanics

Physical States

Every physical system is associated with a Hilbert Space, every unit vector in the space corresponds to a possible pure state of the system, and every possible pure state, to some vector in the space.

Physical Quantities

Hermitian operators in the Hilbert space associated with a system represent physical quantities, and their eigenvalues represent the possible results of measurements of those quantities.

There is an operator, called the Hamiltonian that plays a special role in quantum theory because the dynamics of a system can be conveniently formulated by tracking its evolution. The Hamiltonian – written H, or $\hat{H}$ – stands for the total energy of the system. Its eigenvalues are the possible results that might be obtained in measurements of total energy. It is given by summing over the kinetic and potential energies of the system's components.

Composition

The Hilbert space associated with a complex system is the tensor product of those associated with the simple systems (in the standard, non-relativistic, theory: the individual particles) of which it is composed.

Dynamics

a. *Contexts of type 1:* Given the state of a system at t and the forces and constraints to which it is subject, there is an equation, 'Schrödinger's equation' that gives the state at any other time $U|v_t> \rightarrow |v_{t'}>$. The important properties of U for our purposes are that it is deterministic, which is to say that it takes the state of a system at one time into a unique state at any other, it is unitary, which means that it is an automorphism of the Hilbert space on which it acts (i.e., a mapping of that space onto itself that preserves the linear space structure and inner product), and it is linear, which is to say that if it takes a state |A> onto the state |A′>, and it takes the state |B> onto the state |B′>, then it takes any state of the form $\alpha|A> + \beta|B>$ onto the state $\alpha|A'> + \beta|B'>$.

b. *Contexts of type 2* ("Measurement Contexts"): Carrying out a "measurement" of an observable B on a system in a state |A> has the effect of collapsing the system into a B-eigenstate corresponding to the eigenvalue observed. This is known as the Collapse Postulate. Which *particular* B-eigenstate it collapses into is a matter of probability, and the probabilities are given by a rule known as Born's Rule:

   $prob(b_i) = |<A|B=b_i>|^2$.

There are two important points to note about these two kinds of contexts:

- The distinction between contexts of type 1 and 2 remains to be made out in quantum mechanical terms; nobody has managed to say in a completely satisfactory way, in the terms provided by the theory, which contexts are measurement contexts,
- Even if the distinction is made out, it is an open interpretive question whether there *are* contexts of type 2; i.e., it is an open interpretive question whether there are any contexts in which systems are governed by a dynamical rule *other* than Schrödinger›s equation.

Application of Quantum Mechanics

Decay of the Kaon

The kaon (also called the K^0 meson), discovered in 1947, is produced in high-energy collisions between nuclei and other particles. It has zero electric charge, and its mass is about one-half the mass of the proton. It is unstable and, once formed, rapidly decays into either 2 or 3 pi-mesons. The average lifetime of the kaon is about 10^{-10} second.

In spite of the fact that the kaon is uncharged, quantum theory predicts the existence of an antiparticle with the same mass, decay products, and average lifetime; the antiparticle is denoted by K^0. During the early 1950s, several physicists questioned the justification for postulating the existence of two particles with such similar properties. In 1955, however, Murray Gell-Mann and Abraham Pais made an interesting prediction about the decay of the kaon. Their reasoning provides an excellent illustration of the quantum mechanical axiom that the wave function Ψ can be a superposition of states; in this case, there are two states, the K^0 and K^0 mesons themselves.

A K^0 meson may be represented formally by writing the wave function as $\Psi = K^0$; similarly $\Psi = K^0$ represents a K^0 meson. From the two states, K^0 and K^0, the following two new states are constructed:

$$K_1 = \frac{\left(K^0 + \bar{K}^0\right)}{\sqrt{2}},$$

$$K_2 = \frac{\left(K^0 + \bar{K}^0\right)}{\sqrt{2}}.$$

From these two equations it follows that:

$$K^0 = \frac{\left(K_1 + K_2\right)}{\sqrt{2}},$$

$$\bar{K}^0 = \frac{\left(K_1 + K_2\right)}{\sqrt{2}}.$$

The reason for defining the two states K_1 and K_2 is that, according to quantum theory, when the K^0 decays, it does not do so as an isolated particle; instead, it combines with its antiparticle to form the states K_1 and K_2. The state K_1 (called the K-short [K^0_S]) decays into two pi-mesons with a very short lifetime (about 9×10^{-11} second), while K_2 (called the K-long [K^0_L]) decays into three pi-mesons with a longer lifetime (about 5×10^{-8} second).

The physical consequences of these results may be demonstrated in the following experiment. K^0 particles are produced in a nuclear reaction at the point A. They move to the right in the figure and start to decay. At point A, the wave function is $\Psi = K^0$, which, from equation, can be expressed as the sum of K_1 and K_2. As the particles move to the right, the K_1 state begins to decay rapidly. If the particles reach point B in about 10^{-8} second, nearly all the K_1 component has decayed, although hardly any of the K_2 component has done so. Thus, at point B, the beam has changed from one of pure K^0 to one of almost pure K_2, which equation shows is an equal mixture of K^0 and K^0. In other words, K^0 particles appear in the beam simply because K_1 and K_2decay at different rates. At point B, the beam enters a block of absorbing material. Both the K^0 and K^0 are absorbed by the nuclei in the block, but the K^0 are absorbed more strongly. As a result, even though the beam is an equal mixture of K^0 and K^0 when it enters the absorber, it is almost pure $K^0$when it exits at point C. The beam thus begins and ends as K^0.

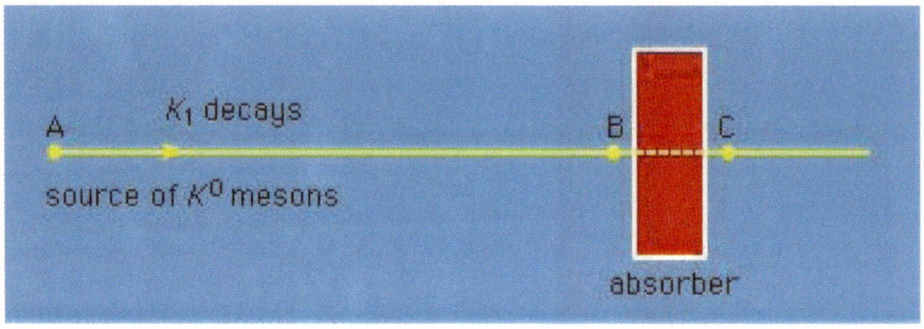

Figure: Decay of the K^o meson

Gell-Mann and Pais predicted all this, and experiments subsequently verified it. The experimental observations are that the decay products are primarily two pi-mesons with a short decay time near A, three pi-mesons with longer decay time near B, and two pi-mesons again near C. (This account exaggerates the changes in the K_1 and K_2 components between A and B and in the K^o and K^o components between B and C; the argument, however, is unchanged.) The phenomenon of generating the K^o and regenerating the K_1 decay is purely quantum. It rests on the quantum axiom of the superposition of states and has no classical counterpart.

Cesium Clock

The cesium clock is the most accurate type of clock yet developed. This device makes use of transitions between the spin states of the cesium nucleus and produces a frequency which is so regular that it has been adopted for establishing the time standard.

Like electrons, many atomic nuclei have spin. The spin of these nuclei produces a set of small effects in the spectra, known as hyperfine structure. (The effects are small because, though the angular momentum of a spinning nucleus is of the same magnitude as that of an electron, its magnetic moment, which governs the energies of the atomic levels, is relatively small.) The nucleus of the cesium atom has spin quantum number $^7/_2$. The total angular momentum of the lowest energy states of the cesium atom is obtained by combining the spin angular momentum of the nucleus with that of the single valence electron in the atom. (Only the valence electron contributes to the angular momentum because the angular momenta of all the other electrons total zero. Another simplifying feature is that the ground states have zero orbital momenta, so only spin angular momenta need to be considered.) When nuclear spin is taken into account, the total angular momentum of the atom is characterized by a quantum number, conventionally denoted by F, which for cesium is 4 or 3. These values come from the spin value $^7/_2$ for the nucleus and $^1/_2$ for the electron. If the nucleus and the electron are visualized as tiny spinning tops, the value $F = 4$ ($^7/_2 + {}^1/_2$) corresponds to the tops spinning in the same sense, and $F = 3$ ($^7/_2 - {}^1/_2$) corresponds to spins in opposite senses. The energy difference ΔE of the states with the two F values is a precise quantity. If electromagnetic radiation of frequency ν_0, is applied to a system of cesium atoms, transitions will occur between the two states. An apparatus that can detect the occurrence of transitions thus provides an extremely precise frequency standard. This is the principle of the cesium clock.

where

$$h\nu_0 = \Delta E,$$

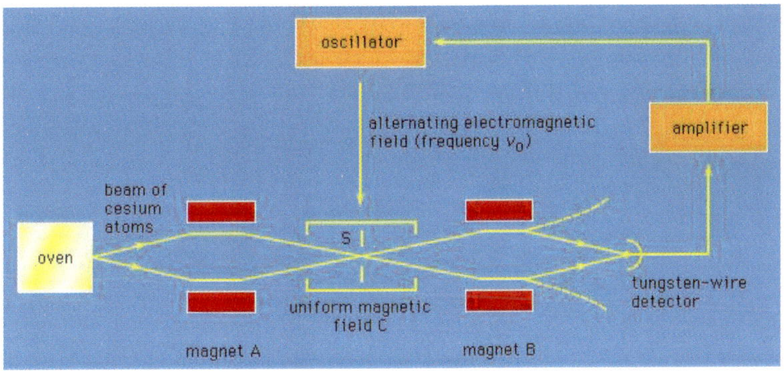

Figure: Cesium clock

The apparatus is shown schematically in figure. A beam of cesium atoms emerges from an oven at a temperature of about 100° C. The atoms pass through an inhomogeneous magnet A, which deflects the atoms in state $F = 4$ downward and those in state $F = 3$ by an equal amount upward. The atoms pass through slit S and continue into a second inhomogeneous magnet B. Magnet B is arranged so that it deflects atoms with an unchanged state in the same direction that magnet A deflected them. The atoms follow the paths indicated by the broken lines in the figure and are lost to the beam. However, if an alternating electromagnetic field of frequency ν_0 is applied to the beam as it traverses the center region C, transitions between states will occur. Some atoms in state $F = 4$ will change to $F = 3$, and vice versa. For such atoms, the deflections in magnet B are reversed. The atoms follow the whole lines in the diagram and strike a tungsten wire, which gives electric signals in proportion to the number of cesium atoms striking the wire. As the frequency ν of the alternating field is varied, the signal has a sharp maximum for $\nu = \nu_0$. The length of the apparatus from the oven to the tungsten detector is about one meter.

Each atomic state is characterized not only by the quantum number F but also by a second quantum number m_F. For $F = 4$, m_F can take integral values from 4 to −4. In the absence of a magnetic field, these states have the same energy. A magnetic field, however, causes a small change in energy proportional to the magnitude of the field and to the m_F value. Similarly, a magnetic field changes the energy for the $F = 3$ states according to the m_F value which, in this case, may vary from 3 to −3. The energy changes are indicated in figure. In the cesium clock, a weak constant magnetic field is superposed on the alternating electromagnetic field in region C. The theory shows that the alternating field can bring about a transition only between pairs of states with m_F values that are the same or that differ by unity. However, as can be seen from the figure, the only transitions occurring at the frequency ν_0 are those between the two states with $m_F = 0$. The apparatus is so sensitive that it can discriminate easily between such transitions and all the others.

If the frequency of the oscillator drifts slightly so that it does not quite equal ν_0, the detector output drops. The change in signal strength produces a signal to the oscillator to bring the frequency back to the correct value. This feedback system keeps the oscillator frequency automatically locked to ν_0.

The cesium clock is exceedingly stable. The frequency of the oscillator remains constant to about one part in 10^{13}. For this reason, the device is used to redefine the second. This base unit of time in the SI system is defined as equal to 9, 192, 631, 770 cycles of the radiation corresponding to the transition between the levels $F = 4$, $m_F = 0$ and $F = 3$, $m_F = 0$ of the ground state of the cesium-133atom.

Prior to 1967, the second was defined in terms of the motion of Earth. The latter, however, is not nearly as stable as the cesium clock. Specifically, the fractional variation of Earth's rotation period is a few hundred times larger than that of the frequency of the cesium clock.

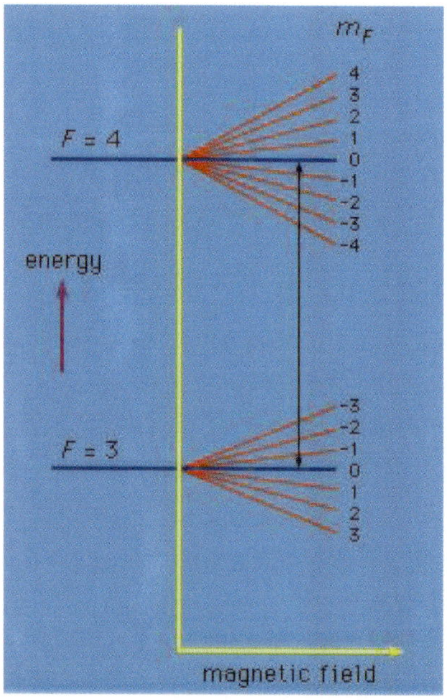

Figure: Variation of energy with magnetic-field strength for the $F = 4$ and $F = 3$ states in cesium-133

A Quantum Voltage Standard

Quantum theory has been used to establish a voltage standard, and this standard has proven to be extraordinarily accurate and consistent from laboratory to laboratory.

If two layers of superconducting material are separated by a thin insulating barrier, a super current (i.e., a current of paired electrons) can pass from one superconductor to the other. This is another example of the tunneling process described earlier. Several effects based on this phenomenon were predicted in 1962 by the British physicist Brian D. Josephson. Demonstrated experimentally soon afterwards, they are now referred to as the Josephson effects.

If a DC (direct-current) voltage V is applied across the two superconductors, the energy of an electron pair changes by an amount of $2eV$ as it crosses the junction. As a result, the super current oscillates with frequency ν given by the Planck relationship ($E = h\nu$). Thus,

$$2eV = h\upsilon.$$

This oscillatory behavior of the super current is known as the AC (alternating-current) Josephson effect. Measurement of V and ν permits a direct verification of the Planck relationship. Although the oscillating super current has been detected directly, it is extremely weak. A more sensitive method of investigating equation is to study effects resulting from the interaction of microwave radiation with the supercurrent.

Several carefully conducted experiments have verified equation to such a high degree of precision that it has been used to determine the value of $2e/h$. This value can in fact be determined more precisely by the AC Josephson effect than by any other method. The result is so reliable that laboratories now employ the AC Josephson effect to set a voltage standard. The numerical relationship between V and v is $\frac{2e}{h} = \frac{v}{V} = 483.597.9 \times 10^9$ hertz per volt.

In this way, measuring a frequency, which can be done with great precision, gives the value of the voltage. Before the Josephson method was used, the voltage standard in metrological laboratories devoted to the maintenance of physical units was based on high-stability Weston cadmium cells. These cells, however, tend to drift and so caused inconsistencies between standards in different laboratories. The Josephson method has provided a standard giving agreement to within a few parts in 10^8 for measurements made at different times and in different laboratories.

The experiments described in the preceding two sections are only two examples of high-precision measurements in physics. The values of the fundamental constants, such as c, h, e and m_e, are determined from a wide variety of experiments based on quantum phenomena. The results are so consistent that the values of the constants are thought to be known in most cases to better than one part in 10^6. Physicists may not know what they are doing when they make a measurement, but they do it extremely well.

Quantum State

Quantum state is a state of a system characterized by a set of quantum numbers and represented by an eigenfunction. The energy of each state is precise within the limits imposed by the uncertainty principle but may be changed by applying a field of force. States that have the same energy are called degenerate.

The quantum state of a system is described by a complex function Ψ, which depends on the coordinate x and on time:

$$\text{quantum state} \sim \Psi(x, t)$$

The wave function does not depend on the momentum of the particle. Compared to classical mechanics, we seem to have lost the symmetry between coordinates and momenta. We shall revisit this issue later. The wave function encodes all the information about the system, albeit in a probabilistic sense. This is a peculiarity of Quantum Mechanics: as postulated by Born, the theory can only predict the probability of the outcome of an experiment. This probability can be computed from the wave function. There are cases where a complicated computation is needed, and there are cases where this probability can be obtained very easily. For instance, $|\Psi(x, t)|^2 \, dx$ is the probability that a measurement of the position of the particle yields a result in the interval $x \rightarrow x + dx$

Thus $|\Psi(x{=}t)|^2$ is a probability per unit length or probability density. The total probability of finding the particle somewhere along the real axis must be unity, thus:

$$\|\Psi\|^2 = \int dx \, |\Psi(x, t)|^2 = 1 .$$

Any function such that its integral along the real axis is finite can be normalized by multiplying by an appropriate constant. In practice two wave functions that differ by an arbitrary factor $c \in C$ describe the same physical system.

Let us discuss an example of a normalizable function. The function $\psi(x) = e^{-x^2/2}$ is clearly normalizable. Its norm is

$$\|\psi\|^2 = \int dx\, e^{-x^2} = \sqrt{\pi},$$

and therefore the normalized wave function is:

$$\psi(x) = \frac{1}{\pi^{1/4}} \exp\left[-x^2/2\right].$$

On the other hand, the function $e^{-x^2/2}$ is non-normalizable and therefore does not represent a physical state.

In general, if $\int dx |\psi(x)|^2 = c$, then the normalized wave function is $\frac{1}{\sqrt{c}}\psi(x)$.

Example: In order to understand better how the information about the system is encoded in the wave function, we shall start with a simpler version of our one-dimensional system. Let us consider a particle in a discretized space. The particle can only be in a finite number of positions along the real axis, as shown in figure. In this particular example, the particle can be in one of six 6 points along the real axis labelled 0,...,5. The lattice spacing (i.e. the distance between two points) is denoted $\in$.

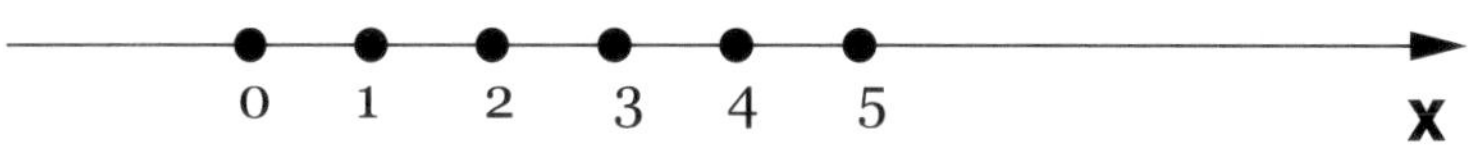

Discretized one-dimensional system with six sites. The particle can only be in one of the six points denoted by 0, ... ,5. The distance between successive points is $\in$,

According to the probabilistic interpretation of the wave function, the probability for the particle to be at x_i at time t is given by $|\Psi(x_i, t)|^2 \in$. Note the factor $\in$ that multiplies the modulo square of the wave function. This is needed because the modulo square is the probability density per unit length. Hence to find the probability we need to multiply by the distance between two points. If we redefine.

$$\psi_i = \sqrt{\in}\Psi(x_i, t),\ i = 0,\ldots,5,$$

then the whole information about this system is encoded in a six-dimensional complex vector:

$$|\Psi\rangle = (\psi 0, \ldots, \psi 5).$$

The x dependence is not explicitly written in the ket $|\Psi\rangle$. The values of Ψ at different spatial points are the components of the state vector. On the other hand, we will use $|\Psi(t)\rangle$ to indicate the dependence of the state on time when necessary.

In equation we have used the *ket* notation introduced by Dirac to denote a state vector. We shall keep using this notation throughout these notes.

Example: What is the state vector, $|2\rangle$, of a particle localized at the site x_2? A particle localized at x_2 is a particle that has a probability one of being at x_2 and probability zero to be at any other site. Hence:

$$|2\rangle = (0,\ 0,\ 1,\ 0,\ 0,\ 0).$$

Actually, from the discussions above, it should be clear that any state of the form:

$$|2\rangle = (0,\ 0,\ z,\ 0,\ 0,\ 0) \text{where } |z|^2 = 1,$$

represents a localized solution.

In the (simple) discretized system that we have discussed, we can see explicitly that the state of the system is represented by a vector. This is an important concept to remember.

The normalization condition encodes the fact that the total probability of finding the particle somewhere on the line must be one, i.e.

$$\sum_{k=0}^{5} |\psi_k|^2 = 1,$$

i.e. $|\Psi\rangle$ is a complex vector of unit norm. We shall frequently use the following convention:

$$\langle\Psi|\Psi\rangle = \sum_{k=0}^{5} |\psi_k|^2,$$

where we have used Dirac's notation to indicate the scalar product of two vectors:

$$\langle\Phi|\Psi\rangle = \sum_k \phi_k^* \psi_k.$$

The wave function for a continuous system can be seen as the limit of the discretized case where the number of points goes to infinity, while the distance becomes infinitesimally small.

In this limit, instead of a finite-dimensional vector, we obtain an infinite number of coordinates, encoded in a continuous function $\psi(x)$.

We shall still refer to the wave function as the state vector, bearing in mind that in this case the vector space is infinite-dimensional. There is one coordinate $\Psi(x,t)$ for each point x on the real axis, and there is an infinity of points along the real axis. We denote the state vector using a ket $|\Psi(t)\rangle$. Note that the time dependence of the state vector is explicit in Dirac's notation. It should be clear from the discussion above that the spatial coordinate x labels the components of the state vector.

The norm of the state vector can be written as the limit of the norm of the finite dimensional vector for $\in \to 0$. Starting from equation and taking the limit:

$$\langle\Psi(t)|\Psi(t)\rangle = \lim_{\in\to 0} \sum_k \in |\Psi(x_k,t)|^2 = \int dx |\Psi(x_k,t)|^2,$$

we recover equation. Similarly the scalar product of two wave functions can be defined as the limit of the discrete case:

$$\langle \Phi(t) | \Psi(t) \rangle = \int dx\, \Phi(x,t)^* \, \Psi(x,t).$$

Two-state Quantum System

Two-state systems are systems in which only two quantum states are of importance. That makes such systems the simplest nontrivial quantum systems.

The classic example is the intrinsic spin of an electron or proton, a purely quantum mechanical label that takes on only two values: + or −. Or it might be that all other degrees of freedom of the system are "frozen out", requiring much more energy to excite. An example is the configuration of the ammonia molecule, NH_3, where the nitrogen sits either above (+) or below (−) the plane formed by the three hydrogens. Excitations that rearrange the molecule in more significant ways or ionize it require much more energy than is required to flip the nitrogen from + to −.

Whatever the reason, we shall assume that our Hilbert space is spanned by two basis states, $|+\rangle$ and$|\ \rangle$, that we assume to be orthonormal, $\langle \pm | \pm \rangle = 1$, $\langle \pm | \mp \rangle = 0$ and complete, $\sum_i = \pm |i\rangle\langle i| = I$, where I is the identity operator.

The unifying concept relating the two bonds is that of two-state systems. Such systems involve two intuitive basic states ψ_1 and ψ_2.

For the hydrogen molecular ion, one state, $\psi_1 = \psi_1$, described that the electron was in the ground state around the left proton. A physically equivalent state, $\psi_2 = \psi_r$, had the electron in the ground state around the right proton. For the hydrogen molecule, $\psi_1 = \psi_1 \psi_r$ had electron 1 around the left proton and electron 2 around the right one. The other state $\psi_2 = \psi_r \psi_1$ was physically the same, but it had the electrons reversed.

There are many other physical situations that may be described as two state systems. Covalent chemical bonds involving atoms other than hydrogen would be an obvious example. Just substitute a positive ion for one or both protons.

As another example of a two-state system, consider the C_6H_6 "benzene molecular ring." This molecule consists of a hexagon of 6 carbon atoms that are held together by 9 covalent bonds. The logical way that 9 bonds can be arranged between the atoms of a 6 atom ring is to make every second bond a double one. However, that still leaves two possibilities; the locations of the single and double bonds can be swapped. So there are once again two different but equivalent states ψ_1 and ψ_2.

The NH 3 there are two intuitive physical states ψ_1 and ψ_2. The peculiarities of the quantum mechanics of two-state systems arise from states that are combinations of these two states, as in

$$\psi = c_1 \psi_1 + c_2 \psi_2$$

Note that according to the ideas of quantum mechanics, the square magnitude of the first coefficient of the combined state, $|c_1|^2$, represents the probability of being in state ψ_1 and $|c_2|^2$ the probability of being in state ψ_2. Of course, the total probability of being in one of the states should be one:

$$|c_1|^2 + |c_2|^2 = 1$$

Consider the simplest possible non-trivial quantum mechanical system. In such a system, there are only *two* independent eigenstates of the unperturbed Hamiltonian: *i.e.*,

$$\begin{aligned} H_0\psi_1 &= E_1\psi_1, \\ H_0\psi_2 &= E_2\psi_2. \end{aligned}$$

It is assumed that these states, and their associated eigenvalues, are known. We also expect the states to be orthonormal, and to form a complete set.

Let us now try to solve the modified energy eigenvalue problem

$$(H_0 + H_1)\psi_E = E\,\psi_E.$$

We can, in fact, solve this problem exactly. Since the eigenstates of H_0 form a complete set, we can write

$$\psi_E = \langle 1|E\rangle\psi_1 + \langle 2|E\rangle\psi_2.$$

It follows from $(H_0 + H_1)\psi_E = E\,\psi_E$ that

$$\langle i|H_0 + H_1|E\rangle = E\langle i|E\rangle,$$

where $i = 1$ or 2. Equations $H_0\psi_1 = E_1\psi_1$, $H_0\psi_2 = E_2\psi_2$, $\psi_E = \langle 1|E\rangle\psi_1 + \langle 2|E\rangle\psi_2$, $\langle i|H_0 + H_1|E\rangle = E\langle i|E\rangle$ and the orthonormality condition

$$\langle i|j\rangle = \delta_{ij},$$

yield two coupled equations which can be written in matrix form:

$$\begin{pmatrix} E_1 - E + e_{11} & e_{12} \\ e_{12}^* & E_2 - E + e_{22} \end{pmatrix} \begin{pmatrix} \langle 1|E\rangle \\ \langle 1|E\rangle \end{pmatrix} = \begin{pmatrix} 0 \\ 0 \end{pmatrix},$$

where

$$\begin{aligned} e_{11} &= \langle 1|H_1|1\rangle, \\ e_{22} &= \langle 2|H_1|2\rangle, \\ e_{12} &= \langle 1|H_1|2\rangle = \langle 2|H_1|1\rangle^*. \end{aligned}$$

Here, use has been made of the fact that H_1 is an Hermitian operator.

Consider the special (but not uncommon) case of a perturbing Hamiltonian whose diagonal matrix elements are zero, so that

$$e_{11} = e_{22} = 0.$$

The solution of equation $\begin{pmatrix} E_1 - E + e_{11} & e_{12} \\ e_{12}^* & E_2 - E + e_{22} \end{pmatrix} \begin{pmatrix} \langle 1|E\rangle \\ \langle 1|E\rangle \end{pmatrix} = \begin{pmatrix} 0 \\ 0 \end{pmatrix}$, (obtained by setting the determinant of the matrix to zero) is

$$E = \frac{(E_1 + E_2) \pm \sqrt{(E_1 - E_2)^2 + 4|e_{12}|^2}}{2}$$

Let us expand in the supposedly small parameter

$$\epsilon = \frac{|e_{12}|}{|E_1 - E_2|}.$$

We obtain

$$E \simeq \frac{1}{2}(E_1 + E_2) \pm \frac{1}{2}(E_1 - E_2)(1 + 2\epsilon^2 + \ldots).$$

The above expression yields the modification of the energy eigenvalues due to the perturbing Hamiltonian:

$$\begin{aligned} E_1' &= E_1 + \frac{|e_{12}|^2}{E_1 - E_2} + \cdots, \\ E_2' &= E_2 - \frac{|e_{12}|^2}{E_1 - E_2} + \cdots. \end{aligned}$$

Note that H_1 causes the upper eigenvalue to rise, and the lower to fall. It is easily demonstrated that the modified eigenstates take the form

$$\begin{aligned} \psi_1' &= \psi_1 + \frac{e_{12}^*}{E_1 - E_2}\psi_2 + \cdots, \\ \psi_2' &= \psi_2 - \frac{e_{12}}{E_1 - E_2}\psi_1 + \cdots. \end{aligned}$$

Thus, the modified energy eigenstates consist of one of the unperturbed eigenstates, plus a slight admixture of the other. Now our expansion procedure is only valid when $\epsilon \ll 1$. This suggests that the condition for the validity of the perturbation method as a whole is

$$|e_{12}| \ll |E_1 - E_2|.$$

In other words, when we say that H_1 needs to be small compared to H_0, what we are really saying is that the above inequality must be satisfied.

Time Dependence of an Isolated Two State System

The time development of an isolated system (one in which the Hamiltonian does not depend explicitly on the time) is generated by the unitary operator,

$$u(t) = e^{-iHt}.$$

Suppose $|\psi\rangle \; (=|\psi\rangle_H = |\psi(0)\rangle s\rangle$ is an arbitrary state defined as a superposition of the fixed basis vectors $|\pm\rangle$ at time $t = 0$. Then in accordance with our discussion of Heisenberg and Schrödinger pictures, we define,

$$\begin{aligned}|\psi\rangle_H &= a_+|\pm\rangle + a_-|-\rangle \\ |\psi\rangle_S &= u(t)|\psi(0)\rangle_S \\ &= a_+(t)|\pm\rangle + a_-(t)|-\rangle\end{aligned}$$

a_+ and a_- define the state at time $t = 0$. In the Schrödinger picture they evolve with time by the action of $u(t)$— $a\pm(t) = e^{\mp i\in t}/2_{a_\pm}$.

It is particularly easy to follow the time dependence of observables in the Heisenberg picture. Let A be an arbitrary observable defined by the c-number constants $a_0, \cdots a_3,$ *a la* The general relation between an operator in Schrödinger picture, A_S and Heisenberg picture, $A_H(t)$ is given by

$$A_H(t) = e^{iHt/\hbar} A_S e^{-iHt/\hbar}$$

or in this case,

$$\begin{aligned}&= e^{\frac{i\in t}{2\hbar}\sigma 3}\left(a_0 I + \vec{a}\cdot\vec{\sigma}\right)e^{-\frac{i\in t}{2\hbar}\sigma 3} \\ &= a_0 I + a_3\sigma_3 + e^{i\lambda\sigma 3}\left(a_1\sigma_1 + a_2\sigma_2\right)e^{-i\lambda\sigma 3},\end{aligned}$$

Where $\lambda = \in t/2\hbar$. It is possible to carry out the operator algebra exactly in this case. We define,

$$\sum_k(\lambda) = e^{i\lambda\sigma 3}\sigma_k e^{-i\lambda\sigma 3},$$

so

$$\begin{aligned}\frac{d\sum_k}{d\lambda} &= ie^{i\lambda\sigma 3}\left[\sigma_3, \sigma_k\right]e^{-i\lambda\sigma 3} \\ &= -2\in_{3kj}\Sigma_j.\end{aligned}$$

Specifically, $d\Sigma_1/d\lambda = -2\Sigma_2$ and $d\Sigma_2/d\lambda = 2\Sigma_1$. Using the initial values $\Sigma_k(0) = \sigma_k$, it is easy to integrate these coupled first order (matrix) equations,

$$\Sigma_1(\lambda) = \sigma_1 \cos 2\lambda - \sigma_2 \sin 2\lambda$$
$$\Sigma_2(\lambda) = \sigma_2 \cos 2\lambda + \sigma_1 \sin 2\lambda$$

Thus we have found the time dependence of the most general operator:

$$A_H(t) = a_0 I + a_3\sigma_3 + a_1\left(\sigma_1 \cos \frac{\in t}{\hbar} - \sigma_2 \sin \frac{\in t}{\hbar}\right) + a_2\left(\sigma_2 \cos \frac{\in t}{\hbar} + \sigma_1 \sin \frac{\in t}{\hbar}\right)$$

Let us analyse this result. The piece of an operator proportional to the identity, a_0, labels both states the same way and is of no interest to us. It may be regarded as an overall property of the system that does not vary in the problem at hand. The piece proportional to σ_3 commutes with H, so it labels the two energy eigenstates differently and distinguishes between them. The pieces of A proportional to σ_1 and σ_2 vary sinusoidally in time with a frequency determined by the energy difference between the two eigenstates.

Analysis of Some Important two-state Systems

Precession in a Field

Consider the case of a spin-1/2 particle in a magnetic field $B = B\hat{n}$. The interaction Hamiltonian for this system is

$$H = -\mu \cdot B = -\mu\sigma \cdot B\,,$$

where μ is the magnitude of the particle's magnetic moment and σ is the vector of Pauli matrices. Solving the time dependent Schrödinger equation $H\psi = i\hbar\partial_t\psi$ yields

$$\psi(t) = e^{i\omega t\acute{o}\cdot\hat{n}}\psi(0),$$

where $\omega = \mu B/\hbar$ and $e^{i\omega t\acute{o}\cdot\hat{n}} = \cos(\omega t)I + i\hat{n}\cdot\sigma\sin(\omega t)$. Physically, this corresponds to the Bloch vector precessing around $\hat{n}$ with angular frequency 2ω. Without loss of generality, assume the field is uniform points in $\hat{z}$, so that the time evolution operator is given as

$$e^{i\omega t\acute{o}\cdot\hat{n}} = \begin{pmatrix} e^{i\omega t} & 0 \\ 0 & e^{-i\omega t} \end{pmatrix}$$

It can be seen that such a time evolution operator acting on a general spin state of a spin-1/2 particle will lead to the precession about the axis defined by the applied magnetic field (this is the quantum mechanical equivalent of Larmor precession).

The above method can be applied to the analysis of any generic two-state system that is interacting with some field (equivalent to the magnetic field in the previous case) if the interaction is given by an appropriate coupling term that is analogous to the magnetic moment. The precession of the state vector (which need not be a physical spinning as in the previous case) can be viewed as the precession of the state vector on the Bloch sphere.

The representation on the Bloch sphere for a state vector $\psi(0)$ will simply be the vector of expectation values $\mathrm{R} = (\langle\sigma_x\rangle, \langle\sigma_y\rangle, \langle\sigma_z\rangle)$. As an example, consider a state vector $\psi(0)$ that is a normalized superposition of $|\uparrow\rangle$ and $|\downarrow\rangle$, that is, a vector that can be represented in the σ_z basis as

$$\psi(0) = \frac{1}{\sqrt{2}}\begin{pmatrix}1\\1\end{pmatrix}$$

The components of $\psi(t)$ on the Bloch sphere will simply be $\mathrm{R} = (\cos 2\omega t, -\sin 2\omega t, 0)$. This is a unit vector that begins pointing along $\hat{\mathrm{x}}$ and precesses around $\hat{\mathrm{z}}$ in a left-handed manner. In general, by a rotation around $\hat{\mathrm{z}}$, any state vector $\psi(0)$ can be represented as $a|\uparrow\rangle + b|\downarrow\rangle$ with real coefficients a and b. Such a state vector corresponds to a Bloch vector in the *xz*-plane making an angle $\tan(\theta/2) = b/a$ with the *z*-axis. This vector will proceed to precess around $\hat{\mathrm{z}}$. In theory, by allowing the system to interact with the field of a particular direction and strength for precise durations, it is possible to obtain any orientation of the Bloch vector, which is equivalent to obtaining any complex superposition. This is the basis for numerous technologies including quantum computing and MRI.

Evolution in a Time-dependent Field: Nuclear magnetic Resonance

Nuclear magnetic resonance (NMR) is an important example in the dynamics of two-state systems because it is involves the exact solution to a time dependent Hamiltonian. The NMR phenomenon is achieved by placing a nucleus in a strong, static field B_0 (the "holding field") and then applying a weak, transverse field B_1 that oscillates at some radiofrequency ω_r. Explicitly, consider a spin-1/2 particle in a holding field $B_0\hat{z}$ and a transverse rf field B_1 rotating in the *xy*-plane in a right-handed fashion around B_0:

$$\mathrm{B} = \begin{pmatrix} B_1 \cos \omega_r t \\ B_1 \sin \omega_r t \\ B_0 \end{pmatrix}$$

As in the free precession case, the Hamiltonian is $H = -\mu\sigma \cdot \mathrm{B}$ and the evolution of a state vector $\psi(t)$ is found by solving the time-dependent Schrödinger equation $H\psi = i\hbar\partial\psi/\partial t$. After some manipulation, it can be shown that the Schrödinger equation becomes

$$\frac{\partial\psi}{\partial t} = i\left(\omega_1\sigma_x + \left(w_0 + \frac{\omega_r}{2}\right)\sigma_z\right)\psi$$

where $\omega_0 = \mu B_0/\hbar$ and $\omega_1 = \mu B_1/\hbar$.

The solution to this equation has the Bloch vector precessing around with a frequency that is twice the magnitude of the vector. If ω_0 is sufficiently strong, some proportion of the spins will be pointing directly down prior to the introduction of the rotating field. If the angular frequency of the rotating magnetic field is chosen such that $\omega_r = -2\omega_0$, in the rotating frame the state vector will precess around $\hat{x}$ with frequency $2\omega_1$, and will thus flip from down to up releasing energy in the form of detectable photons. This is the fundamental basis for NMR, and in practice is accomplished by scanning ω_r until the resonant frequency is found at which point the sample will emit light. Similar calculations are done in atomic physics, and in the case that the field is not rotating, but oscillating with a complex amplitude, use is made of the rotating wave approximation in deriving such results.

Derivation of Above Expression for the NMR Schrödinger Equation

Here the Schrödinger equation reads

$$-\mu\sigma\cdot \mathrm{B}\psi = i\hbar\frac{\partial\psi}{\partial t}.$$

Expanding the dot product and dividing by $i\hbar$ yields

$$\frac{\partial\psi}{\partial t} = i\left(\omega_1\sigma_x\cos\omega_r t + \omega_1\sigma_y\sin\omega_r t + \omega_0\sigma_z\right)\psi.$$

To remove the time dependence from the problem, the wave function is transformed according to $\psi \to e^{-i\sigma_z\omega_r t/2}\psi$. The time dependent Schrödinger equation becomes

$$-i\sigma_z\frac{\omega_r}{2}e^{-i\sigma_z\omega_r t/2}\psi + e^{-i\sigma_z\omega_r t/2}\frac{\partial\psi}{\partial t} = i\left(\omega_1\sigma_x\cos\omega_r t + \omega_1\sigma_y\sin\omega_r t + \omega_0\sigma_z\right)e^{-i\sigma_z\omega_r t/2}\psi,$$

which after some rearrangement yields

$$\frac{\partial\psi}{\partial t} = ie^{i\sigma_z\omega_r t/2}\left(\omega_1\sigma_x\cos\omega_r t + \omega_1\sigma_y\sin\omega_r t + \left(\omega_0 + \frac{\omega_r}{2}\right)\sigma_z\right)e^{-i\sigma_z\omega_r t/2}\psi$$

Evaluating each term on the right hand side of the equation

$$e^{i\sigma_z\omega_r t/2}\sigma_x e^{-i\sigma_z\omega_r t/2} = \begin{pmatrix} e^{i\omega_r t/2} & 0 \\ 0 & e^{-i\omega_r t/2}\end{pmatrix}\begin{pmatrix} 0 & 1 \\ 1 & 0\end{pmatrix}\begin{pmatrix} e^{-i\omega_r t/2} & 0 \\ 0 & e^{i\omega_r t/2}\end{pmatrix} = \begin{pmatrix} 0 & e^{i\omega_r t} \\ e^{-i\omega_r t} & 0\end{pmatrix}$$

$$e^{i\sigma_z\omega_r t/2}\sigma_y e^{-i\sigma_z\omega_r t/2} = \begin{pmatrix} e^{i\omega_r t/2} & 0 \\ 0 & e^{-i\omega_r t/2}\end{pmatrix}\begin{pmatrix} 0 & -i \\ i & 0\end{pmatrix}\begin{pmatrix} e^{-i\omega_r t/2} & 0 \\ 0 & e^{i\omega_r t/2}\end{pmatrix} = \begin{pmatrix} 0 & -ie^{i\omega_r t} \\ ie^{-i\omega_r t} & 0\end{pmatrix}$$

$$e^{i\sigma_z\omega_r t/2}\sigma_z e^{-i\sigma_z\omega_r t/2} = \begin{pmatrix} e^{i\omega_r t/2} & 0 \\ 0 & e^{-i\omega_r t/2}\end{pmatrix}\begin{pmatrix} 1 & 0 \\ 0 & -1\end{pmatrix}\begin{pmatrix} e^{-i\omega_r t/2} & 0 \\ 0 & e^{i\omega_r t/2}\end{pmatrix} = \sigma_z$$

The equation now reads

$$\frac{\partial\psi}{\partial t} = i\left(\omega_1\begin{pmatrix} 0 & e^{i\omega_r t}\left(\cos\omega_r t - i\sin\omega_r t\right) \\ e^{-i\omega_r t}\left(\cos\omega_r t + i\sin\omega_r t\right) & 0\end{pmatrix} + \left(w_0 + \frac{\omega_r}{2}\right)\sigma_z\right)\psi,$$

which by Euler's identity becomes

$$\frac{\partial\psi}{\partial t} = i\left(\omega_1\sigma_x + \left(w_0 + \frac{\omega_r}{2}\right)\sigma_z\right)\psi$$

Relation to Bloch Equations

The optical Bloch equations for a collection of spin-1/2 particles can be derived from the time dependent Schrödinger equation for a two level system. Starting with the previously stated Hamiltonian $i\hbar\partial_t\psi = -\mu\vec{\sigma}\cdot\vec{B}\psi$, it can be written in summation notation after some rearrangement as

$$\frac{\partial\psi}{\partial t} = i\frac{\mu}{\hbar}\sigma_i B_i\psi$$

Multiplying by a Pauli matrix σ_i and the conjugate transpose of the wave function, and subsequently expanding the product of two Pauli matrices yields

$$\psi^{\dagger}\sigma_j\frac{\partial\psi}{\partial t} = i\frac{\mu}{\hbar}\psi^{\dagger}\sigma_j\sigma_i B_i\psi = i\frac{\mu}{\hbar}\psi^{\dagger}\left(I\delta_{ij} - i\sigma_k\varepsilon_{ijk}\right)B_i\psi = \frac{\mu}{\hbar}\psi^{\dagger}\left(iI\delta_{ij} + \sigma_k\varepsilon_{ijk}\right)B_i\psi$$

Adding this equation to its own conjugate transpose yields a left hand side of the form

$$\psi^{\dagger}\sigma_j\frac{\partial\psi}{\partial t} + \frac{\partial\psi^{\dagger}}{\partial t}\sigma_j\psi = \frac{\partial\left(\psi^{\dagger}\sigma_j\psi\right)}{\partial t}$$

And a right hand side of the form

$$\frac{\mu}{\hbar}\psi^{\dagger}\left(iI\delta_{ij} + \sigma_k\varepsilon_{ijk}\right)B_i\psi + \frac{\mu}{\hbar}\psi^{\dagger}\left(-iI\delta_{ij} + \sigma_k\varepsilon_{ijk}\right)B_i\psi = \frac{2\mu}{\hbar}\left(\psi^{\dagger}\sigma_k\psi\right)B_i\varepsilon_{ijk}$$

As previously mentioned, the expectation value of each Pauli matrix is a component of the Bloch vector, $\langle\sigma_i\rangle = \psi^{\dagger}\sigma_i\psi = R_i$. Equating the left and right hand sides, and noting that $\frac{2\mu}{\hbar}$ is the gyromagnetic ratio γ, yields another form for the equations of motion of the Bloch vector

$$\frac{\partial R_j}{\partial t} = \gamma R_k B_i\varepsilon_{kij}$$

where the fact that $\varepsilon_{ijk} = \varepsilon_{kij}$ has been used. In vector form these three equations can be expressed in terms of a cross product

$$\frac{\partial\vec{R}}{\partial t} = \gamma\vec{R}\times\vec{B}$$

Classically, this equation describes the dynamics of a spin in a magnetic field. An ideal magnet consists of a collection of identical spins behaving independently, and thus the total magnetization $\vec{M}$ is proportional to the Bloch vector $\vec{R}$. All that is left to obtain the final form of the optical Bloch equations is the inclusion of the phenomenological relaxation terms.

As a final aside, the above equation can be derived by considering the time evolution of the angular momentum operator in the Heisenberg picture.

$$i\hbar\frac{d\sigma_j}{dt}=[\sigma_j,H]=[\sigma_j,-\mu\sigma_i B_i]=-\mu\left(\sigma_j\sigma_i B_i-\sigma_i\sigma_j B_i\right)=\mu[\sigma_i,\sigma_j]B_i=2\mu i\varepsilon_{ijk}\sigma_k B_i$$

When coupled with the fact that $\vec{R_i}=\langle\sigma_i\rangle$, this equation is the same equation as before.

Validity of the two-state Formalism

Two-state systems are the simplest non-trivial quantum systems that occur in nature, but it should be noted that the above-mentioned methods of analysis are not just valid for simple two-state systems. Any general multi-state quantum system can be effectively treated as two-state system as long as a particular property of is being considered (which behaves as a two-state system). For example, a spin-1/2 particle may in reality have additional translational or even rotational degrees of freedom, but those degrees of freedom are irrelevant to the preceding analysis.

Another case where the effective two-state formalism is valid is when the system under consideration has two levels that are effectively decoupled from the system. This is the case in the analysis of the spontaneous or stimulated emission of light by atoms and that of charge qubits. In this case it should be kept in mind that the perturbations (interactions with an external field) are in the right range and do not cause transitions to states other than the ones of interest.

Other Examples and the Significance of the two-state Formalism

Pedagogically, the two-state formalism is among the simplest of mathematical techniques used for the analysis of quantum systems. It can be used to illustrate fundamental quantum mechanical phenomena such as the interference exhibited by particles of the polarization states of the photon, but also more complex phenomena such as neutrino oscillation or the neutral K-meson oscillation.

Two-state formalism can be used to describe simple mixing of states, which leads to phenomena such as resonance stabilization and other level crossing related symmetries. Such phenomena have a wide variety of application in chemistry. Phenomena with tremendous industrial applications such as the maser and laser can be explained using the two-state formalism.

The two-state formalism also forms the basis of quantum computing. Qubits, which are the building blocks of a quantum computer, are nothing but two-state systems. Any quantum computational operation is a unitary operation that rotates the state vector on the Bloch sphere.

Quantum Theory

Quantum theory is the theoretical basis of modern physics that explains the nature and behavior of matter and energy on the atomic and subatomic level. The nature and behavior of matter and energy at that level is sometimes referred to as quantum physics and quantum mechanics.

In 1900, physicist Max Planck presented his quantum theory to the German Physical Society. Planck had sought to discover the reason that radiation from a glowing body changes in color from red, to orange, and, finally, to blue as its temperature rises. He found that by making the assumption that energy existed in individual units in the same way that matter does, rather than just as a constant electromagnetic wave - as had been formerly assumed - and was therefore *quantifiable*, he could find the answer to his question. The existence of these units became the first assumption of quantum theory.

Planck wrote a mathematical equation involving a figure to represent these individual units of energy, which he called *quanta*. The equation explained the phenomenon very well; Planck found that at certain discrete temperature levels (exact multiples of a basic minimum value), energy from a glowing body will occupy different areas of the color spectrum. Planck assumed there was a theory yet to emerge from the discovery of quanta, but, in fact, their very existence implied a completely new and fundamental understanding of the laws of nature. Planck won the Nobel Prize in Physics for his theory in 1918, but developments by various scientists over a thirty-year period all contributed to the modern understanding of quantum theory.

The Development of Quantum Theory

- In 1900, Planck made the assumption that energy was made of individual units, or quanta.
- In 1905, Albert Einstein theorized that not just the energy, but the radiation itself was *quantized* in the same manner.
- In 1924, Louis de Broglie proposed that there is no fundamental difference in the makeup and behavior of energy and matter; on the atomic and subatomic level either may behave as if made of either particles or waves. This theory became known as the *principle of wave-particle duality*: elementary particles of both energy and matter behave, depending on the conditions, like either particles or waves.
- In 1927, Werner Heisenberg proposed that precise, simultaneous measurement of two complementary values - such as the position and momentum of a subatomic particle - is impossible. Contrary to the principles of classical physics, their simultaneous measurement is inescapably flawed; the more precisely one value is measured, the more flawed will be the measurement of the other value. This theory became known as the uncertainty principle, which prompted Albert Einstein's famous comment, "God does not play dice."

The Copenhagen Interpretation and the Many-worlds Theory

The two major interpretations of quantum theory's implications for the nature of reality are the Copenhagen interpretation and the many-worlds theory. Niels Bohr proposed the Copenhagen interpretation of quantum theory, which asserts that a particle is whatever it is measured to be (for example, a wave or a particle), but that it cannot be assumed to have specific properties, or even to exist, until it is measured. In short, Bohr was saying that objective reality does not exist. This translates to a principle called superposition that claims that while we do not know what the state of any object is, it is actually in all possible states simultaneously, as long as we don›t look to check.

To illustrate this theory, we can use the famous and somewhat cruel analogy of Schrodinger's Cat. First, we have a living cat and place it in a thick lead box. At this stage, there is no question that the cat is alive. We then throw in a vial of cyanide and seal the box. We do not know if the cat is alive or if the cyanide capsule has broken and the cat has died. Since we do not know, the cat is both dead and alive, according to quantum law - in a superposition of states. It is only when we break open the box and see what condition the cat is that the superposition is lost, and the cat must be either alive or dead.

The second interpretation of quantum theory is the *many-worlds* (or *multiverse* theory. It holds that as soon as a potential exists for any object to be in any state, the universe of that object transmutes into a series of parallel universes equal to the number of possible states in which that the object can exist, with each universe containing a unique single possible state of that object. Furthermore, there is a mechanism for interaction between these universes that somehow permits all states to be accessible in some way and for all possible states to be affected in some manner. Stephen Hawking and the late Richard Feynman are among the scientists who have expressed a preference for the many-worlds theory.

Quantum Theory's Influence

Although scientists throughout the past century have balked at the implications of quantum theory - Planck and Einstein among them - the theory's principles have repeatedly been supported by experimentation, even when the scientists were trying to disprove them. Quantum theory and Einstein's theory of relativity form the basis for modern physics. The principles of quantum physics are being applied in an increasing number of areas, including quantum optics, quantum chemistry, quantum computing, and quantum cryptography.

Relativistic Quantum Mechanics

The aim of relativistic quantum mechanics is to describe the finer details of the structure of atoms and molecules, where relativistic effects become non-negligible. It is a sort of intermediate realm, between the familiar nonrelativistic quantum mechanics and fully relativistic quantum field theory.

Four Vector Formalism

Any point of space-time is described by x^u, $\mu = 0, 1, 2, 3$ with $x^0 = ct, x^1 = x, x^2 = y$ and $x^3 = z$. Any mathematical object a^μ is called a four-vector if it transforms in the same fashion as x^μ under Lorentz transformations (LT's).

If a^μ and b^μ are four-vectors

$$a \cdot b \equiv a^0 b^0 - a^1 b^1 - a^2 b^2 - a^3 b^3$$

is a scalar (invariant) under LT's. It has the same value for all observers just like

$$\left(x^0\right)^2 - \left(x^1\right)^2 - \left(x^2\right)^2 - \left(x^3\right)^2 = c^2 t^2 - x^2 - y^2 - z^2$$

An important four-vector is the four-momentum

$$p^{\mu} \equiv \left(\frac{E}{c}, p_x, p_y, p_z\right) \equiv \left(\frac{E}{c}, \vec{p}\right)$$

where $\vec{p}$ is the 3-momentum. The corresponding Lorentz invariant (scalar) quantity is

$$p^2 \equiv p^u\, p_u \equiv \left(p^0\right)^2 - \left(p^1\right)^2 - \left(p^3\right)^2 = \frac{E^2}{c^2} - \vec{p}^2 = m^2c^2$$

$$\Rightarrow E^2 = \vec{p}^2c^2 + m^2c^4$$

In QM the corresponding hermitian operators are

$$\begin{aligned} \hat{p}^u &= \left(\frac{i\hbar}{c}\frac{\partial}{\partial t}, -i\hbar\frac{\partial}{\partial x}, -i\hbar\frac{\partial}{\partial y}, -i\hbar\frac{\partial}{\partial z}\right) \\ &\equiv \left(\frac{i\hbar}{c}\frac{\partial}{\partial t}, -i\hbar\vec{\nabla}\right) \\ \Rightarrow \hat{p}^u &= \left(i\hbar\frac{\partial}{\partial x^0}, -i\hbar\vec{\nabla}\right). \end{aligned}$$

If we define the differential operator 4-vectors, $\nabla^{\mu} = \left(\frac{\partial}{\partial x^0}, -\vec{\nabla}\right) = \left(\frac{1}{c}\frac{\partial}{\partial t}, -\vec{\nabla}\right)$ and $\nabla_{\mu} = \left(\frac{\partial}{\partial x^0}, \vec{\nabla}\right) = \left(\frac{1}{c}\frac{\partial}{\partial t}, \vec{\nabla}\right)$, then the 4-momentum operator is

$$\hat{p}^{\mu} = i\hbar\nabla^{\mu},$$

and the Lorentz invariant (scalar) constructed from ∇^{μ} is

$$\Box \equiv \nabla.\nabla = \frac{1}{c^2}\frac{\partial^2}{\partial t^2} - \vec{\nabla}^2 = \frac{1}{c^2}\frac{\partial^2}{\partial t^2} - \frac{\partial^2}{\partial x^2} - \frac{\partial^2}{\partial y^2} - \frac{\partial^2}{\partial z^2}$$

The Klein-Gordon Equation

In non-relativistic QM, the free Hamiltonian $H = E = \frac{\vec{p}^2}{2m}$ is quantized by the substitution

$$H \to i\hbar\frac{\partial}{\partial t},\ \vec{p} \to -i\hbar\vec{\nabla}$$

to give the Schrodinger equation

$$i\hbar\frac{\partial\Psi}{\partial t} = -\frac{\hbar^2}{2m}\vec{\nabla}^2\Psi.$$

A relativistic free particle has Hamiltonian

$$H = E = \sqrt{\vec{p}^2c^2 + m^2c^4}$$

and hence the same, naive substitution gives

$$i\hbar\frac{\partial\Psi}{\partial t} = \sqrt{m^2c^4 - \hbar^2c^2\vec{\nabla}^2}\Psi.$$

But what to do about the square root of the operator? One interpretation is to make a series expansion, but then we get a Hamiltonian with derivatives of arbitrarily high order.

A more sensible route is to start from $H^2 = \vec{p}^2c^2 + m^2c^4$ to get

$$-\hbar^2\frac{\partial^2}{\partial t^2}\Psi = \left(-\hbar^2c^2\vec{\nabla}^2 + m^2c^4\right)\Psi$$

$$\Rightarrow \left(\frac{1}{c^2}\frac{\partial^2}{\partial t^2} - \vec{\nabla}^2\right)\Psi + \left(\frac{mc}{\hbar}\right)^2\Psi = 0$$

$$\Rightarrow \left(\Box + \left(\frac{mc}{\hbar}\right)^2\right)\Psi = 0$$

By analogy with the Schrodinger equation it is possible to derive a continuity equation for the Klein-Gordon (KG) equation

$$\frac{\partial\rho}{\partial t} + \vec{\nabla}\cdot\vec{J} = 0$$

where ρ is the probability density and $\vec{J}$ is the probability current.

Derivation: Subtracting the following two equations

$$\Psi^*\left(\Box + \left(\frac{mc}{\hbar}\right)^2\right)\Psi = 0$$

$$\Psi\left(\Box + \left(\frac{mc}{\hbar}\right)^2\right)\Psi^* = 0$$

gives

$$\Psi^*\nabla.\nabla\Psi - \Psi\nabla\cdot\nabla\Psi^*.$$

Furthermore, this implies

$$\nabla\cdot\left(\Psi^*\nabla\Psi - \Psi\nabla\Psi^*\right) = 0$$

$$\Rightarrow \frac{\partial}{\partial t}\left(\frac{1}{c^2}\left(\Psi^*\partial_t\Psi - \Psi\partial_t\Psi^*\right)\right) - \vec{\nabla}.\left(\Psi^*\vec{\nabla}\Psi - \Psi\vec{\nabla}\Psi^*\right) = 0.$$

After multiplying through by ic^2, in order to make ρ real, we can write this as desired as a continuity equation with

$$\rho = i\left(\Psi^* \partial_t \Psi \ - \ \Psi \partial_t \Psi^*\right) \text{and } \vec{J} = ic^2\left(\Psi^* \vec{\nabla} \Psi \ - \ \Psi \vec{\nabla} \Psi^*\right)$$

Now because of the negative sign between the two terms in ρ, the probability density can both take positive and negative values (in contrast to non-relativistic QM where $\rho = |\Psi|^2$ is positive definite)! This is an absurd and nonsensical result for a probability density.

Junk the KG equation for the moment and try harder. Schrödinger was the first to write down this relativistic wave equation, but discarded it for a different reason; the spectrum is not bounded from below.

The Dirac Equation

Let us go back to our starting point

$$\hat{H}\Psi = i\hbar \frac{\partial \Psi}{\partial t}$$

and try to give a meaning to the square root in

$$\hat{H} = \sqrt{m^2c^4 + \hat{\vec{p}}c^2} = \sqrt{m^2c^4 - \hbar^2c^2\vec{\nabla}^2}.$$

Since the equation is linear in $\frac{\partial}{\partial t}$, Lorentz covariance suggests it should be linear also in the $\frac{\partial}{\partial x^i}, i = 1, 2, 3.$

So we write

$$\hat{H} - c\vec{\alpha} \,.\, \hat{\vec{p}} + \beta mc^2$$
$$= -i\hbar c\vec{\alpha} \,.\, \vec{\nabla} + \beta mc^2,$$

where α^i and β are coefficients to be determined. More explicitly this equation can be written

$$\hat{H} = -i\hbar c\left(\alpha^1 \frac{\partial}{\partial x^1} + \alpha^2 \frac{\partial}{\partial x^2} + \alpha^3 \frac{\partial}{\partial x^3}\right) + \beta mc^2.$$

Now we determine the coefficients α^1 and β by requiring that this linear operator "squares" to the KG operator

$$\hat{H}^2 = -\hbar^2c^2\vec{\nabla}^2 + m^2c^4.$$

We find

$$\hat{H}^2 = -\hbar^2c^2\left(\left(\alpha^1\right)^2 \frac{\partial^2}{\partial\left(x^1\right)^2} + \left(\alpha^2\right)^2 \frac{\partial^2}{\partial\left(x^2\right)^2} + \left(\alpha^3\right)^2 \frac{\partial^2}{\partial\left(x^3\right)^2}\right) + \beta^2m^2c^4$$
$$-i\hbar mc^3\left(\left(\alpha^1\beta + \beta\alpha^1\right)\frac{\partial}{\partial x^1} + \ldots\right)$$

$$-\hbar^2 c^2 \left(\left(\alpha^1\alpha^2 + \alpha^2\alpha^1\right) \frac{\partial^2}{\partial x^1 \partial x^2} + \ldots \right).$$

Thus we need to solve

$$\left(\alpha^i\right)^2 = \mathbb{I} \quad i = 1, 2, 3$$
$$\alpha^i\alpha^j + \alpha^j\alpha^i = 0, i \neq j$$
$$\beta^2 = \mathbb{I}$$
$$\alpha^i\beta + \beta\alpha^i = 0, i = 1, 2, 3,$$

where $\mathbb{I}$ denotes a unit matrix (if a subscript is added it denotes the dimensionality, e.g. $\mathbb{I}_2$ denotes a 2 × 2 unit matrix).

It is obviously NOT possible to solve those equations if the coefficients are simply complex numbers. So let us assume that they are N ×N matrices. With some (guess) work it can be shown that the smallest value of N for which equation above can be solved is N = 4. This implies that the Dirac wave function is a 4-component column vector

$$\Psi(x) = \begin{pmatrix} \Psi_1(x) \\ \Psi_2(x) \\ \Psi_3(x) \\ \Psi_4(x) \end{pmatrix}$$

where $x \equiv \left(x^o, \vec{x}\right)$ and the Dirac equation becomes a matrix equation

$$i\hbar \frac{\partial \Psi}{\partial t} = \hat{H}\Psi = \left(c\vec{\alpha} \,.\, \hat{\vec{p}} + \beta mc^2\right)\Psi = \left(-i\hbar c\vec{\alpha} \,.\, \vec{\nabla} + \beta mc^2\right)\Psi.$$

This is a set of 4 first order linear differential equations to determine $\Psi_1, \ldots, \Psi_4$.

Spin and Electromagnetically Interacting Particles

Including interactions in RWEs is generally difficult. Minimal coupling is a simple way to include the electromagnetic interaction. For one charged particle of electric charge q in an electromagnetic field, given by the magnetic vector potential A(r, t) defined by the magnetic field B = ∇ × A, and electric scalar potential ϕ(r, t), this is:

$$E - e\phi \approx mc^2, \quad \mathrm{p} \approx mv,$$

where P_μ is the four-momentum that has a corresponding 4-momentum operator, and A_μ the four-potential. In the following, the non-relativistic limit refers to the limiting cases:

$$E - e\phi \approx mc^2, \quad \mathrm{p} \approx mv,$$

that is, the total energy of the particle is approximately the rest energy for small electric potentials, and the momentum is approximately the classical momentum.

Spin 0

In RQM, the KG equation admits the minimal coupling prescription;

$$(\hat{E}-q\phi)^2\psi = c^2(\hat{\mathrm{p}}-q\mathrm{A})^2\psi+(mc^2)^2\psi \rightleftharpoons \left[(\hat{P}_\mu-qA_\mu)(\hat{P}^\mu-qA^\mu)-(mc)^2\right]\psi = 0.$$

In the case where the charge is zero, the equation reduces trivially to the free KG equation so nonzero charge is assumed below. This is a scalar equation that is invariant under the irreducible one-dimensional scalar (0,0) representation of the Lorentz group. This means that all of its solutions will belong to a direct sum of (0,0) representations. Solutions that do not belong to the irreducible (0,0) representation will have two or more *independent* components. Such solutions cannot in general describe particles with nonzero spin since spin components are not independent. Other constraint will have to be imposed for that, e.g. the Dirac equation for spin $\frac{1}{2}$. Thus if a system satisfies the KG equation *only*, it can only be interpreted as a system with zero spin.

The electromagnetic field is treated classically according to Maxwell's equations and the particle is described by a wave function, the solution to the KG equation. The equation is, as it stands, not always very useful, because massive spinless particles, such as the π-mesons, experience the much stronger strong interaction in addition to the electromagnetic interaction. It does, however, correctly describe charged spinless bosons in the absence of other interactions.

The KG equation is applicable to spinless charged bosons in an external electromagnetic potential. As such, the equation cannot be applied to the description of atoms, since the electron is a spin $\frac{1}{2}$ particle. In the non-relativistic limit the equation reduces to the Schrödinger equation for a spinless charged particle in an electromagnetic field:

$$\left(i\hbar\frac{\partial}{\partial t}-q\phi\right)\psi = \frac{1}{2m}(\hat{\mathrm{p}}-q\mathrm{A})^2\psi \Leftrightarrow \hat{H} = \frac{1}{2m}(\hat{\mathrm{p}}-q\mathrm{A})^2+q\phi.$$

Spin $\frac{1}{2}$

Non relativistically, spin was phenomenologically introduced in the Pauli equation by Pauli in 1927 for particles in an electromagnetic field:

$$\left(i\hbar\frac{\partial}{\partial t}-q\phi\right)\psi = \left[\frac{1}{2m}(\sigma\cdot(\mathrm{p}-q\mathrm{A}))^2\right]\psi \Leftrightarrow \hat{H} = \frac{1}{2m}(\sigma\cdot(\mathrm{p}-q\mathrm{A}))^2+q\phi$$

by means of the 2×2 Pauli matrices, and ψ is not just a scalar wave function as in the non-relativistic Schrödinger equation, but a two-component spinor field:

$$\psi = \begin{pmatrix}\psi_\uparrow \\ \psi_\downarrow\end{pmatrix}$$

where the subscripts $\uparrow$ and $\downarrow$ refer to the "spin up" ($\sigma = +\frac{1}{2}$) and "spin down" ($\sigma = -\frac{1}{2}$) states.

In RQM, the Dirac equation can also incorporate minimal coupling, rewritten from above;

$$\left(i\hbar\frac{\partial}{\partial t}-q\phi\right)\psi=\gamma^{0}\left[c\tilde{\mathrm{a}}\cdot(\hat{\mathrm{p}}-q\mathrm{A})-mc^{2}\right]\psi \rightleftharpoons \left[\gamma^{\mu}(\hat{P}_{\mu}-qA_{\mu})-mc^{2}\right]\psi=0$$

and was the first equation to accurately *predict* spin, a consequence of the 4 × 4 gamma matrices $\gamma^0 = \beta$, $\gamma = (\gamma_1, \gamma_2, \gamma_3) = \beta\alpha = (\beta\alpha_1, \beta\alpha_2, \beta\alpha_3)$. There is a 4 × 4 identity matrix pre-multiplying the energy operator (including the potential energy term), conventionally not written for simplicity and clarity (i.e. treated like the number 1). Here ψ is a four-component spinor field, which is conventionally split into two two-component spinors in the form:

$$\psi=\begin{pmatrix}\psi_{+}\\ \psi_{-}\end{pmatrix}=\begin{pmatrix}\psi_{+\uparrow}\\ \psi_{+\downarrow}\\ \psi_{-\uparrow}\\ \psi_{-\downarrow}\end{pmatrix}$$

The 2-spinor ψ_+ corresponds to a particle with 4-momentum (E, p) and charge q and two spin states ($\sigma = \pm\frac{1}{2}$, as before). The other 2-spinor ψ_- corresponds to a similar particle with the same mass and spin states, but *negative* 4-momentum $-(E, p)$ and *negative* charge $-q$, that is, negative energy states, time-reversed momentum, and negated charge. This was the first interpretation and prediction of a particle and *corresponding antiparticle*. When applied a one-electron atom or ion, setting A = 0 and ϕ to the appropriate electrostatic potential, additional relativistic terms include the spin-orbit interaction, electron gyromagnetic ratio, and Darwin term. In ordinary QM these terms have to be put in by hand and treated using perturbation theory. The positive energies do account accurately for the fine structure.

Within RQM, for massless particles the Dirac equation reduces to:

$$\left(\frac{\hat{E}}{c}+\sigma\cdot\hat{\mathrm{p}}\right)\psi_{+}=0,\quad \left(\frac{\hat{E}}{c}-\sigma\cdot\hat{\mathrm{p}}\right)\psi_{-}=0 \rightleftharpoons \sigma^{\mu}\hat{P}_{\mu}\psi_{+}=0,\quad \sigma_{\mu}\hat{P}^{\mu}\psi_{-}=0,$$

the first of which is the Weyl equation, a considerable simplification applicable for massless neutrinos. This time there is a 2 × 2 identity matrix pre-multiplying the energy operator conventionally not written. In RQM it is useful to take this as the zeroth Pauli matrix σ_0 which couples to the energy operator (time derivative), just as the other three matrices couple to the momentum operator (spatial derivatives).

The Pauli and gamma matrices were introduced here, in theoretical physics, rather than pure mathematics itself. They have applications to quaternions and to the SO(2) and SO(3) Lie groups, because they satisfy the important commutator [,] and anticommutator [,]+ relations respectively:

$$\left[\sigma_a,\sigma_b\right]=2i\varepsilon_{abc}\sigma_c,\quad \left[\sigma_a,\sigma_b\right]_{+}=2\delta_{ab}\sigma_0$$

where ε_{abc} is the three-dimensional Levi-Civita symbol. The gamma matrices form bases in Clifford algebra, and have a connection to the components of the flat space time Minkowski metric $\eta^{\alpha\beta}$ in the anti-commutation relation:

$$\left[\gamma^{\alpha},\gamma^{\beta}\right]_{+}=\gamma^{\alpha}\gamma^{\beta}+\gamma^{\beta}\gamma^{\alpha}=2\eta^{\alpha\beta},$$

(This can be extended to curved spacetime by introducing vierbeins, but is not the subject of special relativity).

In 1929, the Breit equation was found to describe two or more electromagnetically interacting massive spin $\frac{1}{2}$ fermions to first-order relativistic corrections; one of the first attempts to describe such a relativistic quantum many-particle system. This is, however, still only an approximation, and the Hamiltonian includes numerous long and complicated sums.

Helicity and Chirality

The helicity operator is defined by;

$$\hat{h}=\hat{S}\cdot\frac{\hat{p}}{|p|}=\hat{S}\cdot\frac{c\hat{p}}{\sqrt{E^2-(m_0c^2)^2}}$$

where p is the momentum operator, S the spin operator for a particle of spin *s*, *E* is the total energy of the particle, and m_0 its rest mass. Helicity indicates the orientations of the spin and translational momentum vectors. Helicity is frame-dependent because of the 3-momentum in the definition, and is quantized due to spin quantization, which has discrete positive values for parallel alignment, and negative values for antiparallel alignment.

A automatic occurrence in the Dirac equation (and the Weyl equation) is the projection of the spin $\frac{1}{2}$ operator on the 3-momentum (times *c*), σ · *c* p, which is the helicity (for the spin $\frac{1}{2}$ case) times $\sqrt{E^2-(m_0c^2)^2}$.

For massless particles the helicity simplifies to:

$$\hat{h}=\hat{S}\cdot\frac{c\hat{p}}{E}$$

Higher Spins

The Dirac equation can only describe particles of spin $\frac{1}{2}$. Beyond the Dirac equation, RWEs have been applied to free particles of various spins. In 1936, Dirac extended his equation to all fermions, three years later Fierz and Pauli rederived the same equation. The Bargmann–Wigner equations were found in 1948 using Lorentz group theory, applicable for all free particles with any spin. Considering the factorization of the KG equation above, and more rigorously by Lorentz group theory, it becomes apparent to introduce spin in the form of matrices.

The wave functions are multicomponent spinor fields, which can be represented as column vectors of functions of space and time:

$$\psi(\mathrm{r},t)=\begin{bmatrix}\psi_{\sigma=s}(\mathrm{r},t)\\ \psi_{\sigma=s-1}(\mathrm{r},t)\\ \vdots\\ \psi_{\sigma=-s+1}(\mathrm{r},t)\\ \psi_{\sigma=-s}(\mathrm{r},t)\end{bmatrix}\rightleftharpoons \psi(\mathrm{r},t)^{\dagger}=\begin{bmatrix}\psi_{\sigma=s}(\mathrm{r},t)^{\star} & \psi_{\sigma=s-1}(\mathrm{r},t)^{\star} & \cdots & \psi_{\sigma=-s+1}(\mathrm{r},t)^{\star} & \psi_{\sigma=-s}(\mathrm{r},t)^{\star}\end{bmatrix}$$

where the expression on the right is the Hermitian conjugate. For a *massive* particle of spin *s*, there are 2*s* + 1 components for the particle, and another 2*s* + 1 for the corresponding antiparticle (there are 2*s* + 1 possible σ values in each case), altogether forming a 2(2*s* + 1)-component spinor field:

$$\psi(\mathrm{r},t)=\begin{bmatrix}\psi_{+,\sigma=s}(\mathrm{r},t)\\ \psi_{\aleph\sigma\ s\ 1}(\mathrm{r},t)\\ \vdots\\ \psi_{+,\sigma=-s+1}(\mathrm{r},t)\\ \psi_{\aleph\sigma\ \ s}(\mathrm{r},t)\\ \psi_{-,\sigma=s}(\mathrm{r},t)\\ \psi_{\aleph\sigma\ s\ 1}(\mathrm{r},t)\\ \vdots\\ \psi_{-,\sigma=-s+1}(\mathrm{r},t)\\ \psi_{\aleph\sigma\ \ s}(\mathrm{r},t)\end{bmatrix}\rightleftharpoons \psi(\mathrm{r},t)^{\dagger}\begin{bmatrix}\psi_{+,\aleph=\ddot{u}}(\mathrm{r},t)^{\star} & \psi_{+,\ =\ -1}(\mathrm{r},t)^{\star} & \cdots & \psi_{-,\ =-}(\mathrm{r},t)^{\star}\end{bmatrix}$$

with the + subscript indicating the particle and – subscript for the antiparticle. However, for *massless* particles of spin *s*, there are only ever two-component spinor fields; one is for the particle in one helicity state corresponding to +*s* and the other for the antiparticle in the opposite helicity state corresponding to −*s*:

$$\psi(\mathrm{r},t)=\begin{pmatrix}\psi_{+}(\mathrm{r},t)\\ \psi_{-}(\mathrm{r},t)\end{pmatrix}$$

According to the relativistic energy-momentum relation, all massless particles travel at the speed of light, so particles traveling at the speed of light are also described by two-component spinors. Historically, Élie Cartan found the most general form of spinors in 1913, prior to the spinors revealed in the RWEs following the year 1927.

For equations describing higher-spin particles, the inclusion of interactions is nowhere near as simple minimal coupling, they lead to incorrect predictions and self-inconsistencies. For spin greater than $\frac{\hbar}{2}$, the RWE is not fixed by the particle's mass, spin, and electric charge; the electromagnetic moments (electric dipole moments and magnetic dipole moments) allowed by the spin quantum number are arbitrary. (Theoretically, magnetic charge would contribute also). For example, the spin $\frac{1}{2}$ case only allows a magnetic dipole, but for spin 1 particles magnetic quadrupoles and electric dipoles are also possible.

Velocity Operator

The Schrödinger/Pauli velocity operator can be defined for a massive particle using the classical definition p = *m* v, and substituting quantum operators in the usual way:

$$\hat{\mathrm{v}} = \frac{1}{m}\hat{\mathrm{p}}$$

which has eigenvalues that take *any* value. In RQM, the Dirac theory, it is:

$$\hat{\mathrm{v}} = \frac{i}{\hbar}\left[\hat{H}, \hat{\mathrm{r}}\right]$$

which must have eigenvalues between $\pm c$.

Relativistic Quantum Lagrangians

The Hamiltonian operators in the Schrödinger picture are one approach to forming the differential equations for ψ. An equivalent alternative is to determine a Lagrangian (really meaning *Lagrangian density*), then generate the differential equation by the field-theoretic Euler–Lagrange equation:

$$\partial_\mu\left(\frac{\partial\mathcal{L}}{\partial(\partial_\mu\psi)}\right) - \frac{\partial\mathcal{L}}{\partial\psi} = 0$$

For some RWEs, a Lagrangian can be found by inspection. For example, the Dirac Lagrangian is:

$$\mathcal{L} = \bar{\psi}(\gamma^\mu P_\mu - mc)\psi$$

and Klein–Gordon Lagrangian is:

$$\mathcal{L} = -\frac{\hbar^2}{m}\eta^{\mu\nu}\partial_\mu\psi^*\partial_\nu\psi - mc^2\psi^*\psi.$$

This is not possible for all RWEs; and is one reason the Lorentz group theoretic approach is important and appealing: fundamental invariance and symmetries in space and time can be used to derive RWEs using appropriate group representations. The Lagrangian approach with field interpretation of ψ is the subject of QFT rather than RQM: Feynman's path integral formulation uses invariant Lagrangians rather than Hamiltonian operators, since the latter can become extremely complicated.

Relativistic Quantum Angular Momentum

In non-relativistic QM, the angular momentum operator is formed from the classical pseudovector definition L = r × p. In RQM, the position and momentum operators are inserted directly where they appear in the orbital relativistic angular momentum tensor defined from the four-dimensional position and momentum of the particle, equivalently a bivector in the exterior algebra formalism:

$$M^{\alpha\beta} = X^{\alpha}P^{\beta} - X^{\beta}P^{\alpha} = 2X^{[\alpha}P^{\beta]} \quad \rightleftharpoons \quad \mathrm{M} = \mathrm{X} \wedge \mathrm{P},$$

which are six components altogether: three are the non-relativistic 3-orbital angular momenta; $M^{12} = L^3$, $M^{23} = L^1$, $M^{31} = L^2$, and the other three M^{01}, M^{02}, M^{03} are boosts of the center of mass of the rotating object. An additional relativistic-quantum term has to be added for particles with spin. For a particle of rest mass m, the *total* angular momentum tensor is:

$$J^{\alpha\beta} = 2X^{[\alpha}P^{\beta]} + \frac{1}{m^2}\varepsilon^{\alpha\beta\gamma\delta}W_{\gamma}p_{\delta} \quad \rightleftharpoons \quad \mathrm{J} = \mathrm{X} \wedge \mathrm{P} + \frac{1}{m^2}\star(\mathrm{W} \wedge \mathrm{P})$$

where the star denotes the Hodge dual, and

$$\ddot{u}_{\alpha} = \frac{1}{2}\varepsilon_{\alpha\beta\gamma\delta} \quad \aleph \quad \rightleftharpoons \quad \mathrm{W} = \star(\mathrm{M} \wedge \mathrm{P})$$

is the Pauli–Lubanski pseudovector. For more on relativistic spin.

Thomas Precession and Spin-orbit Interactions

In 1926 the Thomas precession is discovered: relativistic corrections to the spin of elementary particles with application in the spin–orbit interaction of atoms and rotation of macroscopic objects. In 1939 Wigner derived the Thomas precession.

In classical electromagnetism and special relativity, an electron moving with a velocity v through an electric field E but not a magnetic field B, will in its own frame of reference experience a Lorentz-transformed magnetic field B′:

$$B' = \frac{\mathrm{E} \times \mathrm{v}}{c^2\sqrt{1-(v/c)^2}}.$$

In the non-relativistic limit $v << c$:

$$B' = \frac{\mathrm{E} \times \mathrm{v}}{c^2},$$

so the non-relativistic spin interaction Hamiltonian becomes:

$$\hat{\mathrm{H}} = -\mathrm{B}' \cdot \hat{\mu}_{\mathrm{S}} = -\left(\mathrm{B} + \frac{\mathrm{E} \times \mathrm{v}}{c^2}\right) \cdot \hat{\mu}_{\mathrm{S}},$$

where the first term is already the non-relativistic magnetic moment interaction, and the second term the relativistic correction of order $(v/c)^2$, but this disagrees with experimental atomic spectra by a factor of $1/2$. It was pointed out by L. Thomas that there is a second relativistic effect: An electric field component perpendicular to the electron velocity causes an additional acceleration of the electron perpendicular to its instantaneous velocity, so the electron moves in a curved path. The electron moves in a rotating frame of reference, and this additional precession of the electron is called the *Thomas precession*. It can be shown that the net result of this effect is that the spin–orbit interaction is reduced by half, as if the magnetic field experienced by the electron has only one-half the value, and the relativistic correction in the Hamiltonian is:

$$\hat{H} = -B' \cdot \hat{\mu}_S = -\left(B + \frac{E \times v}{c^2}\right) \cdot \hat{\mu}_S.$$

In the case of RQM, the factor of $\frac{1}{2}$ is predicted by the Dirac equation.

Interpretations of Quantum Mechanics

Interpretations' of quantum mechanics are traditionally characterized by their responses to the measurement problem:

The Copenhagen Interpretation

The earliest consensus concerning the meaning of quantum mechanics formed around the work of Niels Bohr and Werner Heisenberg in Copenhagen during the 1920s, and hence became known as the Copenhagen interpretation. Bohr's position is that our conception of the world is necessarily classical; we think of the world in terms of objects (for example, waves or particles) moving through three-dimensional space, and this is the only way we can think of it. Quantum mechanics doesn't permit such a conceptualization, either in terms of waves or particles, and so the quantum world is in principle unknowable by us. Quantum mechanics shouldn't be taken as a description of the quantum world, and neither should the evolution of the quantum state over time be taken as a causal explanation of the phenomena we observe. Rather, quantum mechanics is an extremely effective tool for predicting measurement results that takes the configuration of the measuring apparatus (described classically) as input, and produces probabilities for the possible measurement outcomes (described classically) as output.

It is sometimes claimed that the Copenhagen interpretation is a product of the logical positivism that flourished in Europe during the same period. The logical positivists held that the meaningful content of a scientific theory is exhausted by its empirical predictions; any further speculation into the nature of the world that produces these measurement outcomes is quite literally meaningless. This certainly has some resonances with the Copenhagen interpretation, particularly as described by Heisenberg. But Bohr's views are importantly different from Heisenberg's, and are more Kantian than positivist. Bohr is happy to say that the micro-world exists, and that it can't be conceived of in causal terms, both of which would be meaningless claims according to positivist scruples. However, Bohr thinks we can say little else about the micro-world. Bohr, like Kant, thinks that we can only conceive of things in certain ways, and that the world as it is in itself is not amenable to such conceptualization. If this is correct, it is inevitable that our fundamental physical theories are unable to describe the world as it is, and the fact that we can make no sense of quantum mechanics as a description of the world should not concern us.

Unless one is convinced of Kant's position concerning our conceptual access to the world, one may not find Bohr's pronouncements concerning what we can conceive compelling. However, the motivation for adopting a Copenhagen-style interpretation can be made independent of any overarching philosophical position. Since the intensity of the wavefunction at a location gives the probability of the particle occupying that location, it is natural to regard the wavefunction as a reflection of our knowledge of the system rather than a description of the system itself. This view, held by Einstein, suggests that quantum mechanics is incomplete, since it gives us only an instrumental recipe for calculating the probabilities of outcomes, rather than a description of the underlying state of the system that gives rise to those probabilities. But it was later proved (as we shall see) that given certain plausible assumptions, it is impossible to construct such a description of the underlying state. Bohr did not know at the time that Einstein's task was impossible, but its evident difficulty provides some motivation for regarding the quantum world as inscrutable.

However, the Copenhagen interpretation has at least two major drawbacks. First, a good deal of the early evidence for quantum mechanics comes from its ability to explain the results of interference experiments involving particles like electrons. Bohr's insistence that quantum mechanics is not descriptive takes away this explanation (although, of course, viewing the wavefunction as descriptive only of our knowledge does no better). Second, Bohr's position requires a "cut" between the macroscopic world described by classical concepts and the microscopic world subsumed under (but not described by) quantum mechanics. Since macroscopic objects are made out of microscopic components, it looks like macroscopic objects must obey the laws of quantum mechanics too; there can be no such "cut", either sharp or vague, delimiting the realm of applicability of quantum mechanics.

The Many-Worlds Interpretation

In 1957 Hugh Everett proposed a radically new way of interpreting the quantum state. His proposal was to take quantum mechanics as descriptive and universal; the quantum state is a genuine description of the physical system concerned, and macroscopic systems are just as well described in this way as microscopic ones. This immediately solves both the above problems; there is no "cut" between the micro and macro worlds, and the explanation of particle interference in terms of waves is retained.

An immediate problem facing such a realist interpretation of the quantum state is the provenance of the outcomes of quantum measurements. Recall that in the case of electron interference, what is detected is not a spread-out wave, but a particle with a well-defined location, where the wavefunction intensity at a location gives the probability that the particle is located there.

How does Everett account for these facts? What he suggests is that we model the measurement process itself quantum mechanically. It is by no means uncontroversial that measuring devices and human observers admit of a quantum mechanical description, but given the assumption that quantum mechanics applies to all material objects, such a description ought to be available at least in principle. So consider for simplicity the situation in which the wavefunction intensity for the electron at the end of the experiment is non-zero in only two regions of space, A and B. The detectors at these locations can be modeled using a wavefunction too, with the result that the electron wavefunction component at A triggers a corresponding change in the wavefunction of the A-detector, and similarly at B. In the same way, we can model the experimenter who observes the detectors using a wavefunction, with the result that the change in the wavefunction of the A-detector causes a change in the wavefunction of the observer corresponding to seeing that the A-detector has fired, and the change in the wavefunction of the B-detector causes a change in the wavefunction of the observer corresponding to seeing that the B-detector has fired. The observer's final state, then, is modeled by two distinct wave structures *superposed*, much in the way two images are superposed in a double-exposure photograph.

In sum, the wave structure of the electron-detector-observer system consists of two distinct branches, the A-outcome branch and the B-outcome branch. Since these two branches are relatively causally isolated from each other, we can describe them as two distinct *worlds*, in one of which the electron hits the detector at A and the observer sees the A-detector fire, and in the other of which the electron hits the detector at B and the observer sees the B-detector fire. This talk of worlds needs to be treated carefully, though; there is just one physical world, described by the quantum state, but because observers (along with all other physical objects) exhibit this branching structure, it is *as if* the world is constantly splitting into multiple copies. It is not clear whether Everett himself endorsed this talk of worlds, but this is the understanding of his work that has become canonical; call it the *many-worlds interpretation*.

According to the many-worlds interpretation, then, every physically possible outcome of a measurement actually occurs in some branch of the quantum state, but as an inhabitant of a particular branch of the state, a particular observer only sees one outcome. This explains why, in the electron interference experiment, the outcome looks like a discrete particle even though the object that passes through the interference device is a wave; each point in the wave generates its own branch of reality when it hits the detectors, so from within each of the resulting branches it looks like the incoming object was a particle.

The main advantage of the many-worlds interpretation is that it is a realist interpretation that takes the physics of standard quantum mechanics literally. It is often met with incredulity, since it entails that people (along with other objects) are constantly branching into innumerable copies, but this by itself is no argument against it. Still, the branching of people leads to philosophical dif ficulties concerning identity and probability, and these (particularly the latter) constitute genuine difficulties facing the approach.

The problem of identity is a philosophically familiar one: if a person splits into two copies, then the copies can't be identical to (that is, the same person as) the original person, or else they would be identical to (the same person as) each other. Various solutions have been developed in the literature. One might follow Derek Parfit and bite the bullet here: what fission cases like this show is that strict identity is not a useful concept for describing the relationship between people and their successors. Or one might follow David Lewis and rescue strict identity by stipulating that a person is a four-dimensional *history* rather than a three dimensional object. According to this picture, there are two people (two complete histories) present both before and after the fission event; they initially overlap but later diverge. Identity over time is preserved, since each of the pre-split people is identical with exactly one of the post-split people. Both of these positions have been proposed as potential solutions to the problem of personal identity in a many-worlds universe. A third solution that is sometimes mentioned is to stipulate that a person is the *whole* of the branching entity, so that the pre-split person is identical to *both* her successors, and (despite our initial intuition otherwise) the successors are identical to each other.

So the problem of identity admits of a number of possible solutions, and the only question is how one should try to decide between them. Indeed, one might argue that there is no need to decide between them, since the choice is a pragmatic one about the most useful language to use to describe branching persons.

The problem of probability, though, is potentially more serious. As noted above, quantum mechanics makes its predictions in the form of probabilities: the square of the wavefunction amplitude in a region tells us the probability of the particle being located there. The striking agreement of the observed distribution of outcomes with these probabilities is what underwrites our confidence in quantum mechanics. But according to the many-worlds interpretation, every outcome of a measurement actually occurs in some branch of reality, and the well-informed observer knows this. It is hard to see how to square this with the concept of probability; at first glance, it looks like every outcome has probability 1, both objectively and epistemically. In particular, if a measurement results in two branches, one with a large squared amplitude and one with a small squared amplitude, it is hard to see why we should regard the former as more probable than the latter. But unless we can do so, the empirical success of quantum mechanics evaporates.

It is worth noting, however, that the foundations of probability are poorly understood. When we roll two dice, the chance of rolling 7 is higher than the chance of rolling 12. But there is no consensus concerning the meaning of chance claims, or concerning why the higher chance of 7 should constrain our expectations or behavior. So perhaps a quantum branching world is in no worse shape than a classical linear world when it comes to understanding probability. We may not understand how squared wavefunction amplitude could function as chance in guiding our expectations, but perhaps that is no barrier to postulating that it does so function.

A more positive approach has been developed by David Deutsch and David Wallace, arguing that given some plausible constraints on rational behavior, rational individuals should behave as if squared wavefunction amplitudes are chances. If one combines this with a functionalist attitude towards chance—that whatever functions as chance in guiding behavior *is* chance—then this program promises to underwrite the contention that squared wave amplitudes are chances. However, the assumptions on which the Deutsch-Wallace argument is based can be challenged. In particular, they assume that it is irrational to care about branching *per se*: having two successors experiencing a given outcome is neither better nor worse than having one successor experiencing that outcome. But it is not clear that this is a matter of *rationality* any more than the question of whether having several happy children is better than having one happy child.

A further worry about the many-words theory that has been largely put to rest concerns the ontological status of the worlds. It has been argued that the postulation of many worlds is ontologically profligate. However, the current consensus is that worlds are emergent entities just like tables and chairs, and talk of worlds is just a convenient way of talking about the features of the quantum state. On this view, the many-worlds interpretation involves no entities over and above those represented by the quantum state, and as such is ontologically parsimonious. There remains the residual worry that the number of branches depends sensitively on mathematical choices about how to represent the quantum state. Wallace, however, embraces this indeterminacy, arguing that even though the many-worlds universe is a branching one, there is no well-defined number of branches that it has. If tenable, this goes some way towards resolving the above concern about the rationality of caring about branching *per se*: if there is no number of branches, then it is irrational to care about it.

Hidden Variable Theories

The many-worlds interpretation would have us believe that we are mistaken when we think that a quantum measurement results in a unique outcome; in fact such a measurement results in multiple outcomes occurring on multiple branches of reality. But perhaps that is too much to swallow, or perhaps the problems concerning identity and probability mentioned above are insuperable. In that case, one is led to the conclusion that quantum mechanics is incomplete, since there is nothing in the quantum state that picks out one of the many possible measurement results as the single *actual* measurement result. As mentioned above, this was Einstein's view. If this view is correct, then quantum mechanics stands in need of completion via the addition of extra variables describing the *actual* state of the world. These additional variables are commonly known as *hidden*variables.

However, a theorem proved by John Bell in 1964 shows that, subject to certain plausible assumptions, no such hidden-variable completion of quantum mechanics is possible. One version of the proof concerns the properties of a pair of particles. Each particle has a property called spin: when the spin of the particle is measured in some direction, one either gets the result *up* or *down*. Suppose that the spin of each particle can be measured along one of three directions 120° apart. What quantum mechanics predicts is that if the spins of the particles are measured along the same direction, they always agree (both *up* or both *down*), but if they are measured along different directions they agree 25% of the time and disagree 75% of the time. According to the hidden variable approach, the particles have determinate spin values for each of the three measurement directions prior to measurement. The question is how to ascribe spin values to particles to reproduce the predictions of quantum mechanics. And what Bell proved is that there is no way to do this; the task is impossible.

Many physicists concluded on the basis of Bell's theorem that no hidden-variable completion of quantum mechanics is possible. However, this was not Bell's conclusion. Bell concluded instead that one of the assumptions he relied on in his proof must be false. First, Bell assumed *locality*—that the result of a measurement performed on one particle cannot influence the properties of the other particle. This seems secure because the measurements on the two particles can be widely separated, so that a signal carrying such an influence would have to travel faster than light. Second, Bell assumed *independence*—that the properties of the particles are independent of which measurements will be performed on them. This assumption too seems secure, because the choice of measurement can be made using a randomizing device or the free will of the experimenter.

Despite the apparent security of his assumptions, Bell knew when he proved his theorem that a hidden-variable completion of quantum mechanics had been explicitly constructed by David Bohm in 1952. Bohm assumed that in addition to the wave described by the quantum state, there is also a set of particles whose positions are given by the hidden variables. The wave pushes the particles around according to a new dynamical law formulated by Bohm, and the law is such that if the particle positions are initially statistically distributed according to the squared amplitude of the wave, then they are always distributed in this way. In an electron interference experiment, then, the existence of the wave explains the interference effect, the existence of the particles explains why each electron is observed at a precise location, and the new Bohmian law explains why the probability of observing an electron at a given location is given by the squared amplitude of the wave. As Bell often pointed out, to call Bohm's theory a hidden variable theory is something of a misnomer, since it is the values of the hidden variables—the positions of the particles—that are directly observed on measurement. Nevertheless, the name has stuck.

Bohm's theory, then, provides a concrete example of a hidden variable theory of quantum mechanics. However, it is not a counterexample to Bell's theorem, because it violates Bell's locality assumption. The new law introduced by Bohm is explicitly non-local: the motion of each particle is determined in part by the positions of all the other particles at that instant. In the case of Bell's spin experiment, a measurement on one particle instantaneously affects the motion of the other particle, even if the particles are widely separated. This is a prima facie violation of special relativity, since according to special relativity simultaneity is dependent on one's choice of coordinates, making it impossible to define "instantaneous" in any objective way. However, this does not mean that Bohm's theory is immediately *refuted* by special relativity, since one can instead take Bohm's theory to show the need to *add* a universal standard of simultaneity to special relativity. Bell recognized this possibility. It is worth noting that even though Bohm's theory requires instantaneous action at a distance, it also prevents these influences from being controlled so as to send a signal; there is no "Bell telephone".

Bohm chooses positions as the properties described by the hidden variables of his theory. His reason for this is that it is plausible that it is the positions of things that we directly observe, and hence completing quantum mechanics via positions suffices to ensure that measurements have unique outcomes. But it is possible to construct measurements in which the outcome is recorded in some property other than position. As a response to this possibility, one might suggest adding hidden variables describing *every* property of the particles simultaneously, rather than just their positions. However, a theorem proved by Kochen and Specker in 1967 shows that no such theory can reproduce the predictions of quantum mechanics. A second response is to stick with Bohm's theory as it is, and argue that while such measurements may initially lack a unique outcome, they will rapidly acquire a unique outcome as the recording device becomes correlated with the positions of the surrounding objects in the environment.

A final way to accommodate such measurements within a hidden variable theory is to make it a contingent matter *which* properties of a system are ascribed determinate values at a particular time. That is, rather than supplementing the wavefunction with variables describing a fixed property (the positions of things), one can let the wavefunction state itself determine which properties of the system are described by the hidden variables at that time. The idea is that the algorithm for ascribing hidden variables to a system is such that whenever a measurement is performed, the algorithm ascribes a determinate value to the property recording the outcome of the measurement. Such theories are known as modal theories. But while Bohm's theory provides an explicit dynamical law describing the motion of the particles over time, modal theories generally do not provide a dynamical law governing their hidden variables, and this is regarded as a weakness of the approach.

Modal theories, like Bohm's theory, evade Bell's theorem by violating Bell's locality assumption. In the modal case, the rule for deciding which properties of the system are made determinate depends on the complete wavefunction state at a particular instant, and this allows a measurement on one particle to affect the properties ascribed to another particle, however distant. As mentioned above, one can solve this problem by supplementing special relativity with a preferred standard of simultaneity. But this is widely regarded as an ad hoc and unwarranted addition to an otherwise elegant and well-confirmed physical theory. Indeed, the same charge is often levelled at the hidden variables themselves; they are an ad hoc and unwarranted addition to quantum mechanics.

If hidden variable theories turn out to be the only viable interpretations of quantum mechanics, though, the force of this charge is reduced considerably.

Nevertheless, it may be possible to construct a hidden variable theory that does not violate locality. In order to evade Bell's theorem, then, it will have to violate the independence assumption—the assumption that the properties of the particles are independent of which measurements will be performed on them. Since one can choose the measurements however one likes, it is initially hard to see how this assumption *could* be violated. But there are a couple of ways it might be done. First, one could simply accept that there are brute, uncaused correlations in the world. There is no causal link (in either direction) between my choice of which measurement to perform on a (currently distant) particle and its properties, but nevertheless there is a correlation between them. This approach requires giving up on the common cause principle—the principle that a correlation between two events indicates either that one causes the other or that they share a cause. However, there is little consensus concerning this principle anyway.

A second approach is to postulate a common cause for the correlation—a past event that causally influences both the choice of measurement and the properties of the particle. But absent some massive unseen conspiracy on the part of the universe, one can frequently ensure that there is no common cause in the past by isolating the measuring device from external influences. However, the measuring device and the particle to be measured will certainly interact in the *future*, namely when the measurement occurs. It has been proposed that this future event can constitute the causal link explaining the correlation between the particle properties and the measurements to be performed on them. This requires that later events can cause earlier events—that causation can operate backwards in time as well as forwards in time. For this reason, the approach is known as the retrocausal approach.

The retrocausal approach allows correlations between distant events to be explained without instantaneous action at a distance, since a combination of ordinary causal links and retrocausal links can amount to a causal chain that carries an influence between simultaneous distant events. No absolute standard of simultaneity is required by such explanations, and hence retrocausal hidden variable theories are more easily reconciled with special relativity than non-local hidden variable theories.

Bohm's theory operates with a two-element ontology—a wave steering a set of particles. Retrocausal theories vary in their ontological presuppositions. Some—retrocausal Bohmian theories—incorporate two waves steering a set of particles; one wave carries the "forward-causal" influences on the particles from the initial state of the system, and the other carries the "backward-causal" influences on the particles from the final state of the system. But it may be possible to make do with the particles alone, with the wavefunction representing our knowledge of the particle positions rather than the state of a real object. The idea is that the interaction between the causal influences on the particles from the past and from the future can explain all the quantum phenomena we observe, including interference. However, at present this is just a promising research program; no explicit dynamical laws for such a theory have been formulated.

Spontaneous Collapse Theories

Hidden variable theories attempt to complete quantum mechanics by positing extra ontology in addition to (or perhaps instead of) the wavefunction. Spontaneous collapse theories, on the other hand, (at least initially) take the wavefunction to be a complete representation of the state of a system, and posit instead that the dynamical law of standard quantum mechanics—the Schrödinger equation—is not exactly right. The Schrödinger equation is linear; this means that if initial state A leads to final state A' and initial state B leads to final state B', then initial state A + B leads to final state A' + B'. For example, if a measuring device fed a spin-up particle leads to a spin-up reading, and a measuring device fed a spin-down particle leads to a spin-down reading, then a measuring device fed a particle whose state is a sum of spin-up and spin-down states will end up in a state which is a sum of reading spin-up and reading spin-down. This is the multiplicity of measurement outcomes embraced by the many-worlds interpretation.

To avoid sums of distinct measurement outcomes, one needs to modify the basic dynamical equation of the quantum mechanics equation so that it is non-linear. The first proposal along these lines was made by Gian Carlo Ghirardi, Alberto Rimini, and Tullio Weber in 1986; it has become known as the GRW theory. The GRW theory adds an irreducibly probabilistic "collapse" term to the otherwise deterministic Schrödinger dynamics. In particular, for each particle in a system there is a small chance per unit time of the wavefunction undergoing a process in which it is instantly and discontinuously localized in the coordinates of that particle. The localization process multiplies the wave state by a narrow Gaussian (bell curve), so that if the wave was initially spread out in the coordinates of the particle in question, it ends up concentrated around a particular point. The point on which this collapse process is centered is random, with a probability distribution given by the square of the pre-collapse wave amplitude (averaged over the Gaussian collapse curve).

The way this works is as follows. The collapse rate for a single particle is very low—about one collapse per hundred million years. So for individual particles (and systems consisting of small numbers of individual particles), we should expect that they obey the Schrödinger equation. And this is exactly what we observe; there are no known exceptions to the Schrödinger equation at the microscopic level. But macroscopic objects contain on the order of a trillion trillion particles, so we should expect about ten million collapses per second for such an object. Furthermore, in solid objects the positions of those particles are strongly correlated with each other, so a collapse in the coordinates of any particle in the object has the effect of localizing the wavefunction in the coordinates of every particle in the object. This means that if the wavefunction of a macroscopic object is spread over a number of distinct locations, it very quickly collapses to a state in which its wavefunction is highly localized around one location.

In the case of electron interference, then, each electron passes through the apparatus in the form of a spread-out wave. The collapse process is vanishingly unlikely to affect this wave, which is important, as its spread-out nature is essential to the explanation of interference: wave components traveling distinct paths must be able to come together and either reinforce each other or cancel each other out. But when the electron is detected, its position is indicated by something we can directly observe, for example, by the location of a macroscopic pointer. To measure the location of the electron, then, the position of the pointer must become correlated with the position of the electron. Since the wave representing the electron is spread out, the wave representing the pointer will initially be spread out too. But within a fraction of a second, the spontaneous collapse process will localize the pointer (and the electron) to a well-defined position, producing the unique measurement outcome we observe.

The spontaneous collapse approach is related to earlier proposals (for example, by John von Neumann) that the measurement process itself causes the collapse that reduces the multitude of pre-measurement wave branches to the single observed outcome. However, unlike previous proposals, it provides a physical *mechanism* for the collapse process in the form of a deviation from the standard Schrödinger dynamics. This mechanism is crucial; without it, as we have seen, there is no way for the measurement process to generate a unique outcome.

Note that, unlike in Bohm's theory, there are no particles at the fundamental level in the GRW theory. In the electron interference case, particle behavior emerges during measurement; the measured system exhibits only wave-like behavior prior to measurement. Strictly speaking, to say that a system contains n particles is just to say that its wave representation has $3n$ dimensions, and to single out one of those particles is really just to focus attention on the form of the wave in three of those dimensions.

An immediate difficulty that faces the GRW theory is that the localization of the wave induced by collapse is not perfect. The collapse process multiplies the wave by a Gaussian, a function which is strongly peaked around its center but which is non-zero everywhere. No part of the pre-collapse wavefunction is driven to zero by this process; if the wavefunction represents a set of possible measurement results, the wave component corresponding to one result becomes large and the wave component corresponding to the others become small, but they do not disappear. Since one motivation for adopting a spontaneous collapse theory is the perceived failure of the many-worlds interpretation to recover probability claims, it cannot be argued that the small terms are intrinsically improbable. Instead, it looks like the GRW spontaneous collapse process fails to ensure that measurements have unique outcomes.

A second difficulty with the GRW theory is that the wavefunction is not an object in a three-dimensional space, but an object occupying a high-dimensional space with three dimensions for each "particle" in the system concerned. David Albert has argued that this makes the three dimensional world of experience *illusory*.

A third difficulty with the GRW theory is that the collapse process acts instantaneously on spatially separated parts of the system; it instantly multiplies the wavefunction *everywhere* by a Gaussian. Like Bohm's theory, the GRW theory violates Bells' locality assumption, since a measurement performed on one particle can instantaneously affect the state of a distant particle (although in the case of the GRW theory talk of "particles" has to be cashed out in terms of the coordinates of the wavefunction). As discussed in relation to Bohm's theory, this requires an objective conception of simultaneity that is absent from special relativity, and hence it is hard to see how to reconcile the GRW theory with relativity.

One way of responding to these difficulties, advocated by Ghirardi, is to postulate a three-dimensional mass distribution in addition to and determined by the wavefunction, such that our experience is determined directly by the mass distribution rather than the wavefunction. This responds to the second difficulty, since the mass distribution that we directly experience is three-dimensional, and hence our experience of a three-dimensional world is veridical. It may also go some way towards resolving the first difficulty, since the mass density corresponding to non-actual measurement outcomes is likely to be negligible relative to the background mass density surrounding the actual measurement outcome (the mass density of air, for example). Ghirardi's mass density is not intended to address the third difficulty; this requires modifying the collapse process itself, and several proposals for constructing a relativistic collapse process based on the GRW theory have been developed.

An alternative approach to the difficulties facing the GRW theory is to adapt a suggestion made by John Bell that the center of each collapse event should be regarded as a "flash of determinacy" out of which everyday objects and everyday experience are built. Roderich Tumulka has developed this suggestion into a "flashy" spontaneous collapse theory, in which the wavefunction is regarded instrumentally as that which connects the distribution of flashes at one time with the probability distribution of flashes at a later time. On this proposal, the small wave terms corresponding to non-actual measurement outcomes can be understood in a straightforwardly probabilistic way: there is only a small chance that a flash will be associated with such a term, and so only a small chance that the non-actual measurement outcome will be realized. The flashes are located in three-dimensional space, so there is no worry that three-dimensionality is an illusion. And since the flashes, unlike the wavefunction, are located at space-time *points*, it is easier to envision a reconciliation between the flashy theory and special relativity.

Consistent Histories

The consistent histories interpretation generalizes the conventional Copenhagen interpretation and attempts to provide a natural interpretation of quantum cosmology. The theory is based on a consistency criterion that allows the history of a system to be described so that the probabilities for each history obey the additive rules of classical probability. It is claimed to be consistent with the Schrödinger equation.

According to this interpretation, the purpose of a quantum-mechanical theory is to predict the relative probabilities of various alternative histories (for example, of a particle).

Ensemble Interpretation

The ensemble interpretation, also called the statistical interpretation, can be viewed as a minimalist interpretation. That is, it claims to make the fewest assumptions associated with the standard mathematics. It takes the statistical interpretation of Born to the fullest extent. The interpretation states that the wave function does not apply to an individual system – for example, a single particle – but is an abstract statistical quantity that only applies to an ensemble (a vast multitude) of similarly prepared systems or particles. Probably the most notable supporter of such an interpretation was Einstein:

> The attempt to conceive the quantum-theoretical description as the complete description of the individual systems leads to unnatural theoretical interpretations, which become immediately unnecessary if one accepts the interpretation that the description refers to ensembles of systems and not to individual systems.
>
> —*Einstein in Albert Einstein*

It is evident from this double-slit experiment with an ensemble of individual electrons that, since the quantum mechanical wave function (absolutely squared) describes the *completed* interference pattern, it must describe an ensemble. A new version of the ensemble interpretation that relies on a reformulation of probability theory was introduced by Raed Shaiia.

De Broglie–Bohm Theory

The de Broglie–Bohm theory of quantum mechanics (also known as the pilot wave theory) is a theory by Louis de Broglie and extended later by David Bohm to include measurements. Particles, which always have positions, are guided by the wave function. The wave function evolves according to the Schrödinger wave equation, and the wave function never collapses. The theory takes place in a single space-time, is non-local, and is deterministic. The simultaneous determination of a particle's position and velocity is subject to the usual uncertainty principle constraint. The theory is considered to be a hidden variable theory, and by embracing non-locality it satisfies Bell's inequality. The measurement problem is resolved, since the particles have definite positions at all times. Collapse is explained as phenomenological.

Relational Quantum Mechanics

The essential idea behind relational quantum mechanics, following the precedent of special relativity, is that different observers may give different accounts of the same series of events: for example, to one observer at a given point in time, a system may be in a single, "collapsed" eigenstate, while to another observer at the same time, it may be in a superposition of two or more states. Consequently, if quantum mechanics is to be a complete theory, relational quantum mechanics argues that the notion of "state" describes not the observed system itself, but the relationship, or correlation, between the system and its observers. The state vector of conventional quantum mechanics becomes a description of the correlation of some degrees of freedom in the observer, with respect to the observed system. However, it is held by relational quantum mechanics that this applies to all physical objects, whether or not they are conscious or macroscopic. Any "measurement event" is seen simply as an ordinary physical interaction, an establishment of the sort of correlation. Thus the physical content of the theory has to do not with objects themselves, but the relations between them.

An independent relational approach to quantum mechanics was developed in analogy with David Bohm's elucidation of special relativity, in which a detection event is regarded as establishing a relationship between the quantized field and the detector. The inherent ambiguity associated with applying Heisenberg's uncertainty principle is subsequently avoided.

Transactional Interpretation

The transactional interpretation of quantum mechanics (TIQM) by John G. Cramer is an interpretation of quantum mechanics inspired by the Wheeler–Feynman absorber theory. It describes the collapse of the wave function as resulting from a time-symmetric transaction between a possibility wave from the source to the receiver (the wave function) and a possibility wave from the receiver to source (the complex conjugate of the wave function). Since the possibility wave is collapsed by interaction with the receiver, consciousness plays no role in the theory, eliminating Schrödinger's cat paradox. This interpretation of quantum mechanics is unique in that it not only views the wave function as a real entity, but the complex conjugate of the wave function, which appears in the Born rule for calculating the expected value for an observable, as also real.

Stochastic Mechanics

An entirely classical derivation and interpretation of Schrödinger's wave equation by analogy with Brownian motion was suggested by Princeton University professor Edward Nelson in 1966. Similar considerations had previously been published.

Objective Collapse Theories

Objective collapse theories differ from the Copenhagen interpretation in regarding both the wave function and the process of collapse as ontologically objective. In objective theories, collapse occurs randomly ("spontaneous localization"), or when some physical threshold is reached, with observers having no special role. Thus, they are realistic, in deterministic, no-hidden-variables theories. The mechanism of collapse is not specified by standard quantum mechanics, which needs to be extended if this approach is correct, meaning that Objective Collapse is more of a theory than an interpretation. Examples include the Ghirardi-Rimini-Weber theory and the Penrose interpretation. Another example is the deterministic variant of an objective collapse theory.

Consciousness Causes Collapse (Von Neumann–Wigner Interpretation)

He concluded that the entire physical universe could be made subject to the Schrödinger equation (the universal wave function). He also described how measurement could cause a collapse of the wave function. This point of view was prominently expanded on by Eugene Wigner, who argued that human experimenter consciousness (or maybe even dog consciousness) was critical for the collapse, but he later abandoned this interpretation.

Variations of the consciousness causes collapse interpretation include:

- Subjective reduction research

This principle, that consciousness causes the collapse, is the point of intersection between quantum mechanics and the mind/body problem; and researchers are working to detect conscious events correlated with physical events that, according to quantum theory, should involve a wave function collapse; but, thus far, results are inconclusive.

- Participatory anthropic principle (PAP)

John Archibald Wheeler's participatory anthropic principle says that consciousness plays some role in bringing the universe into existence.

Other physicists have elaborated their own variations of the consciousness causes collapse interpretation; including:

- Henry P. Stapp (*Mindful Universe: Quantum Mechanics and the Participating Observer*)
- Bruce Rosenblum and Fred Kuttner (*Quantum Enigma: Physics Encounters Consciousness*)
- Amit Goswami (*The Self-Aware Universe*).

Many Minds

The many-minds interpretation of quantum mechanics extends the many-worlds interpretation by proposing that the distinction between worlds should be made at the level of the mind of an individual observer.

Quantum Logic

Quantum logic can be regarded as a kind of propositional logic suitable for understanding the apparent anomalies regarding quantum measurement, most notably those concerning composition of measurement operations of complementary variables. This research area and its name originated in the 1936 paper by Garrett Birkhoff and John von Neumann, who attempted to reconcile some of the apparent inconsistencies of classical boolean logic with the facts related to measurement and observation in quantum mechanics.

Quantum Information Theories

Quantum informational approaches have attracted growing support. They subdivide into two kinds:

- Information ontologies, such as J. A. Wheeler's "it from bit". These approaches have been described as a revival of immaterialism.
- Interpretations where quantum mechanics is said to describe an observer's knowledge of the world, rather than the world itself. This approach has some similarity with Bohr's thinking. Collapse (also known as reduction) is often interpreted as an observer acquiring information from a measurement, rather than as an objective event. These approaches have been appraised as similar to instrumentalism.

The state is not an objective property of an individual system but is that information, obtained from a knowledge of how a system was prepared, which can be used for making predictions about future measurements. A quantum mechanical state being a summary of the observer's information about an individual physical system changes both by dynamical laws, and whenever the observer acquires new information about the system through the process of measurement. The existence of two laws for the evolution of the state vector becomes problematical only if it is believed that the state vector is an objective property of the system. The "reduction of the wave packet" does take place in the consciousness of the observer, not because of any unique physical process which takes place there, but only because the state is a construct of the observer and not an objective property of the physical system.

Modal Interpretations of Quantum Theory

Modal interpretations of quantum mechanics were first conceived of in 1972 by B. van Fraassen, in his paper "A formal approach to the philosophy of science." However, this term now is used to describe a larger set of models that grew out of this approach.

- The Copenhagen variant.
- Kochen-Dieks-Healey Interpretations.

- Motivating Early Modal Interpretations, based on the work of R. Clifton, M. Dickson and J. Bub.

Time-symmetric Theories

Several theories have been proposed which modify the equations of quantum mechanics to be symmetric with respect to time reversal. This creates retrocausality: events in the future can affect ones in the past, exactly as events in the past can affect ones in the future. In these theories, a single measurement cannot fully determine the state of a system (making them a type of hidden variables theory), but given two measurements performed at different times, it is possible to calculate the exact state of the system at all intermediate times. The collapse of the wave function is therefore not a physical change to the system, just a change in our knowledge of it due to the second measurement. Similarly, they explain entanglement as not being a true physical state but just an illusion created by ignoring retro causality. The point where two particles appear to "become entangled" is simply a point where each particle is being influenced by events that occur to the other particle in the future.

Not all advocates of time-symmetric causality favor modifying the unitary dynamics of standard quantum mechanics. Thus a leading exponent of the two-state vector formalism, Lev Vaidman, highlights how well the two-state vector formalism dovetails with Hugh Everett's many-worlds interpretation.

Branching Space-time Theories

BST theories resemble the many worlds interpretation; however, "the main difference is that the BST interpretation takes the branching of history to be a feature of the topology of the set of events with their causal relationship rather than a consequence of the separate evolution of different components of a state vector." In MWI, it is the wave functions that branches, whereas in BST, the space-time topology itself branches. BST has applications to Bell's theorem, quantum computation and quantum gravity. It also has some resemblance to hidden variable theories and the ensemble interpretation: particles in BST have multiple well defined trajectories at the microscopic level. These can only be treated stochastically at a coarse grained level, in line with the ensemble interpretation.

Interpretations

There are several other interpretations of quantum mechanics available that don't fit neatly into one of the categories discussed above. Here are some prominent ones.

The consistent histories (or decoherent histories) interpretation developed by Robert Griffiths, Murray Gell-Mann and James Hartle, and defended by Roland Omnès, is mathematically something of a hybrid between collapse theories and hidden variable theories. Like spontaneous collapse theories, the consistent histories approach incorporates successive localizations of the wavefunction. But unlike spontaneous collapse theories, these localizations are not regarded as physical events, but just as a means of picking out a particular history of the system in question as actual, much as hidden variables pick out a particular history as actual. If the localizations all constrain the position of a particle, then the history picked out resembles a Bohmian trajectory. But the consistent histories approach also allows localizations to constrain properties other than position, resulting in a more general class of possible histories.

However, not all such sets of histories can be ascribed consistent probabilities: notably, interference effects often prevent the assignment of probabilities obeying the standard axioms to histories. However, for systems that interact strongly with their environment, interference effects are rapidly suppressed; this phenomenon is called decoherence. Decoherent histories can be ascribed consistent probabilities—hence the two alternative names of this approach. It is assumed that only consistent sets of histories can describe the world, but other than this consistency requirement, there is no restriction on the kinds of histories that are allowed. Indeed, Griffiths maintains that there is no unique set of possible histories: there are many ways of constructing sets of possible histories, where one among each set is actual, even if the alternative actualities so produced describe the world in mutually incompatible ways. Absent a many worlds ontology, however, some have worried about how such a plurality of true descriptions of the world could be coherent. Gell-Mann and Hartle respond to such concerns by arguing that organisms evolve to exploit the relative predictability of one among the competing sets of histories.

The transactional interpretation, initially developed by John Cramer, also incorporates elements of both collapse and hidden variable approaches. It starts from the observation that some versions of the dynamical equation of quantum mechanics admit wave-like solutions traveling backward in time as well as forward in time. Typically the former solutions are ignored, but the transactional interpretation retains them. Just as in retrocausal hidden variable theories, the backward-travelling waves can transmit information about the measurements to be performed on a system, and hence allow the transactional interpretation to evade the conclusion of Bell's theorem.

The transactional interpretation posits rules according to which the backward and forward waves generate "transactions" between preparation events and measurement events, and one of these transactions is taken to represent the actual history of the system in question, where probabilities are assigned to transactions via a version of the Born rule. The formation of a transaction is somewhat reminiscent of the spontaneous collapse of the wavefunction, but due to the retrocausal nature of the theory, one might conclude that the wavefunction never exists in a pre-collapse form, since the completed transaction exists as a timeless element in the history of the universe. Hence some have questioned the extent to which the story involving forwards and backwards waves constitutes a genuine explanation of transaction formation, raising questions about the tenability of the transactional interpretation as a description of the quantum world. Ruth Kastner responds to these challenges by developing a possibilist transactional interpretation, embedding the transactional interpretation in a dynamic picture of time in which multiple future possibilities evolve to a single present actuality.

Relational interpretations, such as those developed by David Mermin and by Carlo Rovelli, take quantum mechanics to be about the relations between systems rather than the properties of the individual systems themselves. According to such an interpretation, there is no need to assign properties to individual particles to explain the correlations exhibited by Bell's experiment, and hence one can evade Bells' theorem without violating either locality or independence. Superficially, this approach resembles Everett's, according to which systems have properties only relative to a given branch of the wavefunction. But whereas Everettians typically say that a relation such as an observer seeing a particular measurement result holds on the basis of the properties of the observer and of the measured system within a branch, Mermin denies that there are such relata; rather, the relation itself is fundamental. Hence this is not a many worlds interpretation, since world-relative properties provide the relata that relational interpretations deny. Without such relata, though, it is hard to understand relational quantum mechanics as a description of a *single*world either. However, citing analogies with spatiotemporal properties in relativistic theories, Rovelli insists that it is enough that quantum mechanics ascribe properties to a system relative to the state of a second system (for example, an observer).

Informational interpretations, such as those developed by Jeffrey Bub and by Carlton Caves, Christopher Fuchs and Rüdiger Schack, interpret quantum mechanics as describing constraints on our degrees of belief. They develop rules of quantum credence by analogy with the rules of classical information theory, expressing the difference between quantum systems and classical systems in informational terms, for example in terms of an unavoidable loss of information associated with a quantum measurement. Some proponents of an informational interpretation take an explicitly instrumentalist stance: quantum mechanics is just about the beliefs of observers, treated as external to the quantum systems under consideration. Others take their informational interpretation to be a realist one, in the sense that it can in principle be applied to the whole universe, with "information" serving as a new physical primitive. However, the adequacy of the informational approach as realist can be challenged, for example, on the basis that it does not provide a dynamics for the evolution of the actual state of the world over time. Bub responds that an account of the information-theoretic properties of our measurement results may be the deepest explanation we can hope for.

References

- P. Mohn (2003). Magnetism in the Solid State: An Introduction. Springer Series in Solid-State Sciences Series. 134. Springer. p. 6. ISBN 3-540-43183-7
- Bargmann, V.; Wigner, E. P. (1948). "Group theoretical discussion of relativistic wave equations" (PDF). Proc. Natl. Acad. Sci. U.S.A. 34 (5): 211–23. Bibcode:1948PNAS...34..211B. doi:10.1073/pnas.34.5.211. PMC 1079095. PMID 16578292
- Applications-of-quantum-mechanics, quantum-mechanics-physics, science: britannica.com, Retrieved 13 April 2018
- "Review of Penrose's Shadows of the Mind". Thymos.com. 1999. Archived from the original on 2001-02-09. Retrieved 2011-01-24
- Nunn, C. M. H.; et al. (1994). "Collapse of a Quantum Field may Affect Brain Function. '". Journal of Consciousness Studies'. 1 (1): 127–139
- Quantum-theory: whatis.techtarget.com, Retrieved 30 March 2018
- Y. Peleg; R. Pnini; E. Zaarur; E. Hecht (2010). Quantum Mechanics. Shaum's outlines (2nd ed.). McGraw–Hill. p. 181. ISBN 978-0-07-162358-2

2

Schrödinger Equation

The Schrödinger equation is a mathematical equation used for the systems which have significant quantum effects. It describes the change in the system occurring over time. It can be represented in a time dependent and time independent form. The aim of this chapter is to explore the fundamentals of the time-dependent and time-independent Schrödinger equation and the non-linear Schrödinger equation.

Schrödinger equation is the fundamental equation of the science of submicroscopic phenomena known as quantum mechanics.

The equation, developed (1926) by the Austrian physicist Erwin Schrödinger, has the same central importance to quantum mechanics as Newton's laws of motion have for the large-scale phenomena of classical mechanics.

Essentially a wave equation, the Schrödinger equation describes the form of the probability waves (or wave functions that govern the motion of small particles, and it specifies how these waves are altered by external influences. Schrödinger established the correctness of the equation by applying it to the hydrogen atom, predicting many of its properties with remarkable accuracy. The equation is used extensively in atomic, nuclear, and solid-state physics.

The term "Schrödinger equation" actually refers to two separate equations, often called the time-dependent and time-independent Schrödinger equations. The time-dependent Schrödinger equation is a partial differential equation that describes how the wave function evolves over time, while the time-independent Schrödinger equation is an equation of state for wave functions of definite energy.

Derivation from Wave Mechanics

Consider a "free particle state" corresponding to $V = 0$: a particle traveling in space with no potential holding it down. Since microscopic particles behave as waves in quantum mechanics, a good model for the particle is a "wave packet" formed by a superposition of plane waves at different momenta $\hbar k$, corresponding to the wave function:

$$\Psi(x,t) = \frac{1}{\sqrt{2\pi}} \int_{-\infty}^{\infty} \phi(k) e^{i(kx-wt)} dk.$$

Notably, since the form of $\phi(k)$ is not given, the above statement is just that $\Psi(x, t)$ is the inverse Fourier transform of some function $\phi(k)$.

Using de Broglie's equations of momentum $p = \hbar k = \frac{h}{v}$ and energy $E = \hbar\omega = \frac{p^3}{2m}$, the equation for the wave-packet equation can be rewritten as

$$\Psi(x,t) = \frac{1}{\sqrt{2\pi\hbar}} \int_{-\infty}^{\infty} \phi(p) e^{i(px-Et)/\hbar} dp.$$

To attempt to write down an equation of motion for $\Psi(x,t)$, it is instructive to look at the different derivatives of $\Psi(x,t)$ that are possible. Physicists typically believe the equation of state of an arbitrary system should be given in terms of first or second derivatives at the most, so that the initial configuration and motion of a system are enough to specify the state at all future times.

Taking the time derivative and two derivatives in space yields

$$\frac{\partial\Psi}{\partial t} = \frac{1}{\sqrt{2\pi\hbar}} \int_{-\infty}^{\infty} \phi(p) \left(\frac{-iE}{\hbar}\right) e^{i(px-Et)/\hbar} dp.$$

$$\frac{\partial\Psi}{\partial x} = \frac{1}{\sqrt{2\pi\hbar}} \int_{-\infty}^{\infty} \phi(p) \left(\frac{-ip}{\hbar}\right) e^{i(px-Et)/\hbar} dp$$

$$\frac{\partial^2\Psi}{\partial x^2} = \frac{1}{\sqrt{2\pi\hbar}} \int_{-\infty}^{\infty} \phi(p) \left(\frac{-p^2}{\hbar^2}\right) e^{i(px-Et)/\hbar} dp.$$

Multiplying the time derivative by $i\hbar$ and the second spatial derivative by $\frac{-\hbar^2}{2m}$ and using the de Broglie relations, one has the following interesting equality:

$$i\hbar\frac{\partial\Psi}{\partial t} = \frac{-\hbar^2}{2m}\frac{\partial^2\Psi}{\partial x^2}.$$

The right-hand size has units of energy; it is the kinetic energy of the quantum-mechanical state (a statement which is further justified below). To generalize this free-particle derivation to a particle in an arbitrary potential, one needs only to add $V\Psi$ to the right hand side. This is sufficient to obtain the result, the time-dependent Schrödinger equation:

$$i\hbar\frac{\partial\Psi}{\partial t} = \frac{-\hbar^2}{2m}\frac{\partial^2\Psi}{\partial x^2} + V\Psi.$$

Operator Formulation of the Schrödinger Equation

Quantum mechanics is inherently linear, which means linear algebra is the language of QM. Thus, it is most appropriate to write the Schrödinger equation in operator form. For now, note that it is useful to define the energy operator $\hat{E}$ and momentum operator $\hat{p}$:

$$\hat{E} = i\hbar\partial_t, \qquad \hat{p} = -i\hbar\partial_x.$$

This discussion is motivated by the description of a particle as a superposition of plane wave states. Considering a plane wave $\psi = e^{\frac{i}{\hbar}(px-Et)}$, applying these operators yields.

$$\hat{E}\psi = i\hbar\partial_t e^{\frac{i}{\hbar}(px-Et)} = i\hbar\left(-\frac{i}{\hbar}E\right)\psi = E\psi$$

$$\hat{p}\psi = i\hbar\partial_x e^{\frac{i}{\hbar}(px-Et)} = -i\hbar\left(\frac{i}{\hbar}p\right)\psi = p\psi.$$

Therefore, the eigenvalues of these operators correspond to the energy and momentum of a wave function, respectively.

Using this fact, it is straightforward to formulate the Schrödinger equation directly from operators. Note that the energy of a classical particle is

$$E = \frac{p^2}{2m} = V,$$

where V is the potential energy. Considering the system in quantum mechanics requires "promoting" all variables to operators acting on a wave function. The energy becomes the Hamiltonian operator $\hat{H}$:

$$\hat{H}\psi = \frac{\hat{p}^2}{2m}\psi + V\psi.$$

Equating the Hamiltonian operator $\hat{H}$ with the energy operator $\hat{E}$, one obtains the time-dependent Schrödinger equation again from the operator formulation:

$$i\hbar\partial_t\psi = -\frac{\hbar^2}{2m}\partial_x^2\psi + V\psi.$$

Recalling the form of the energy operator $\hat{E} = i\hbar\partial_t$, the equation can also be rewritten in the extremely compact form:

$$\hat{E}\psi = \hat{H}\psi,$$

where $\hat{E}$ generates time evolution $\hat{H}$ and depends on the kinetic and potential energies of the quantum particle.

Example

Find the energy of the state given at some time by

$$\psi(x) = \left(\frac{m\omega}{\pi\hbar}\right)^{1/4} e^{-\frac{m\omega x^2}{2\hbar}}$$

in the harmonic oscillator potential, $V(x) = \frac{1}{2}m\omega^2 x^2$, given that this state is an eigenfunction of the corresponding Hamiltonian.

Since the state is an eigenfunction of the given Hamiltonian, one only needs to apply the Hamiltonian operator to this state:

$$\hat{H}\psi = -\frac{\hbar^2}{2m}\partial_x^2\psi + \frac{1}{2}m\omega^2 x^2\psi$$

$$= -\frac{\hbar^2}{2m}\partial_x\left(-\frac{m\omega x}{\hbar}\psi\right)+\frac{1}{2}m\omega^2x^2\psi$$

$$= -\frac{\hbar^2}{2m}\left(\left(-\frac{m\omega x}{\hbar}\right)^2\psi-\frac{m\omega}{\hbar}\psi\right)+\frac{1}{2}m\omega^2x^2\psi$$

$$= -\frac{1}{2}m\omega^2x^2\psi+\frac{1}{2}\hbar\omega\psi+\frac{1}{2}m\omega^2x^2\psi$$

$$= \frac{1}{2}\hbar\omega\psi.$$

Reading off the eigenvalue, the energy is $E=\frac{1}{2}\hbar\omega$. This is the ground state of the quantum harmonic oscillator.

Formulation from Unitary Evolution

A last formulation of the time-dependent Schrödinger equation is the method from unitary evolution. Recall that a unitary operator is one that preserves the norm of a state. To this end, suppose that states evolve in time unitarily so that the total probability of a wave function is always one:

$$\psi(x,t)=\hat{U}(t)\psi(x,0),$$

with $\hat{U}^{\dagger}\hat{U}=\hat{1}$ at all times.

Plugging this state into the Schrödinger equation, one can verify that it is a solution at all times:

$$i\hbar\frac{\partial\hat{U}\psi(x,0)}{\partial t}\aleph\hat{H}\hat{U}\psi(x\ddot{u}\ )\quad i\hbar\frac{\partial\hat{U}}{\partial t}\quad\hat{H}\hat{U}$$

where the second equation is purely an equation for the operator $\hat{U}(t)$. Assuming the Hamiltonian is time-independent, the solution to this first-order ODE is

$$\hat{U}=e^{-i\hat{H}t/\hbar}.$$

Therefore, the state

$$\psi(x,t)=e^{-i\hat{H}t/\hbar}\psi(x,0)$$

obeys the Schrödinger equation at all times, and it has been shown that unitary evolution is equivalent to Schrödinger evolution.

Since any state $\psi(x,0)$ can be written in a superposition of energy eigenstates as

$$\psi(x,t)=\sum_n e^{-iE_nt/\hbar}\phi_n(x).$$

A particle in quantum mechanics has the non-normalized wave function at time $t=0$:

$$\psi(x)=2\phi_2+i\phi_5,$$

where the ϕ_n are the orthonormal eigenstates of some Hamiltonian with energies $E_n = n^3$.

Find the normalized wave function at all times.

First, normalize the wave function at time $t = 0$. Since Schrödinger evolution is unitary, the state will stay normalized at all times if it starts normalized. The normalized wave function is:

$$\psi(x) = \frac{2}{\sqrt{5}}\phi_2 + \frac{i}{\sqrt{5}}\phi_5.$$

To check this, compute the square of $\psi(x)$ and use the orthonormality of the ϕ_n :

$$\left|\frac{2}{\sqrt{5}}\right|^2 + \left|\frac{i}{\sqrt{5}}\right|^2 = \frac{4+1}{5} = 1.$$

Now, since the ϕ_n are stationary states, each can be time-evolved individually, and the answer is

$$\psi(x,t) = \frac{2}{\sqrt{5}} e^{-8it/\hbar}\phi_2(x) + \frac{i}{\sqrt{5}} e^{-125it/\hbar}\phi_5(x).$$

Schrödinger Group

The Schrödinger group is a non-relativistic conformal group and it is the symmetry group for the free Schrödinger equation.

The Schrödinger group Sch(d) is defined as the following set of space-time transformations in d space dimensions

$$t \mapsto t' = \frac{\alpha t + \beta}{\gamma t + \delta},\ r \mapsto r' = \frac{Rr + vt + a}{\gamma t + \delta};\ \alpha\delta - \beta\gamma = 1$$

where R is a rotation matrix. Consider the free Schrödinger equation

$$2\,\mathrm{mi}\frac{\partial\phi}{\partial t} - \frac{\partial^2\phi}{\partial r^2} = 2M\frac{\partial\phi}{\partial t} - \frac{\partial^2\phi}{\partial r^2} = 0$$

where the mass m is a constant. In 1972, Niederer showed that the maximal kinematic invariance group of is the Schrödinger group Sch(d), while Lie had already shown in 1882 that the free diffusion equation is invariant under Sch(d). The action of Sch(d) on the space of solutions φ is projective, that is, the wave function φ = φ(t, r) transforms into

$$\phi(t,r) \mapsto (T_g\phi)(t,r) = f_g\left[g^{-1}(t,r)\phi\left[g^{-1}(t,r)\right]\right]$$

(the companion function f_g is explicitly known). Independently, Hagen showed that the freefield action from which can be derived as equation of motion is Schrödinger-invariant. In this paper, we shall mainly consider the Lie algebra schd of the Schrödinger group Sch(d). In particular, one has the following set of generators for sch_1:

$X_{-1} = -\partial_t$, $Y_{-1/2} = -\partial_r$ time and space translations

$Y_{1/2} = -t\partial_r - Mr$ Galilei transformation

$X_0 = -t\partial_t - \frac{1}{2} r\partial_r - \frac{x}{2}$ dilatation

$X_1 = -t^2\partial_t - tr\partial_r - \frac{M}{2} r^2 - 2xt$ special Schrödinger transformation

$M_0 = -M$ phase shift

Here, M = im and x is the scaling dimension of the wave function φ on which the generators of sch1 act. For a solution of the free Schrödinger equation one has x = d/2. The non-vanishing commutators of sch_1 are

$$[X_n, X_{n'}] = (n-n')X_{n+n'} \quad , [X_n, Y_m] = \left(\frac{n}{2} - m\right) Y_{n+m} \quad , [Y_{1/2,} Y_{-1/2,}] = M_0$$

$(n, n' \in \{\pm 1, 0\}, m \in \{\pm 1/2\})$. The Schrödinger group has been introduced as a non-relativistic analogue of the conformal group in d dimensions (whose Lie algebra will be denoted here confd). Indeed, it was argued by Barut that the Schrödinger Lie algebra schd could be obtained from the conformal algebra $conf_{d+1}$ "by a combined process of group contraction and a 'transfer' of the transformation of mass to the co-ordinates". The projective unitary irreducible representations of the Schrödinger group Sch(d) are classified. In particular, Sch(1) is isomorphic to the semi-direct product $Sl(2, \mathbb{R}) \propto H(1)$, where $Sl(2, \mathbb{R})$ comes from the exponentiation of the X-generators, and the 1-dimensional Heisenberg group H(1) from the exponentiation of $Y_{\pm 1/2}$ and M_0.

Schrödinger Algebra

The Schrödinger algebra is the Lie algebra of the Schrödinger group. It is not semi-simple. In one space dimension, it can be obtained as a semi-direct sum of the Lie algebra sl(2,R) and the Heisenberg algebra; similar constructions apply to higher spatial dimensions.

It contains a Galilei algebra with central extension.

$$[J_a, J_b] = i\epsilon_{abc} J_c,$$

$$[J_a, P_b] = i\epsilon_{abc} P_c,$$

$$[J_a, K_b] = i\epsilon_{abc} K_c,$$

$$[P_a, P_b] = 0, [K_a, K_b] = 0, [K_a, P_b] = i\delta_{ab} M,$$

$$[H, J_a] = 0, [H, P_a] = 0, [H, K_a] = iP_a.$$

Where J_a, P_a, K_a, H are generators of rotations (angular momentum operator), spatial translations (momentum operator), Galilean boosts and time translation (Hamiltonian) correspondingly (i is the imaginary unit, $i^2 = -1$. The specific form of the commutators of the generators of rotation

J_a is the one of three-dimensional space, then $\ddot{u}$, , $=1,\ldots,3)$.The central extension M has an interpretation as non-relativistic mass and corresponds to the symmetry of Schrödinger equation under phase transformation (and to the conservation of probability).

There are two more generators which we shall denote by D and C. They have the following commutation relations:

$$[H,C]=iD,[C,D]=-2iC,[H,D]=2iH,$$

$$[P_a,D]=iP_a,[K_i,D]=-iK_a,$$

$$[P_a,C]=-iK_a,[K_a,C]=0,$$

$$[J_a,C]=[J_a,D]=0.$$

The generators H, C and D form the sl(2,R) algebra.

A more systematic notation allows to cast these generators into the four (infinite) families $X_n, Y_m^{(j)}, M_n$ and $R_n^{(jk)}=-R_n^{(kj)}$, where $n\in Z$ is an integer and $m\in Z+1/2$ is a half-integer and $j,k=1,\ldots,d$ label the spatial direction, in d spatial dimensions. The non-vanishing commutators of the Schrödinger algebra become (euclidean form)

$$[X_n,X_{n'}]=(n-n')X_{n+n'}$$

$$[X_n,Y_m^{(j)}]=\left(\frac{n}{2}-m\right)Y_{n+m}^{(j)}$$

$$[X_n,M_{n'}]=-n'M_{n+n'}$$

$$[X_n,R_{n'}^{(jk)}]=-n'R_{n'}^{(jk)}$$

$$[Y_m^{(j)},Y_{m'}^{(k)}]=\delta_{j,k}(m-m')M_{m+m'}$$

$$[R_n^{(ij)},Y_m^{(k)}]=\delta_{i,k}Y_{n+m}^{(j)}-\delta_{j,k}Y_{n+m}^{(i)}$$

$$[R_n^{(ij)},R_{n'}^{(kl)}]=\delta_{i,k}R_{n+n'}^{(jl)}+\delta_{j,l}R_{n+n'}^{(ik)}-\delta_{i,l}R_{n+n'}^{(jk)}-\delta_{j,k}R_{n+n'}^{(il)}$$

The Schrödinger algebra is finite-dimensional and contains the generators $X_{-1,0,1}, Y_{-1/2,1/2}^{(j)}, M_0, R_0^{(jk)}$. In particular, the three generators $X_{-1}=H, X_0=D, X_1=C$ span the sl(2, R) sub-algebra. Space-translations are generated by $Y_{-1/2}^{(j)}$ and the Galilei-transformations by $Y_{1/2}^{(j)}$.

In the chosen notation, one clearly sees that an infinite-dimensional extension exists, which is called the Schrödinger-Virasoro algebra. Then, the generators X_n with n integer span a loop-Virasoro algebra. An explicit representation as time-space transformations is given by, with $n\in Z$ and $m\in Z+1/2$

$$X_n=-t^{n+1}\partial_t-\frac{n+1}{2}t^n\vec{r}\cdot\partial_{\vec{r}}-\frac{n(n+1)}{4}Mt^{n-1}\vec{r}\cdot\vec{r}-\frac{x}{2}(n+1)t^n$$

$$Y_m^{(j)} = -t^{m+1/2}\partial_{r_j} - \left(m + \frac{1}{2}\right)Mt^{m-1/2}r_j$$

$$M_n = -t^n M$$

$$R_n^{(jk)} = -t^n\left(r_j\partial_{r_k} - r_k\partial_{r_j}\right)$$

This shows how the central extension M_0 of the non-semi-simple and finite-dimensional Schrödinger algebra becomes a component of an infinite family in the Schrödinger-Virasoro algebra. In addition, and in analogy with either the Virasoro algebra or the Kac-Moody algebra, further central extensions are possible. However, a non-vanishing result only exists for the commutator $[X_n, X_{n'}]$, where it must be of the familiar Virasoro form, namely

$$[X_n, X_{n'}] = (n-n')X_{n+n'} + \frac{c}{12}(n^3 - n)\delta_{n+n',0}$$

Or for the commutator between the rotations $R_n^{(jk)}$, where it must have a Kac-Moody form. Any other possible central extension can be absorbed into the Lie algebra generators.

The Role of the Schrödinger Group in Mathematical Physics

Though the Schrödinger group is defined as symmetry group of the free particle Schrödinger equation, it is realised in some interacting non-relativistic systems (for example cold atoms at criticality).

The Schrödinger group in d spatial dimensions can be embedded into relativistic conformal group in d+1 dimensions SO(2,d+2). This embedding is connected with the fact that one can get the Schrödinger equation from the massless Klein–Gordon equation through Kaluza–Klein compactification along null-like dimensions and Bargmann lift of Newton–Cartan theory. This embedding can be also be viewed as the extension of the Schrödinger algebra to the maximal parabolic sub-algebra of SO(2,d+2).

The Schrödinger group also arises as dynamical symmetry in condensed-matter applications: it is the dynamical symmetry of the Edwards-Wilkinson model of kinetic interface growth. It also describes the kinetics of phase-ordering, after a temperature quench from the disordered to the ordered phase, in magnetic systems.

Theoretical and Experimental Justification for the Schrödinger Equation

The theoretical and experimental justification for the Schrödinger equation motivates the discovery of the Schrödinger equation, the equation that describes the dynamics of nonrelativistic particles. The motivation uses photons, which are relativistic particles with dynamics determined by Maxwell's equations, as an analogue for all types of particles.

Classical Electromagnetic Waves

Nature of Light

The quantum particle of light is called a photon. Light has both a wave-like and a particle-like nature. In other words, light can appear to be made of photons (particles) in some experiments and light can act like waves in other experiments. The dynamics of classical electromagnetic waves are completely determined by Maxwell's equations, the classical description of electrodynamics. In the absence of sources, Maxwell's equations can be written as wave equations in the electric and magnetic field vectors. Maxwell's equations thus describe, among other things, the wave-like properties of light. When "classical" (coherent or thermal) light is incident on a photographic plate or CCD, the average number of "hits", "dots", or "clicks" per unit time that result is approximately proportional to the square of the electromagnetic fields of the light. By *formal analogy*, the wave function of a material particle can be used to find the probability density by taking its absolute-value squared. Unlike electromagnetic fields, quantum-mechanical wave functions are complex. (Often in the case of EM fields complex notation is used for convenience, but it is understood that in fact the fields are real. On the contrary, wave functions are genuinely complex.)

Maxwell's equations were completely known by the latter part of the nineteenth century. The dynamical equations for light were, therefore, well-known long before the discovery of the photon. This is not true for other particles such as the electron. It was surmised from the interaction of light with atoms that electrons also had both a particle-like and a wave-like nature. Newtonian mechanics, a description of the particle-like behavior of macroscopic objects, failed to describe very small objects such as electrons. Abductive reasoning was performed to obtain the dynamics of massive objects (particles with mass) such as electrons. The electromagnetic wave equation, the equation that described the dynamics of light, was used as a prototype for discovering the Schrödinger equation, the equation that describes the wave-like and particle-like dynamics of nonrelativistic massive particles.

Electromagnetic Wave Equation

The electromagnetic wave equation describes the propagation of electromagnetic waves through a medium or in a vacuum. The homogeneous form of the equation, written in terms of either the electric field E or the magnetic field B, takes the form:

$$\nabla^2 \mathbf{E} - \frac{1}{c^2}\frac{\partial^2 \mathbf{E}}{\partial t^2} = 0$$

$$\nabla^2 \mathbf{B} - \frac{1}{c^2}\frac{\partial^2 \mathbf{B}}{\partial t^2} = 0$$

where c is the speed of light in the medium. In a vacuum, $c = 2.998 \times 10^8$ meters per second, which is the speed of light in free space.

The magnetic field is related to the electric field through Faraday's law (cgs units)

$$\nabla \times \mathbf{E} = -\frac{1}{c}\frac{\partial \mathbf{B}}{\partial t}.$$

Plane Wave Solution of the Electromagnetic Wave Equation

The plane sinusoidal solution for an electromagnetic wave traveling in the z direction is (cgs units and SI units).

$$\mathrm{E}(\mathrm{r},t)=|\mathrm{E}|\mathrm{Re}\{|\zeta\rangle\exp[i(kz-\omega t)]\}\equiv|\mathrm{E}|\mathrm{Re}\{|\phi\rangle\}$$

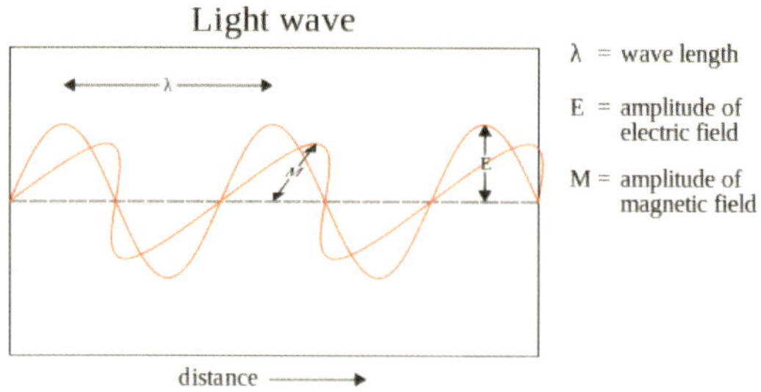

Electromagnetic radiation can be imagined as a self-propagating transverse oscillating wave of electric and magnetic fields. This diagram shows a plane linearly polarised wave propagating from left to right.

for the electric field and

$$\mathrm{B}(\mathrm{r},t)=\hat{\mathrm{z}}\times\mathrm{E}(\mathrm{r},t)$$

for the magnetic field, where k is the wavenumber,

$$\omega=ck$$

is the angular frequency of the wave, and c is the speed of light. The hats on the vectors indicate unit vectors in the x, y, and z directions. In complex notation, the quantity $|\mathrm{E}|$ is the amplitude of the wave.

Here

$$|\zeta\rangle\equiv\begin{pmatrix}\zeta_x\\ \zeta_y\end{pmatrix}=\begin{pmatrix}\cos\theta\exp(i\alpha_x)\\ \sin\theta\exp(i\alpha_y)\end{pmatrix}$$

is the Jones vector in the x-y plane. The notation for this vector is the bra–ket notation of Dirac, which is normally used in a quantum context. The quantum notation is used here in anticipation of the interpretation of the Jones vector as a quantum state vector. The angles θ, α_x, and α_y are the angle the electric field makes with the x axis and the two initial phases of the wave, respectively.

The quantity

$$|\phi\rangle=\exp[i(kz-\omega t)]|\zeta\rangle$$

is the state vector of the wave. It describes the polarization of the wave and the spatial and temporal functionality of the wave. For a coherent state light beam so dim that its average photon number is much less than 1, this is approximately equivalent to the quantum state of a single photon.

Energy Density of Classical Electromagnetic Waves

Energy in a Plane Wave

The energy per unit volume in classical electromagnetic fields is (cgs units)

$$\mathcal{E}_c = \frac{1}{8\pi}\left[\mathrm{E}^2(\mathrm{r}, t) + \mathrm{B}^2(\mathrm{r}, t)\right].$$

For a plane wave, converting to complex notation (and hence dividing by a factor of 2), this becomes:

$$\mathcal{E}_c = \frac{|\mathrm{E}|^2}{8\pi}$$

where the energy has been averaged over a wavelength of the wave.

Fraction of Energy in Each Component

The fraction of energy in the x component of the plane wave (assuming linear polarization) is

$$f_x = \frac{|\mathrm{E}|^2 \cos^2\theta}{|\mathrm{E}|^2} = \phi_x^* \phi_x$$

with a similar expression for the y component.

The fraction in both components is:

$$\phi_x^* \phi_x + \phi_y^* \phi_y = \langle \phi | \phi \rangle = 1.$$

Momentum Density of Classical Electromagnetic Waves

The momentum density is given by the Poynting vector :

$$\mathcal{P} = \frac{1}{4\pi c} \mathrm{E}(\mathrm{r}, t) \times \mathrm{B}(\mathrm{r}, t)$$

For a sinusoidal plane wave traveling in the z direction, the momentum is in the z direction and is related to the energy density:

$$\mathcal{P}c = \mathcal{E}_c.$$

The momentum density has been averaged over a wavelength.

Angular Momentum Density of Classical Electromagnetic Waves

The angular momentum density is:

$$\mathcal{L} = \mathrm{r} \times \mathcal{P} = \frac{1}{4\pi c} \mathrm{r} \times \left[\mathrm{E}(\mathrm{r}, t) \times \mathrm{B}(\mathrm{r}, t)\right].$$

For a sinusoidal plane wave the angular momentum is in the z direction and is given by (going over to complex notation):

$$\mathcal{L} = \frac{|\mathrm{E}|^2}{8\pi\omega}\left(|\langle R|\phi\rangle|^2 - |\langle L|\phi\rangle|^2\right) = \frac{1}{\omega}\mathcal{E}_c\left(|\phi_R|^2 - |\phi_L|^2\right)$$

where again the density is averaged over a wavelength. Here right and left circularly polarized unit vectors are defined as

$$|R\rangle \equiv \frac{1}{\sqrt{2}}\begin{pmatrix}1\\ i\end{pmatrix}$$

and

$$|L\rangle \equiv \frac{1}{\sqrt{2}}\begin{pmatrix}1\\ -i\end{pmatrix}.$$

Unitary Operators and energy Conservation

A wave can be transformed by, for example, passing through a bire fringent crystal or through slits in a diffraction grating. We can define the transformation of the state from the state at time t to the state at time $(t+\tau)$ as

$$|\phi(t+\tau)\rangle = \hat{U}(\tau)|\phi(t)\rangle.$$

To conserve energy in the wave we require

$$\langle\phi(t+\tau)|\phi(t+\tau)\rangle = \langle\phi(t)|\hat{U}^\dagger(\tau)\hat{U}(\tau)|\phi(t)\rangle = \langle\phi(t)|\phi(t)\rangle = 1$$

where $U^\dagger$ is the adjoint of U, the complex conjugate transpose of the matrix.

This implies that a transformation that conserves energy must obey

$$\hat{U}^\dagger\hat{U} = I$$

where I is the identity operator and U is called a unitary operator. The unitary property is necessary to ensure energy conservation in state transformations.

Hermitian Operators and Energy Conservation

If τ is an infinitesimal real quantity dt, then the unitary transformation is very close to the identity matrix (the final state is very close to the initial state) and can be written

$$\hat{U} \approx I - i\hat{H}\tau$$

and the adjoint by

$$\hat{U}^\dagger \approx I + i\hat{H}^\dagger\tau.$$

The factor of i is introduced for convenience. With this convention, it will be shown that energy conservation requires H to be a Hermitian operator and that H is related to the energy of a particle.

Energy conservation requires

$$I = \hat{U}^{\dagger}\hat{U} \approx \left(I + i\hat{H}^{\dagger}\tau\right)\left(I - i\hat{H}\tau\right) \approx I + i\hat{H}^{\dagger}\tau - i\hat{H}\tau + \hat{H}^{\dagger}\hat{H}\tau^{2}.$$

Since τ is infinitesimal, which means that τ^2 may be neglected with respect to τ, the last term can be omitted. Further, if H is equal to its adjoint:

$$\hat{H} = \hat{H}^{\dagger},$$

it follows that (for infinitesimal translations in time $\tau = dt$)

$$\hat{U}^{\dagger}\hat{U} = I,$$

so that, indeed, energy is conserved.

Operators that are equal to their adjoints are called Hermitian or self-adjoint.

The infinitesimal translation of the polarization state is

$$|\phi(t+dt)\rangle - |\phi(t)\rangle = -i\hat{H}dt\,|\phi(t)\rangle.$$

Thus, energy conservation requires that infinitesimal transformations of a polarization state occur through the action of a Hermitian operator. While this derivation is classical, the concept of a Hermitian operator generating energy-conserving infinitesimal transformations forms an important basis for quantum mechanics. The derivation of the Schrödinger equation follows directly from this concept.

Quantum Analogy of Classical Electrodynamics

The treatment to this point has been classical. However, the quantum mechanical treatment of particles follows along lines *formally analogous* however, to Maxwell's equations for electrodynamics. The analog of the classical "state vectors"

$$|\phi\rangle$$

in the classical description is quantum state vectors in the description of photons.

Energy, Momentum, and Angular Momentum of Photons

Energy

The early interpretation is based on the experiments of Max Planck and the interpretation of those experiments by Albert Einstein, which was that electromagnetic radiation is composed of irreducible packets of energy, known as photons. The energy of each packet is related to the angular frequency of the wave by the relation

$$\epsilon = \hbar\omega$$

where $\hbar$ is an experimentally determined quantity known as the reduced Planck's constant. If there are N photons in a box of volume V, the energy (neglecting zero point energy) in the electromagnetic field is

$$N\hbar\omega$$

and the energy density is

$$\frac{N\hbar\omega}{V}$$

The energy of a photon can be related to classical fields through the correspondence principle which states that for a large number of photons, the quantum and classical treatments must agree. Thus, for very large N, the quantum energy density must be the same as the classical energy density

$$\frac{N\hbar\omega}{V} = \mathcal{E}_c = \frac{|\mathrm{E}|^2}{8\pi}.$$

The average number of photons in the box in a coherent state is then

$$N = \frac{V}{8\pi\hbar\omega}|\mathrm{E}|^2.$$

Momentum

The correspondence principle also determines the momentum and angular momentum of the photon. For momentum

$$\mathcal{P}_c = \frac{N\hbar\omega}{cV} = \frac{N\hbar k}{V}$$

which implies that the momentum of a photon is

$$\hbar k \text{ (or equivalently } \frac{h}{\lambda}\text{)}.$$

Angular Momentum and Spin

Similarly for the angular momentum

$$\mathcal{L} = \frac{1}{\omega}\mathcal{E}_c\left(|\psi_R|^2 - |\psi_L|^2\right) = \frac{\hbar}{V}\left(|\psi_R|^2 - |\psi_L|^2\right)$$

which implies that the angular momentum of the photon is

$$l_z = \hbar\left(|\psi_R|^2 - |\psi_L|^2\right).$$

The quantum interpretation of this expression is that the photon has a probability of $|\psi_R|^2$ of having an angular momentum of $\hbar$ and a probability of $|\psi_L|^2$ of having an angular momentum of $-\hbar$. We can therefore think of the angular momentum of the photon being quantized as well as the energy. This has indeed been experimentally verified. Photons have only been observed to have angular momenta of $\pm\hbar$.

Spin Operator

The spin of the photon is defined as the coefficient of $\hbar$ in the angular momentum calculation. A photon has spin 1 if it is in the $|R\rangle$ state and -1 if it is in the $|L\rangle$ state. The spin operator is defined as the outer product

$$\hat{S} \equiv |R\rangle\langle R| - |L\rangle\langle L| = \begin{pmatrix} 0 & -i \\ i & 0 \end{pmatrix}.$$

The eigenvectors of the spin operator are $|R\rangle$ and $|L\rangle$ with eigenvalues 1 and -1, respectively.

The expected value of a spin measurement on a photon is then

$$\langle\psi|\hat{S}|\psi\rangle = |\psi_R|^2 - |\psi_L|^2.$$

An operator S has been associated with an observable quantity, the angular momentum. The eigenvalues of the operator are the allowed observable values. This has been demonstrated for angular momentum, but it is in general true for any observable quantity.

Probability for a single photon

There are two ways in which probability can be applied to the behavior of photons; probability can be used to calculate the probable number of photons in a particular state, or probability can be used to calculate the likelihood of a single photon to be in a particular state. The former interpretation is applicable to thermal or to coherent light. The latter interpretation is the option for a single-photon Fock state. Dirac explains this in the context of the double-slit experiment:

> Sometime before the discovery of quantum mechanics people realized that the connection between light waves and photons must be of a statistical character. What they did not clearly realize, however, was that the "wave function" gives information about the probability of one photon being in a particular place and not the probable number of photons in that place. The importance of the distinction can be made clear in the following way. Suppose we have a beam of light consisting of a large number of photons split up into two components of equal intensity. On the assumption that the beam is connected with the probable number of photons in it, we should have half the total number going into each component. If the two components are now made to interfere, we should require a photon in one component to be able to interfere with one in the other. Sometimes these two photons would have to annihilate one another and other times they would have to produce four photons. This would contradict the conservation of energy. The new theory, which connects the wave function with probabilities for one photon gets over the difficulty by making each photon

go partly into each of the two components. Each photon then interferes only with itself. Interference between two different photons never occurs.

—Paul Dirac

Probability Amplitudes

The probability for a photon to be in a particular polarization state depends on the probability distribution over the fields as calculated by the classical Maxwell's equations (in the Glauber-Sudarshan P-representation of a one-photon Fock state.) The expectation value of the photon number in a coherent state in a limited region of space is quadratic in the fields. In quantum mechanics, by analogy, the state or probability amplitude of a single particle contains the basic probability information. In general, the rules for combining probability amplitudes look very much like the classical rules for composition of probabilities:

1. The probability amplitude for two successive probabilities is the product of amplitudes for the individual possibilities.
2. The amplitude for a process that can take place in one of several indistinguishable ways is the sum of amplitudes for each of the individual ways.
3. The total probability for the process to occur is the absolute value squared of the total amplitude calculated by 1 and 2.

de Broglie Waves

Louis de Broglie. De Broglie received the Nobel Prize in Physics in 1929 for his identification of waves with particles.

In 1923 Louis de Broglie addressed the question of whether all particles can have both a wave and a particle nature similar to the photon. Photons differ from many other particles in that they are massless and travel at the speed of light. Specifically de Broglie asked the question of whether a particle that has both a wave and a particle associated with it is consistent with Einstein's two great 1905 contributions, the special theory of relativity and the quantization of energy and momentum. The answer turned out to be positive. The wave and particle nature of electrons was experimentally observed in 1927, two years after the discovery of the Schrödinger equation.

de Broglie Hypothesis

de Broglie supposed that every particle was associated with both a particle and a wave. The angular frequency ω and wave number k of the wave was related to the energy E and momentum p of the particle by,

$$E = \hbar\omega$$

and

$$p = \hbar k .$$

The question reduces to whether every observer in every inertial reference frame can agree on the phase of the wave. If so, then a wave-like description of particles may be consistent with special relativity.

Rest Frame

First consider the rest frame of the particle. In that case the frequency and wavenumber of the wave are related to the energy and momentum of the particles properties by,

$$E_0 = mc^2 = \hbar\omega_0$$

and

$$p_0 = 0 = \hbar k_0$$

where m is the rest mass of the particle.

This describes a wave of infinite wavelength and infinite phase velocity

$$v_\phi = \frac{\omega_0}{k_0} .$$

The wave may be written as proportional to

$$\cos(\omega_0 t) .$$

This, however, is also the solution for a simple harmonic oscillator, which can be thought of as a clock in the rest frame of the particle. We can imagine a clock ticking at the same frequency as the wave is oscillating. The phases of the wave and the clock can be synchronized.

Frame of the Observer

It is shown that the phase of the wave in an observer frame is the same as the phase of the wave in a particle frame, and also the same as clocks in the two frames. There is, therefore, consistency of both a wave-like and a particle-like picture in special relativity.

Phase of the Observer Clock

In the frame of an observer moving at relative speed v with respect to the particle, the particle clock is observed to tick at a frequency

$$\omega_c = \frac{\omega_0}{\gamma}$$

where

$$\gamma = \frac{1}{\sqrt{1 - \frac{v^2}{c^2}}}$$

is a Lorentz factor that describes time dilation of the particle clock as observed by the observer.

The phase of the observer clock is

$$\omega_c t = \frac{\omega_0}{\gamma}(\gamma t_0) = \omega_0 t_0$$

where t_0 is time measured in the particle frame. Both the observer clock and the particle clock agree on the phase.

Phase of the Observer Wave

The frequency and wave number of the wave in the observer frame is given by

$$E = \gamma m_0 c^2 = \hbar\omega = \gamma\hbar\omega_0$$

and

$$\vec{p} = \gamma m_0 \vec{v} = \hbar\vec{k} = \frac{\gamma\hbar\omega_0\vec{v}}{c^2}$$

with a phase velocity

$$v_\phi = \frac{\omega}{k} = \frac{E}{p} = \frac{c^2}{v}.$$

The phase of the wave in the observer frame is

$$\omega t - kx = \omega t - \frac{\omega}{v_\phi} vt = \omega t\left(1 - \frac{v^2}{c^2}\right) = \frac{\omega t}{\gamma^2} = \frac{1}{\gamma^2}\frac{\gamma m_0 c^2}{\hbar}(\gamma t_0) = \omega_0 t_0 = \omega_c t$$

The phase of the wave in the observer frame is the same as the phase in the particle frame, as the clock in the particle frame, and the clock in the observer frame. A wave-like picture of particles is thus consistent with special relativity.

In fact, we now know that these relations can be succinctly written using special relativistic 4-vector notation:

The relevant four-vectors are:

Four-position $\mathrm{X} = (ct, \vec{x})$

Four-velocity $\mathrm{U} = \gamma(c, \vec{u})$

Four-momentum $\mathrm{P} = \left(\frac{E}{c}, \vec{p}\right)$

Four-wavevector $\mathrm{K} = \left(\frac{\omega}{c}, \vec{k}\right) = \left(\frac{\omega}{c}, \frac{\omega}{v_\phi}\hat{n}\right)$

The relations between the four-vectors are as follows:

$$\mathrm{U} = \frac{d\mathrm{X}}{d\tau}$$

$$\mathrm{P} = \hbar\mathrm{K} = m_o\mathrm{U}$$

$$\mathrm{K} - \left(\frac{m_o}{\hbar}\right)\mathrm{U} - \left(\frac{\omega_o}{c^2}\right)\mathrm{U}$$

The phase of the wave is the relativistic invariant:

$$\mathrm{K}\cdot\mathrm{X} = \omega t - \vec{k}\cdot\vec{x} = \omega_o t_o = \omega_o \tau$$

Bohr atom

Niels Bohr. In 1922 the Nobel Prize in Physics was awarded to
Niels Bohr for his contributions to the understanding of quantum mechanics.

Inconsistency of Observation with Classical Physics

The de Broglie hypothesis helped resolve outstanding issues in atomic physics. Classical physics was unable to explain the observed behavior of electrons in atoms. Specifically, accelerating electrons emit electromagnetic radiation according to the Larmor formula. Electrons orbiting a nucleus should lose energy to radiation and eventually spiral into the nucleus. This is not observed. Atoms are stable on timescales much longer than predicted by the classical Larmor formula.

Also, it was noted that excited atoms emit radiation with discrete frequencies. Einstein used this fact to interpret discrete energy packets of light as, in fact, real particles. If these real particles are emitted from atoms in discrete energy packets, however, must the emitters, the electrons, also change energy in discrete energy packets? There is nothing in Newtonian mechanics that explains this.

The de Broglie hypothesis helped explain these phenomena by noting that the only allowed states for an electron orbiting an atom are those that allow for standing waves associated with each electron.

Balmer Series

The Balmer series identifies those frequencies of light that can be emitted from an excited hydrogen atom:

$$\hbar\omega_n \; \aleph R\left(\frac{1}{2^2} \quad \frac{1}{n^2}\right) \quad n \quad 3, 4, 5, \ldots$$

where R is known as the Rydberg constant and is equal to 13.6 electron volts.

Assumptions of the Bohr model

The Bohr model, introduced in 1913, was an attempt to provide a theoretical basis for the Balmer series. The assumptions of the model are:

1. The orbiting electrons existed in circular orbits that had discrete quantized energies. That is, not every orbit is possible but only certain specific ones.
2. The laws of classical mechanics do not apply when electrons make the jump from one allowed orbit to another.
3. When an electron makes a jump from one orbit to another the energy difference is carried off (or supplied) by a single quantum of light (called a photon) which has an energy equal to the energy difference between the two orbitals.
4. The allowed orbits depend on quantized (discrete) values of orbital angular momentum, L according to the equation $L = n\hbar$

   Where n = 1, 2, 3, ... and is called the principal quantum number.

Implications of the Bohr model

In a circular orbit the centrifugal force balances the attractive force of the electron

$$\frac{mv^2}{r} = \frac{e_M^{\;2}}{r^2}$$

where m is the mass of the electron, v is the speed of the electron, r is the radius of the orbit and

$$e_M = \frac{e}{4\pi\epsilon_0}$$

where e is the charge on the electron or proton.

The energy of the orbiting electron is

$$E = \frac{1}{2}mv^2 - \frac{e_M{}^2}{r} = -\frac{1}{2}\frac{e_M{}^2}{r}$$

which follows from the centrifugal force expression.

The angular momentum assumption of the Bohr model implies

$$L = mvr = n\hbar$$

which implies that, when combined with the centrifugal force equation, the radius of the orbit is given by

$$r = \frac{n^2\hbar^2}{me_M{}^2}.$$

This implies, from the energy equation,

$$E_n = -\frac{1}{2}\frac{e_M{}^2}{r} = -\frac{1}{2}\left(\frac{me_M{}^4}{\hbar^2}\right)\frac{1}{n^2}.$$

The difference between energy levels recovers the Balmer series.

De Broglie's Contribution to the Bohr Model

The Bohr assumptions recover the observed Balmer series. The Bohr assumptions themselves, however, are not based on any more general theory. Why, for instance, should the allowed orbits depend on the angular momentum? The de Broglie hypothesis provides some insight.

If we assume that the electron has a momentum given by

$$p = mv = \hbar k$$

as postulated by the de Broglie hypothesis, then the angular momentum is given by

$$L = mvr = \hbar kr = \hbar\left(\frac{2\pi}{\lambda}\right)r$$

where λ is the wavelength of the electron wave.

If only standing electron waves are permitted in the atom then only orbits with perimeters equal to integral numbers of wavelengths are allowed:

$$\lambda = \frac{2\pi r}{n}.$$

This implies that allowed orbits have angular momentum

$$L = n\hbar$$

which is Bohr's fourth assumption.

Assumptions one and two immediately follow. Assumption three follows from energy conservation, which de Broglie showed was consistent with the wave interpretation of particles.

Need for Dynamical Equations

The problem with the de Broglie hypothesis as applied to the Bohr atom is that we have forced a plane wave solution valid in empty space to a situation in which there is a strong attractive potential. We have not yet discovered the general dynamic equation for the evolution of electron waves. The Schrödinger equation is the immediate generalization of the de Broglie hypothesis and the dynamics of the photon.

Schrödinger Equation

Analogy with Photon Dynamics

The dynamics of a photon are given by

$$|\phi(t+dt)\rangle - |\phi(t)\rangle = -i\hat{H}dt\,|\phi(t)\rangle$$

where H is a Hermitian operator determined by Maxwell's equations. The Hermiticity of the operator ensures that energy is conserved.

Erwin Schrödinger assumed that the dynamics for massive particles were of the same form as the energy-conserving photon dynamics.

$$|\psi(t+dt)\rangle - |\psi(t)\rangle = -i\hat{H}dt\,|\psi(t)\rangle$$

where $|\psi(t)\rangle$ is the state vector for the particle and H is now an unknown Hermitian operator to be determined.

Particle State Vector

Rather than polarization states as in the photon case, Schrödinger assumed the state of the vector depended on the position of the particle. If a particle lives in one spatial dimension, then he divided the line up into an infinite number of small bins of length λ and assigned a component of the state vector to each bin

$$|\psi(t)\rangle \equiv \begin{pmatrix} \vdots \\ \psi_{j-1}(t) \\ \psi_j(t) \\ \psi_{j+1}(t) \\ \vdots \end{pmatrix}.$$

The subscript j identifies the bin.

Matrix Form and Transition Amplitudes

The transition equation can be written in matrix form as

$$\psi_j(t+dt) - \psi_j(t) = -i\sum_k H_{jk}\, dt\, \psi_k(t).$$

The Hermitian condition requires

$$H_{jk} = H^*_{kj}.$$

Schrödinger assumed that probability could only leak into adjacent bins during the small time step dt. In other words, all components of H are zero except for transitions between neighboring bins

$$H_{j\pm1,j} \neq 0,$$

$$H_{j,j} \neq 0.$$

Moreover, it is assumed that space is uniform in that all transitions to the right are equal

$$H_{j+1,j} = H_{j,j-1} \equiv H_R.$$

The same is true for transitions to the left

$$H_{j-1,j} = H_{j,j+1} \equiv H_L.$$

The transition equation becomes

$$i\frac{\partial \psi_j(t)}{\partial t} = H_L\psi_{j+1}(t) - H_R\psi_j(t) + H_R\psi_{j-1}(t) - H_L\psi_j(t) + H_{jj}\psi_j(t).$$

The first term on the right side represents the movement of probability amplitude into bin j from the right. The second term represents leakage of probability from bin j to the right. The third term represents leakage of probability into bin j from the left. The fourth term represents leakage from bin j to the left. The final term represents any change of phase in the probability amplitude in bin j.

If we expand the probability amplitude to second order in the bin size λ and assume space is isotropic, $H_R = H_L \equiv H_0$ the transition equation reduces to

$$i\frac{\partial \psi_j(t)}{\partial t} = H_0\lambda^2 \frac{\partial^2 \psi_j(t)}{\partial x^2} + H_{jj}\psi_j(t).$$

Schrödinger Equation in One Dimension

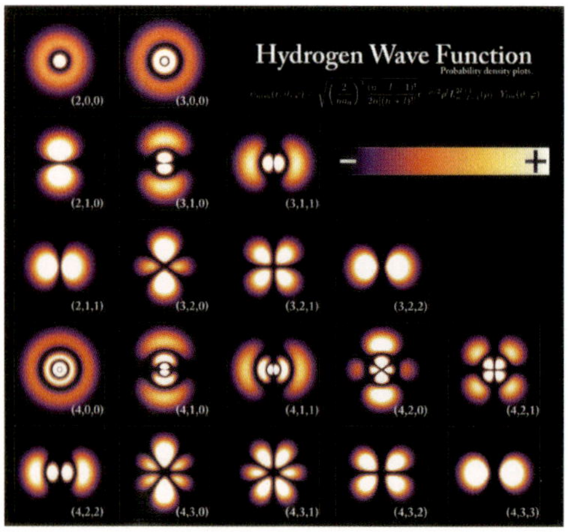

Probability densities for the electron at different quantum numbers in the hydrogen atom.

The transition equation must be consistent with the de Broglie hypothesis. In free space the probability amplitude for the de Broglie wave is proportional to

$$\exp\left[i(kx - \omega t)\right]$$

where

$$E = \hbar\omega = \frac{p^2}{2m} = \frac{\hbar^2 k^2}{2m}$$

in the non-relativistic limit.

The de Broglie solution for free space is a solution of the transition equation if we require

$$H_0\lambda^2 = -\frac{\hbar}{2m}$$

and

$$H_{jj} = 0.$$

The time derivative term in the transition equation can be identified with the energy of the de Broglie wave. The spatial derivative term can be identified with the kinetic energy. This suggests that the term containing H_{jj} is proportional to the potential energy. This yields the Schrödinger equation

$$i\hbar\frac{\partial \psi(x,t)}{\partial t} = -\frac{\hbar^2}{2m}\frac{\partial^2 \psi(x,t)}{\partial x^2} + U(x)\psi(x,t)$$

where U is the classical potential energy and

$$\psi(x,t) \equiv \frac{1}{\sqrt{\lambda}}\psi_j(t)$$

and

$$1 = \int_{-\infty}^{\infty} \psi^*(x,t)\psi(x,t)dx.$$

Schrödinger Equation in Three Dimensions

In three dimensions the Schrödinger equation becomes

$$-\frac{\hbar^2}{2m}\nabla^2\psi + U\psi = i\hbar\frac{\partial}{\partial t}\psi$$

Hydrogen Atom

The solution for the hydrogen atom describes standing waves of energy exactly given by the Balmer series. This was a spectacular validation of the Schrödinger equation and of the wave-like behavior of matter.

Logarithmic Schrödinger Equation

The logarithmic Schrödinger equation (log SE) is one of the nonlinear modifications of Schrödinger's equation, with applications in quantum optics, nuclear physics, diffusion phenomena, stochastic quantum mechanics, effective quantum gravity and Bose–Einstein condensation.

The logarithmic Schrödinger equation

$$i\partial t\phi + \Delta\phi + \phi\log|\phi|^2 = 0,\ \ \phi:[0,\infty]\times\mathbb{R}^n \to \mathbb{C},\ \ n \ge 3,$$

admits applications to quantum mechanics, quantum optics, nuclear physics, transport and diffusion phenomena, open quantum systems, effective quantum gravity, theory of superfluidity and Bose–Einstein condensation.

We are interested in the existence, multiplicity and qualitative properties of the standing waves solution of $i\partial\phi + \Delta\phi + \phi\log|\phi|^2 = 0,\ \phi:[0,\infty]\times\mathbb{R}^n \to \mathbb{C},\ n \ge 3$, i.e. solution in the form $\left(\varphi = e^{i\omega t}u(x)\right)$, where $(\omega \in \mathbb{R})$ and (u) is a real-valued function which has to solve the following semi-linear elliptic problem

$$-\Delta u + \omega u = u\ log\ u^2,\ \ u \in H^1(\mathbb{R}^n).$$

It is well known that the Gausson

$$g(x) = e^{-|x|^2/2}$$

solves $-\Delta u + \omega u = u \log u^2,\ u \in H^1(\mathbb{R}^n)$ for $\omega = -n$. We emphasize that if u is a solution of $-\Delta u + \omega u = u \log u^2,\ u \in H^1(\mathbb{R}^n)$ then $\lambda u, \lambda \neq 0$, is a solution of $-\Delta v + \omega' v = v \log v^2$ with $\omega' = \omega + \log \lambda^2$. This fact allows us to name the solution $\exp\{(\omega + n - |x|^2)/2\}$, Gausson for $-\Delta u + \omega u = u \log u^2,\ u \in H^1(\mathbb{R}^n)$. Moreover, without loss of generality, we can restrict to the case ω > 0, even if our results hold for every $\omega \in \mathbb{R}$. From a variational point of view, the search of nontrivial solutions to $-\Delta u + \omega u = u \log u^2,\ u \in H^1(\mathbb{R}^n)$ can be formally associated with the study of critical points of the functional on $H^1(\mathbb{R}^n)$ defined by

$$J(u) = \frac{1}{2}\int |\nabla u|^2 + \frac{\omega + 1}{2}\int u^2 - \frac{1}{2}\int u^2 \log u^2.$$

Due to the logarithmic Sobolev inequality

$$\int u^2 \log u^2 \leq \frac{a^2}{\pi}\|\nabla u\|_2^2 + \left(\log\|u\|_2^2 - n(1 + \log a)\right)\|u\|_2^2,$$

for $u \in H^1(\mathbb{R}^n)$ *and* $a > 0$,

we have $J(u) > -\infty$ for all $u \in H^1(\mathbb{R}^n)$, but there are elements $u \in H^1(\mathbb{R}^n)$, such that $\int u^2 \log u^2 = -\infty$. Thus, in general, J fails to be finite and C^1 *on* $H^1(\mathbb{R}^n)$. Due to this loss of smoothness, in order to study existence of solutions to $-\Delta u + \omega u = u \log u^2,\ u \in H^1(\mathbb{R}^n)$, to the best of our knowledge, two indirect approaches were followed so far in the literature. On the one hand, the idea is to work on the Banach space

$$W = \left\{u \in H^1(\mathbb{R}^n) \,\middle|\, \int u^2 |\log u^2| < \infty\right\},$$

$$\|u\| W = \|u\| H^1 + \inf\left\{\gamma > 0 \,\middle|\, \int A(\gamma^{-1}|u|) \leq 1\right\},$$

Where $A(s) = -s^2 \log s^2$ on $[0, e^{-3}]$ and $A(s) = 3s^2 + 4e^{-3}s - e^{-6}$ on $[e^{-3}, \infty)$. In fact, it turns out that, in this framework $J : W \rightarrow \mathbb{R}$ is well defined and C^1 smooth. On the other hand, the authors penalize the nonlinearity around the origin and try to obtain a priori estimates to get a nontrivial solution at the limit. However, the drawback of these indirect approaches is that the Palais–Smale condition cannot be obtained, due to a loss of coercivity of the functional J, and, in general, no multiplicity result can be obtained by the Lusternik–Schnirelmann category theory. In fact, it is easy to see that the functional $J:\ H^1_{rad}(\mathbb{R}^n) \rightarrow \mathbb{R} \cup \{+\infty\}$ is lower semicontinuous and that it satisfies the Palais–Smale condition in the sense of weak slope. More precisely, we shall prove the following theorem.

Theorem: Problem $-\Delta u + \omega u = u \log u^2,\ u \in H^1(\mathbb{R}^n)$ has a sequence of solutions $u_k \in H^1_{rad}(\mathbb{R}^n)$ with $J(\ddot{u}_k) \rightarrow +\infty \quad k \rightarrow +\infty$.

To the best of our knowledge this is the first multiplicity result for $-\Delta u + \omega u = u \log u^2,\ u \in H^1(\mathbb{R}^n)$ on the space $H^1(\mathbb{R}^n)$ and it is obtained directly without penalizing the functional and without changing the topology of the space. We also point out that, even without working in the restricted space of radial functions $H^1_{rad}(\mathbb{R}^n)$ since J decreases under polarization of nonnegative functions of $H^1(\mathbb{R}^n)$ we can obtain the existence of a Palais–Smale sequence $\{u_k\} \subset H^1(\mathbb{R}^n)$ with the additional information that $\left\| u_k - |u_k|^* \right\| L^{2^*}(\mathbb{R}^n) \to 0$ as $h \to \infty$, namely $\{u_k\}$ is almost radially symmetric and decreasing. An analogous multiplicity result is expected to hold both for the problem

$$-\Delta_p u + \omega |u|^{p-2} u = |u|^{p-2} u \log u^2, \; u \in W^{1,p}(\mathbb{R}^n),$$

by applying the L^p-logarithmic Sobolev inequality and for the fractional logarithmic Schrödinger equation.

We are able to prove that the nonnegative solutions are strictly positive and that they are smooth. By exploiting the moving plane method (we outline that our nonlinearity is not C^1 in $[0, \infty)$), we show the following theorem.

Theorem: Up to translations, the Gausson for $-\Delta u + \omega u = u \log u^2, \; u \in H^1(\mathbb{R}^n)$ is the unique strictly positive C^2-solution such that $u(x) \to 0$ as $|x| \to \infty$.

Then we get that the first solution u_1 in theorem is the Gausson for $-\Delta u + \omega u = u \log u^2, \; u \in H^1(\mathbb{R}^n)$.

Theorem: The Gausson g is nondegenerate, that is $\mathrm{Ker}(L) = \mathrm{span}\{\partial_{xh} g\}$, where $Lu = -\Delta u + (|x|^2 - n - 2)u$ is the linearized operator for $-\Delta u - nu = u \log u^2 \; at \; g$.

Finally, in theorem, we also obtain a variational characterization of ground state solutions (namely minima of J on the Nehari manifold) of the problem as minima on the L^2-sphere. We believe that the nondegeneracy of g and the connection between the minimization on the Nehari manifold and on the L^2-sphere can be useful in the study of the stability properties of the logarithmic Schrödinger equation $i\partial\phi + \Delta\phi + \phi \log|\phi|^2 = 0, \; \phi : [0, \infty] \times \mathbb{R}^n \to \mathbb{C}, \; n \geq 3,$ possibly in presence of an external driving potential.

Notations

1. $L_c^\infty(\mathbb{R}^n)$ denotes the space of functions in $L^\infty(\mathbb{R}^n)$ with compact support;
2. $H_{rad}^1(\mathbb{R}^n)$ denotes the space of $H^1(\mathbb{R}^n)$ functions that are radially symmetric;
3. C denotes a generic positive constant which can change from line to line.

Schrödinger–Newton Equation

The Schrödinger–Newton equation

$$i\hbar\partial_t \psi(t, r) = \left(-\frac{\hbar^2}{2m}\nabla^2 - Gm^2 \int d^3 r' \frac{|\psi(t, r')|^2}{|r - r'|} \right) \psi(t, r),$$

has been brought into play by Diósi and Penrose to provide a dynamical description of the collapse of the quantum wave-function. It has regained attention in recent times, mainly due to its connection with the question whether gravity should really be quantised, and due to its falsifiability in envisaged experiments.

There has been quite a debate about how the Schrödinger–Newton equation relates to established fundamental principles of physics. Here we wish to clarify the motivations underlying the Schrödinger–Newton equation, in particular, in which sense it follows from semiclassical gravity. We will also discuss its present limitations as a fundamental description of physical phenomena.

First of all one should avoid confusion about what is actually meant by the term semiclassical gravity. Most physicists assume that the gravitational field must be quantised in some way or another, and that semi-classical gravity is only an effective theory, which holds in situations where matter must be treated quantum mechanically, but gravity can be treated classically (although it is fundamentally quantum). In this case, of course, we have a pure quantum theory where everything remains linear. From such a theory we would never expect any nonlinear interaction terms in the Schrödinger equation. Under these assumptions, the Schrödinger–Newton equation can be derived within quantum gravity as a mean-field limit—but only as such—whose validity is restricted to the case of large numbers of particles.

The point taken by the proponents of equation $i\hbar\partial_t\psi(t,r) = \left(-\frac{\hbar^2}{2m}\nabla^2 - Gm^2\int d^3 r' \frac{|\psi(t,r')|^2}{|r-r'|}\right)\psi(t,r)$, however, is different. The Schrödinger–Newton equation does follow from a theory in which only matter fields are quantised, while the gravitational field remains classical even at the fundamental level. This is what we will refer to as semi-classical gravity in the following, thereby adopting the notation. One possible candidate for such a theory is a coupling of gravity to matter by means of the semi-classical Einstein equations

$$R_{\mu\nu} + \frac{1}{2} g_{\mu\nu} R = \frac{8\pi G}{c^4}\left\langle \Psi \middle| \hat{T}_{\mu\nu} \middle| \Psi \right\rangle,$$

that is by replacing the classical energy-momentum tensor in Einstein's equations by the expectation value of the corresponding quantum operator in a given quantum state Ψ. This idea has a long history, dating back to the works of Møller and Rosenfeld. It has been commented repeatedly that such a theory would be incompatible with established principles of physics but these arguments turn out to be inconclusive.

The Schrödinger–Newton Equation from Semi-classical Gravity

As anticipated, here we take the point of view that at the fundamental level quantum theory is coupled to classical General Relativity via the semi-classical Einstein equations $R_{\mu\nu} + \frac{1}{2} g_{\mu\nu} R = \frac{8\pi G}{c^4}\left\langle \Psi \middle| \hat{T}_{\mu\nu} \middle| \Psi \right\rangle$. To treat the full equation in the framework of quantum fields on a curved space-time can be a difficult endeavor. But in the linearized theory of gravity, where the space-time metric is written as

$$g_{\mu\nu} = \eta_{\mu\nu} + h_{\mu\nu},$$

the expansion in $h_{\mu\nu}$ is well-known to yield the gravitational wave equations at leading order. This remains right for the semi-classical equation $R_{\mu\nu} + \frac{1}{2} g_{\mu\nu} R = \frac{8\pi G}{c^4}\left\langle \Psi \middle| \hat{T}_{\mu\nu} \middle| \Psi \right\rangle$ where one obtains

$$\Box h_{\mu\nu} = -\frac{16\pi G}{c^4}\left(\left\langle \Psi \right| \hat{T}_{\mu\nu} \left| \Psi \right\rangle - \frac{1}{2}\eta_{\mu\nu}\left\langle \Psi \right| \eta^{\rho\sigma}\hat{T}_{\rho\sigma} \left| \Psi \right\rangle\right),$$

imposing the de Donder gauge-condition $\partial^{\mu}\left(h_{\mu\nu} - \frac{1}{2}\eta_{\mu\nu}\eta^{\rho\sigma}h_{\rho\sigma}\right) = 0$. Here $\Box$ denotes the d'Alembert operator. Note that the energy-momentum tensor at this order of the linear approximation is that for flat space-time, while it was still in curved space-time. In the Newtonian limit, where $\left\langle \Psi \right| \hat{T}_{00} \left| \Psi \right\rangle$ is large compared to the other nine components of the energy-momentum tensor, equation $\Box h_{\mu\nu} = -\frac{16\pi G}{c^4}\left(\left\langle \Psi \right| \hat{T}_{\mu\nu} \left| \Psi \right\rangle - \frac{1}{2}\eta_{\mu\nu}\left\langle \Psi \right| \eta^{\rho\sigma}\hat{T}_{\rho\sigma} \left| \Psi \right\rangle\right)$ becomes the Poisson equation

$$\nabla^2 V = \frac{4\pi G}{c^2}\left\langle \Psi \right| \hat{T}_{00} \left| \Psi \right\rangle$$

for the potential $V = -\frac{c^2}{2}h_{00}$. This is simply the usual behavior of General Relativity in the Newtonian limit: space-time curvature becomes a Newtonian potential sourced by the energy density term of the energy-momentum tensor.

This potential term now contributes to the Hamiltonian of the matter fields, which in turn yields the dynamics in the Schrödinger equation. To be more specific, in the linearized theory of gravity, the interaction between gravity and matter is given by the Hamiltonian

$$H_{\text{int}} = -\frac{1}{2}\int d^3r\, h_{\mu\nu} T^{\mu\nu}.$$

The quantisation of the matter fields then provides us with the corresponding operator:

$$\hat{H}_{\text{int}} = -\frac{1}{2}\int d^3r\, h_{\mu\nu} T^{\mu\nu}.$$

It is important to point out the difference to a quantised theory of gravity. In the latter, $h_{\mu\nu}$ becomes an operator as well, simply by applying the correspondence principle to the perturbation $h_{\mu\nu}$ of the metric—and thereby treating the classical $h_{\mu\nu}$ like a field living on flat space-time rather than a property of space-time. In contrast to this, $h_{\mu\nu}$ here remains fundamentally classical. It is determined by the wave equations $\Box h_{\mu\nu} = -\frac{16\pi G}{c^4}\left(\left\langle \Psi \right| \hat{T}_{\mu\nu} \left| \Psi \right\rangle - \frac{1}{2}\eta_{\mu\nu}\left\langle \Psi \right| \eta^{\rho\sigma}\hat{T}_{\rho\sigma} \left| \Psi \right\rangle\right)$, which are meant as classical equations of motion.

In the Newtonian limit, where $\hat{T}_{00}$ is the dominant term of the energy-momentum tensor, the interaction Hamiltonian then becomes

$$\hat{H}_{\text{int}} = \int d^3r\, V\, \hat{T}^{00} = -G\int d^3r\, d^3r' \frac{\langle \Psi | \hat{\varrho}(r') | \Psi \rangle}{|\mathrm{r} - \mathrm{r}'|}\hat{\varrho}(\mathrm{r}),$$

where we already integrated equation $\nabla^2 V = \frac{4\pi G}{c^2}\left\langle \Psi \right| \hat{T}_{00} \left| \Psi \right\rangle$ and used $\hat{T}^{00} = c^2\,\hat{\varrho}$ in the non-relativistic limit. The mass density operator $\hat{\varrho}$ is simply $m\hat{\psi}^{\dagger}\hat{\psi}$ when only one kind of particle is present. Therefore, following the standard procedure we end up with the Schrödinger–Newton equation in Fock space:

$$i\hbar\partial_t|\psi\rangle\Big[\int \mathrm{d}^3 r\,\hat{\psi}^\dagger(r)\left(-\frac{\hbar^2}{2m}\nabla^2\right)\hat{\psi}(r)$$
$$-Gm^2\int \mathrm{d}^3r\,\mathrm{d}^3\mathrm{r}'\frac{\langle\Psi|\hat{\psi}^\dagger(\mathrm{r}')\hat{\psi}(\mathrm{r}')|\Psi\rangle}{|\mathrm{r}-\mathrm{r}'|}\hat{\psi}^\dagger(\mathrm{r})\hat{\psi}(\mathrm{r})\Big]|\Psi\rangle.$$

In the non-relativistic limit the number of particles is conserved and we can without further assumptions go over to the first-quantised form. For an N-particle state

$$|\Psi_N\rangle=\frac{1}{\sqrt{N!}}\left\{\int\prod_{i=1}^{N}d^3r_i\right\}\Psi_N(t,r_1,\ldots,r_N)\hat{\psi}^\dagger(r_1)\ldots\hat{\psi}^\dagger(r_N)|0\rangle,$$

Where $|\Psi_N\rangle(t,r_1,\ldots,r_N)$ is the N-particle wave-function, the expectation value is

$$\langle\Psi_N|\hat{\psi}^\dagger(\mathrm{r})\hat{\psi}(\mathrm{r})|\Psi_N\rangle=\sum_{j-1}^{N}\left\{\int\prod_{\substack{i=1\\ i\neq j}}^{N}d^3r_i\right\}\left|\Psi_N\left(t,\mathrm{r}_1,\ldots,\mathrm{r}_{j-1},\,\mathrm{r},\,\mathrm{r}_{j+1}\ldots,a\,\mathrm{r}_N\right)\right|^2.$$

By using, Schrödinger–Newton equation in Fock space with the state $|\Psi_N\rangle=\frac{1}{\sqrt{N!}}\left\{\int\prod_{i=1}^{N}d^3r_i\right\}\Psi_N(t,r_1,\ldots,r_N)\hat{\psi}^\dagger(r_1)\ldots\hat{\psi}^\dagger(r_N)|0\rangle$, inserted therefore yields the N-particle Schrödinger–Newton equation

$$i\hbar\partial_t\Psi_N(t;\mathrm{r}_1,\ldots,\mathrm{r}_N)=\Bigg(-\sum_{i=1}^{N}\frac{\hbar^2}{2m}\nabla_i^2$$
$$-Gm^2\sum_{i=1}^{N}\sum_{j=1}^{N}\left\{\int\prod_{k=1}^{N}d^3r'_k\right\}\frac{\left|\Psi_N(t;\mathrm{r}'_1,\ldots,\mathrm{r}'_N)\right|^2}{|r_i-\mathrm{r}'_j|}\Bigg)\Psi_N(t;\mathrm{r}_1,\ldots,\mathrm{r}_N),$$

and in the one-particle case the Schrödinger–Newton equation $i\hbar\partial_t\psi(t,r)=\left(-\frac{\hbar^2}{2m}\nabla^2-Gm^2\int \mathrm{d}^3\mathrm{r}'\frac{|\psi(t,\mathrm{r}')|^2}{|\mathrm{r}-\mathrm{r}'|}\right)\psi(t,\mathrm{r})$, follows immediately. We therefore unavoidably obtain the Schrödinger–Newton equation for non-relativistic quantum matter if the initial assumptions are correct: that gravity is fundamentally classical, and that the semi-classical Einstein equations $R_{\mu\nu}+\frac{1}{2}g_{\mu\nu}R=\frac{8\pi G}{c^4}\langle\Psi|\hat{T}_{\mu\nu}|\Psi\rangle$, describe its coupling to matter. In this precise sense, the Schrödinger–Newton equation does follow from fundamental principles. Whether or not these principles and the underlying assumptions are correct, is a different story which, eventually, will be decided by experiments.

The Schrödinger–Newton Equation as a Hartree Approximation

It is important to stress that the Schrödinger–Newton equation also appears in a different context, however with a totally different meaning. Assuming that gravity is fundamentally described by a quantum theory in which the metric perturbation $h_{\mu\nu}$ turns into a linear operator, then very similar to what we have in equation $\Box h_{\mu\nu} = -\frac{16\pi G}{c^4}\left(\langle\Psi|\hat{T}_{\mu\nu}|\Psi\rangle - \frac{1}{2}\eta_{\mu\nu}\langle\Psi|\eta^{\rho\sigma}\hat{T}_{\rho\sigma}|\Psi\rangle\right)$ in the weak-field limit one gets the linearised equation:

$$\Box\hat{h}_{\mu\nu} = \frac{16\pi G}{c^4}\left(\hat{T}^{\mu\nu} - \frac{1}{2}\eta_{\mu\nu}\eta^{\rho\sigma}\hat{T}_{\rho\sigma}\right),$$

where $\hat{h}_{\mu\nu}$ now is a linear quantum operator. In complete analogy to equation $\nabla^2 V = \frac{4\pi G}{c^2}\langle\Psi|\hat{T}_{00}|\Psi\rangle$, one obtains the potential $\hat{V}$ in the Newtonian limit where:

$$\nabla^2\hat{V} = \frac{4\pi G}{c^2}\hat{T}_{00}.$$

Note that, contrary to equation $\nabla^2 V = \frac{4\pi G}{c^2}\langle\Psi|\hat{T}_{00}|\Psi\rangle$, $\hat{V}$ also carries a hat now, i. e. it is considered a quantum operator. Then, the corresponding interaction Hamiltonian reads as:

$$\hat{H}_{\text{int}} = \int d^3r\,\hat{V}\,\hat{T}^{00} = -G\int d^3r\,d^3r'\frac{\langle\Psi|(r')|\Psi\rangle}{|\mathrm{r}-\mathrm{r}'|}.$$

Note that the interaction term in equation $\hat{H}_{\text{int}} = \int d^3r\,\hat{V}\,\hat{T}^{00} = -G\int d^3r\,d^3r'\frac{\langle\Psi|(r')|\Psi\rangle}{|\mathrm{r}-\mathrm{r}'|}$ is in the second-quantised formalism, and we can again write down the full Schrödinger equation in Fock space:

$$i\hbar\partial_t|\Psi\rangle = \Big[\int d^3r\,\hat{\psi}^\dagger(r)\left(-\frac{\hbar^2}{2m}\nabla^2\right)\hat{\psi}(r) - Gm^2\int d^3r\,d^3r'\frac{\hat{\psi}^\dagger(\mathrm{r}')\hat{\psi}(\mathrm{r}')\hat{\psi}^\dagger(\mathrm{r})\hat{\psi}(\mathrm{r})}{|\mathrm{r}-\mathrm{r}'|}\Big]|\Psi\rangle.$$

The corresponding N-body Schrödinger equation in first-quantised formalism is given by

$$i\hbar\frac{\partial}{\partial t}\Psi_N(t,\mathrm{r}_1,\ldots,\mathrm{r}_N) = \hat{H}_N\Psi_N(t,\mathrm{r}_1,\ldots,\mathrm{r}_N),$$

Where

$$\hat{H}_N = -\sum_{j=1}^{N}\frac{\hbar^2}{2m_j}\nabla_j^2 - Gm^2\sum_{i\neq j}\frac{1}{|\hat{\mathrm{r}}_i - \hat{\mathrm{r}}_j|}.$$

The last term is the contribution of $\hat{H}_{\text{int}}$ given in equation $\hat{H}_{\text{int}} = \int d^3r\,\hat{V}\,\hat{T}^{00} = -G\int d^3r\,d^3r'\frac{\langle\Psi|(r')|\Psi\rangle}{|\mathrm{r}-\mathrm{r}'|}$.

In the last term on the right-hand side of the above equation, the infinite terms with $i = j$ are treated by standard renormalisation and regularisation techniques at the second-quantised level and they appear as a mass renormalisation.

Then, in the case of many-particle systems for $N \to \infty$, and assuming that all particles have the same mass m, one can obtain the nonlinear Hartree equation as the mean-field limit of equation $i\hbar\frac{\partial}{\partial t}\Psi_N(t, \mathrm{r}_1, \ldots, \mathrm{r}_N) = \hat{H}_N\Psi_N(t, \mathrm{r}_1, \ldots, \mathrm{r}_N),$:

$$i\hbar\frac{\partial}{\partial t}\psi(t,r) = \left(-\frac{\hbar^2}{2m}\nabla^2 - Gm^2\int d^3\mathrm{r}'\frac{|\psi(t,\mathrm{r}')|^2}{|\mathrm{r}-\mathrm{r}'|}\right)\psi(t,\mathrm{r}).$$

Note that the precise mathematical derivation is not a trivial endeavour and involves implementing the quantum Bogoliubov–Born–Green–Kirkwood–Yvon (BBGKY) hierarchy in the limit $N \to \infty$.

However, one should keep in mind that equation $i\hbar\frac{\partial}{\partial t}\psi(t,r) = \left(-\frac{\hbar^2}{\text{üm}}\nabla^{\text{ü}} - Gm\int d\ \mathrm{r}'\frac{|\psi(t,\mathrm{r}')|^2}{|\ -\ '|}\right)\psi(t,\mathrm{r}).$

is derived as the mean-field limit of an N-body linear Schrödinger equation and it is only an effective description to the zero-th order of the dynamics of the N-body system. At the level of the full N-body system the dynamics are still linear, as given in equation $i\hbar\frac{\partial}{\partial t}\Psi_N(t, \mathrm{r}_1, \ldots, \mathrm{r}_N) = \hat{H}_N\Psi_N(t, \mathrm{r}_1, \ldots, \mathrm{r}_N),$. The centre-of-mass wave-function is described by a free Schrödinger equation.

In contrast to this, in semi-classical gravity we have nonlinearity even at the level of the Nbody dynamics, yielding contributions to the center-of-mass motion. Therefore, while in the relative motion the non-gravitational interactions dominate the dynamics, in the center-of-mass dynamics gravitation is the only interaction and can therefore lead to observable effects.

Time-dependent Schrödinger Equation

The time dependent form of the Schrödinger equation depends on the physical situation. The most general form is the time-dependent Schrödinger equation, which gives a description of a system evolving with time.

We are now ready to consider the time-dependent Schrödinger equation. Although we were able to derive the single-particle time-independent Schrödinger equation starting from the classical wave equation and the de Broglie relation, the time-dependent Schrödinger equation cannot be derived using elementary methods and is generally given as a postulate of quantum mechanics. It is possible to show that the time-dependent equation is at least *reasonable* if not derivable, but the arguments are rather involved.

The single-particle three-dimensional time-dependent Schrödinger equation is

$$i\hbar\frac{\partial\psi(\mathrm{r},t)}{\partial t} = -\frac{\hbar^2}{2m}\nabla^2\psi(\mathrm{r},t) + V(\mathrm{r})\psi(\mathrm{r},t)$$

where V is assumed to be a real function and represents the potential energy of the system (a complex function V will act as a source or sink for probability. *Wave Mechanics* is the branch of quantum mechanics with $i\hbar\frac{\partial\psi(\mathrm{r},t)}{\partial t}=-\frac{\hbar^2}{2m}\nabla^2\psi(\mathrm{r},t)+V(\mathrm{r})\psi(\mathrm{r},t)$ as its dynamical law. Note that equation $i\hbar\frac{\partial\psi(\mathrm{r},t)}{\partial t}=-\frac{\hbar^2}{2m}\nabla^2\psi(\mathrm{r},t)+V(\mathrm{r})\psi(\mathrm{r},t)$ does not yet account for spin or relativistic effects.

Of course the time-dependent equation can be used to derive the time-independent equation. If we write the wave function as a product of spatial and temporal terms, $\psi(\mathrm{r},t)=\psi(\mathrm{r})f(t)$, then equation $i\hbar\frac{\partial\psi(\mathrm{r},t)}{\partial t}=-\frac{\hbar^2}{2m}\nabla^2\psi(\mathrm{r},t)+V(\mathrm{r})\psi(\mathrm{r},t)$ becomes

$$\psi(\mathrm{r})i\hbar\frac{df(t)}{dt}=f(t)\left[-\frac{\hbar^2}{2m}\nabla^2+V(\mathrm{r})\right]\psi(\mathrm{r})$$

$$\frac{i\hbar}{f(t)}\frac{df}{dt}=\frac{}{(\mathrm{r})}\left[-\frac{\hbar}{2m}\nabla \;+\; (\mathrm{r})\right]\;(\mathrm{r})$$

Since the left-hand side is a function of $\underline{t}$ only and the right hand side is a function of $\underline{r}$ only, the two sides must equal a constant. If we tentatively designate this constant E (since the right-hand side clearly must have the dimensions of energy), then we extract two ordinary differential equations, namely

$$\frac{1}{f(t)}\frac{df(t)}{dt}=-\frac{iE}{\hbar}$$

and

$$-\frac{\hbar^2}{2m}\nabla^2\psi(\mathrm{r})+V(\mathrm{r})\psi(\mathrm{r})=E\psi(\mathrm{r})$$

The latter equation is once again the time-independent Schrödinger equation. The former equation is easily solved to yield

$$f(t)=e^{-iEt/\hbar}$$

The Hamiltonian in equation $-\frac{\hbar^2}{2m}\nabla^2\psi(\mathrm{r})+V(\mathrm{r})\psi(\mathrm{r})=E\psi(\mathrm{r})$ is a Hermitian operator, and the eigenvalues of a Hermitian operator must be real, so E is real. This means that the solutions $f(t)$ are purely oscillatory, since $f(t)$ never changes in magnitude (recall Euler›s formula $e^{\pm i\theta}=cos\theta\pm i\, sin\theta$). Thus if

$$\psi(\mathrm{r},t)=\psi(\mathrm{r})e^{-iEt/\hbar}$$

then the total wave function $\psi(\mathrm{r},t)$ differs from $\psi(\mathrm{r},t)$ only by a phase factor of constant magnitude. There are some interesting consequences of this. First of all, the quantity $|\psi(\mathrm{r},t)|^2$ is time independent, as we can easily show:

$$|\psi(\mathrm{r},t)|^2 = \psi^*(\mathrm{r},t)\psi(\mathrm{r},t) = e^{iEt/\hbar}\psi^*(\mathrm{r})e^{-iEt/\hbar}\psi(\mathrm{r}) = \psi^*(\mathrm{r})\psi(\mathrm{r})$$

Secondly, the expectation value for any time-independent operator is also time-independent, if $\psi(\mathrm{r},t)$ satisfies equation $\psi(\mathrm{r},t)=\psi(\mathrm{r})e^{-iEt/\hbar}$. By the same reasoning applied above,

$$<A> = \int \psi^*(\mathrm{r},t)\hat{A}\psi(\mathrm{r},t) = \int \psi^*(\mathrm{r})\hat{A}\psi(\mathrm{r})$$

For these reasons, wave functions of the form $\psi(\mathrm{r},t)=\psi(\mathrm{r})e^{-iEt/\hbar}$ are called *stationary states*. The state $\psi(\mathrm{r},t)$ is "stationary," but the particle it describes is not.

Of course equation $\psi(\mathrm{r},t)=\psi(\mathrm{r})e^{-iEt/\hbar}$ represents a particular solution to equation $i\hbar\frac{\partial\psi(\mathrm{r},t)}{\partial t} = -\frac{\hbar^2}{2m}\nabla^2\psi(\mathrm{r},t)+V(\mathrm{r})\psi(\mathrm{r},t)$. The general solution to equation $i\hbar\frac{\partial\psi(\mathrm{r},t)}{\partial t} = -\frac{\hbar^2}{2m}\nabla^2\psi(\mathrm{r},t)+V(\mathrm{r})\psi(\mathrm{r},t)$ will be a linear combination of these particular solutions, i.e.

$$\psi(\mathrm{r},t) = \sum_i c_i e^{-iE_it/\hbar}\psi_i(\mathrm{r})$$

Time-independent Schrödinger Equation

The time-independent Schrödinger equation is the equation describing stationary states. The time-independent Schrödinger equation predicts that wave functions can form standing waves, called stationary states (also called "orbitals", as in atomic orbitals or molecular orbitals). A very important special case of the Schrödinger equation is the situation when the potential energy term does not depend on time. As the name implies, this is the situation when the potential depends only on position. (Could be a constant)).

The Time-Independent Schrödinger Equation

We start with the one-dimensional classical wave equation,

$$\frac{\partial^2 u}{\partial x^2} = \frac{1}{v^2}\frac{\partial^2 u}{\partial t^2}$$

By introducing the separation of variables

$$u(x,t) = \psi(x)f(t)$$

we obtain

$$f(t)\frac{d^2\psi(x)}{dx^2} = \frac{1}{v^2}\psi(x)\frac{d^2 f(t)}{dt^2}$$

If we introduce one of the standard wave equation solutions for $f(t)$ such as $e^{i\omega t}$ (the constant can be taken care of later in the normalization), we obtain

$$\frac{d^2\psi(x)}{dx^2}=\frac{-\omega^2}{v^2}\psi(x)$$

Now we have an ordinary differential equation describing the spatial amplitude of the matter wave as a function of position. The energy of a particle is the sum of kinetic and potential parts

$$E=\frac{p^2}{2m}+V(x)$$

which can be solved for the momentum, p, to obtain

$$p=\left\{2m\left[E-V(x)\right]\right\}^{1/2}$$

Now we can use the de Broglie formula to get an expression for the wavelength

$$\lambda=\frac{h}{p}=\frac{h}{\{2m[E-V(x)]\}^{1/2}}$$

The term ω^2/v^2 in equation $\frac{d^2\psi(x)}{dx^2}=\frac{-\omega^2}{v^2}\psi(x)$ can be rewritten in terms of λ if we recall that $\omega=2\pi\nu$ and $\nu\lambda=v$.

$$\frac{\omega^2}{v^2}=\frac{4\pi^2\nu^2}{v^2}=\frac{4\pi^2}{\lambda^2}=\frac{2m[E-V(x)]}{\hbar^2}$$

When this result is substituted into equation $\frac{d^2\psi(x)}{dx^2}=\frac{-\omega^2}{v^2}\psi(x)$ we obtain the famous *time-independent Schrödinger equation*

$$\frac{d^2\psi(x)}{dx^2}+\frac{2m}{\hbar^2}[E-V(x)]\psi(x)=0$$

which is almost always written in the form

$$-\frac{\hbar^2}{2m}\frac{d^2\psi(x)}{dx^2}+V(x)\psi(x)=E\psi(x)$$

This single-particle one-dimensional equation can easily be extended to the case of three dimensions, where it becomes

$$-\frac{\hbar^2}{2m}\nabla^2\psi(\mathrm{r})+V(\mathrm{r})\psi(\mathrm{r})=E\psi(\mathrm{r}).$$

Non-Linear Schrödinger Equation

The Nonlinear Schrodinger (NLS) equation is a prototypical dispersive nonlinear partial differential equation (PDE) that has been derived in many areas of physics and analyzed mathematically for over 40 years. Historically the essence of NLS equations can be found in the early work of Ginzburg and Landau and Ginzburg in their study of the macroscopic theory of superconductivity, and also of Ginzburg and Pitaevskii who subsequently investigated the theory of superfluidity. Nonetheless, it was not until the works of Chiao and Talanov that the wider physical importance of NLS equation became evident, especially in connection with the phenomenon of self-focusing and the conditions under which an electromagnetic beam can propagate without spreading in nonlinear media. In the general situation, an optical beam in a dielectric broadens due to diffraction. However, in materials whose dielectric constant increases with the field intensity, the critical angle for internal reflection at the beam›s boundary can become greater than the angular divergence due to diffraction and as a consequence the beam does not spread and can, in some situations, continue to focus into extremely high intensity spots.

Starting from the electromagnetic wave equation in the presence of nonlinearities and assuming a linearly polarized wave propagating along the *z*-axis, after a suitable rescaling of the dependent and independent variables one can derive for the propagation of the electromagnetic field the NLS equation in standard non-dimensional form

$$i\partial_z\psi+\Delta_{\perp\psi}+2|\psi|^2\psi = 0$$

where ψ is proportional to the slowly varying complex envelope of the electromagnetic field, z is the propagation variable, and $\Delta_\perp$ denotes the Laplacian with respect to the transverse coordinates. Subscripts x,y,z,t will denote partial differentiation throughout this entry.

Beside the fact that NLS systems have direct applications in many physical problems, the importance of the NLS equation is also due to its universal character. Generically speaking, most weakly nonlinear, dispersive, energy-preserving systems give rise, in an appropriate limit, to the NLS equation. Specifically, the NLS equation provides a "canonical" description for the envelope dynamics of a quasi-monocromatic plane wave propagating in a weakly nonlinear dispersive medium when dissipation can be neglected.

Mathematically, the NLS equation attains broad significance since, in one transverse dimension, it is integrable via the Inverse Scattering Transform (IST, for brevity) -- which is a nonlinear Fourier Transform -- it admits multisoliton solutions, it has an infinite number of conserved quantities, and it possesses many other interesting properties.

There has been a vast amount of literature involving the NLS equation over the years, but recently there has been additional interest, mainly due to the developments in nonlinear optics and soft-condensed matter physics. In the optical context, the experimental developments involving localized pulses in arrays of coupled optical waveguides have drawn attention to discrete NLS models (where the fields are substituted by appropriate finite differences). Related problems involving NLS equations on a lattice background have also generated considerable interest. The vector generalization of the NLS equation has also proved to be particularly valuable from the point of view of nonlinear optics. On the other hand, the experimental realization of Bose-Einstein condensates(BECs) and their mean field modeling by the so-called Gross-Pitaevskii equation which, like optical pulses on a lattice background, is an NLS equation with an external potential, has opened new avenues for the study of NLS-type equations.

The following sections elucidate some of the physical and the mathematical aspects of NLS systems, both continuous and discrete, scalar and vector, in one or more spatial dimensions.

Continuous Models

Scalar (1+1)-Dimensional Systems

The nonlinear propagation of wave packets is governed by NLS-type systems in such diverse fields as fluid, nonlinear optics, magnetic spin waves, plasma physics, etc.

For example, the NLS equation describes self-compression and self-modulation of electromagnetic wave packets in weakly nonlinear media. Hasegawa and Tappert first derived the NLS equation in fiber optics, taking into account both dispersion and nonlinearity.

Dispersion originates from the frequency dependence of the refractive index of the fiber and leads to frequency dependence of the group velocity; this is usually called group velocity dispersion or simply GVD. Due to GVD, different spectral components of an optical pulse propagate at different group velocities and thus arrive at different times. This leads to pulse broadening, resulting in signal distortion.

Fiber nonlinearity is due to the so-called Kerr effect, i.e. the dependence of the refractive index on the intensity of the optical pulse. In the presence of GVD and Kerr nonlinearity, and neglecting polarization-dependent effects, the refractive index is expressed as

$$n(\omega, E) = n_0(\omega) + n_2 |E|^2$$

where ω and E represent the frequency and electric field of the light wave, $n_0^{(\omega)}$ is the frequency dependent linear refractive index and the constant n_2, the Kerr coefficient, has a value of approximately $10^{-22} m^2/W$. Even though fiber nonlinearity is small, the nonlinear effects accumulate over long distances and can have a significant impact due to the high intensity of the light wave over the small fiber cross section. By itself, the Kerr nonlinearity produces as intensity dependent phase shift which results is spectral broadening during propagation.

In the usual transmission process with light waves, the electric field is modulated into a slowly varying amplitude of a carrier wave. Concretely, a modulated electromagnetic light wave is written as

$$E(z,t) = \epsilon(z,t) e^{i(k_0 z - \omega_0 t)} + c.c.$$

where c.c. denotes complex conjugation, z the distance along the fiber, t the time, $k_{0=k_0(\omega_0)}$ the wavenumber, ω_0 the frequency and $\in(z,t)$ is the complex envelope of the electromagnetic field.

A simplified derivation can be conveniently obtained from the nonlinear dispersion relation:

$$k(\omega, E) = \frac{\omega}{c}(n_0(\omega) + n_2 |E|^2)$$

where c denotes the speed of light.

A Taylor series expansion of $k(\omega, E)$ around the carrier frequency $\omega = \omega_0$ yields

$$k - k_0 = k'(\omega_0)(\omega - \omega_0) + \frac{k''(\omega_0)}{2}(\omega - \omega_0)^2 + \frac{\omega_0 n_2}{c} | E |^2$$

where $'$ represents derivative with respect to ω and $k_0 = k(\omega_0)$. Replacing $k - k_0$ and $\omega - \omega_0$ by their Fourier operator equivalents $i\partial/\partial z$ and $i\partial/\partial t$ resp., using $k - k_0 = \frac{\omega}{c} n_0(\omega)$ and letting equation $k - k_0 = k'(\omega_0)(\omega - \omega_0) + \frac{k''(\omega_0)}{2}(\omega - \omega_0)^2 + \frac{\omega_0 n_2}{c} | E |^2$ operate on $\in$ yields

$$i\left(\frac{\partial \epsilon}{\partial z} + k'_0(\omega_0)\frac{\partial \epsilon}{\partial t}\right) - \frac{k''_0(\omega_0)}{2}\frac{\partial^2 \epsilon}{\partial t^2} + \nu | \epsilon |^2 \epsilon = 0$$

Where $\nu = \frac{\omega_0 n_2}{3_c A_{eff}}$, with *Aeff* being the effective cross section area of the fiber (the factor 1/*Aeff* comes from a more detailed derivation that takes into account the transverse dimensions; it is used in order to take into account the variation of field intensity in the cross section of fiber). Note that $k'_0(\omega_0) = 1/v_g$ where v_g represents the group velocity of the wave train. In order to obtain a dimensionless equation, it is standard to introduce a retarded time coordinate $t_{ret} = t - k'_0(\omega_0)z$ and dimensionless variables $t' = t_{ret}/t_*$, $z' = z/z_*$ and $q = \epsilon/\sqrt{P_*}$ where t_*, z_*, P_* are the characteristic time, distance and power respectively. Substitution of this coordinate transformation and choosing the dimensionless variables as $z_* = 1/\nu P_*$, $t_{2*} = z_* |-k''(\omega_0)|$ [and for convenience dropping the $'$] yields the NLS equation

$$i\frac{\partial q}{\partial z} + \frac{\text{sgn}(-k''_0(\omega_0))}{2}\frac{\partial^2 q}{\partial t^2} + | q |^2 q = 0.$$

There are two cases of physical interest depending on the sign of $-k''_0$.

Replacing $z \to 2\text{sgn}(-k''_0(\omega_0))z$ the NLS equation in "standard" form is given by

$$iq_z + q_{tt} \pm 2 | q |^2 q = 0.$$

In these notations, the focusing case is given by the (+) sign in equation $iq_z + q_{tt} \pm 2 | q |^2 q = 0$ and it corresponds to anomalous dispersion. The defocusing case obtains when the dispersion is normal, and it corresponds to the (-) sign in equation $iq_z + q_{tt} \pm 2 | q |^2 q = 0$.

The NLS equation possesses soliton solutions, which are exact solutions decaying to a background state. The focusing (+) NLS equation admits so-called "bright" solitons (namely, solutions that are localized travelling "humps"). A pure one-soliton solution of the focusing NLS has the form

$$q(z,t) = \eta\, sech\, [\eta(t + 2\xi z - t_0)]e^{-i\theta(z,t)}$$

where $\theta(z, t) = \xi t + (\xi^2 - \eta^2)z + \theta_0$. A typical bright soliton is depicted in figure (with $\eta = 1/\sqrt{2}$).

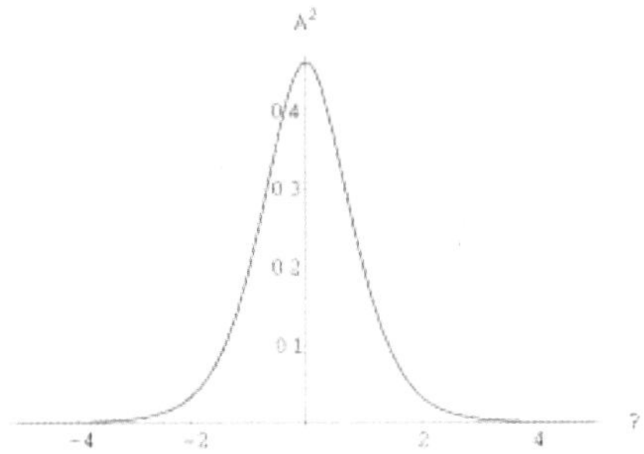

Figure: A bright soliton. A^2 is the square modulus of the solution, ζ is the coordinate in the moving frame.

It is worth noting that in nonlinear optics and many other areas of physics solitary waves are usually called solitons, despite the fact that they generally do not interact elastically. Indeed today most physicists and engineers use the word soliton in this broader sense.

The defocusing (-) NLS equation does not admit solitons that vanish at infinity. However, it does admit soliton solutions on a nontrivial background, called "dark" and "gray" solitons. A dark soliton is a solution of the form

$$q(z,t) = q_0 tanh(q_0 t)e^{2iq_0^2 z}.$$

A gray soliton solution is

$$q(z,t) = q_0 e^{2iq_0^2 z}[\cos\alpha + i\sin\alpha \tanh[\sin\alpha\, q_0(t - 2q_0 \cos\alpha\, z - t_0)]]$$

with $q_{0,}$ α and t_0 arbitrary real parameters. Such solutions satisfy the boundary conditions

$$q(z,t) \to q_\pm(z) = q_0 e^{2iq_0^2 z + i\alpha} \text{ as } x \to \pm\infty$$

and appear as localized dips of intensity $q_0^2 \sin^2\alpha$ on the background field q_0. As $\cos\alpha \to 0^+$, the gray soliton becomes a dark soliton, which is stationary. In figure we illustrate a typical gray soliton (with $\alpha = \pi/3$).

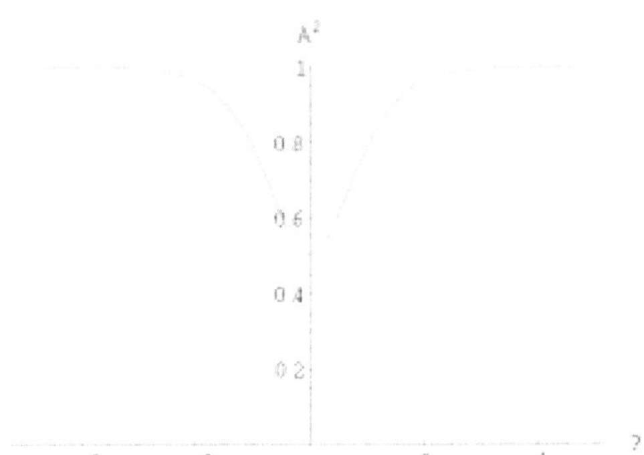

Figure: A gray solito. A^2 is the square modulus of the solution, ζ is the coordinate in the moving frame.

Importantly, the solution of the NLS equation for both decaying initial data and for data which tend to constant amplitude at infinity were obtained by the method of the Inverse Scattering Transform by Zakharov and Shabat.

One of the most remarkable properties of soliton solutions is that interacting scalar solitons affect each other only by a phase shift, that depends only on the soliton powers and velocities, both of which are conserved quantities. Thus, when two soliton collisions occur sequentially, the outcome of the first collision does not affect the second collision, except for a uniform phase shift.

In the context of small-amplitude water waves, the NLS equation was derived by Zakharov for the case of infinite depth and Benney and Roskes for finite depth (the difference between standard NLS equations and Benney-Roskes/Davey-Stewartson type equations). Basically, the NLS equation is obtained from the Euler-Bernoulli equations for the dynamics of an ideal (i.e., incompressible, irrotational and inviscid) fluid under the assumption of a small amplitude quasi-monochromatic wave expansion.

Finally, it should also be mentioned that Ablowitz have shown that, in quadratically nonlinear optical materials, more complicated NLS-type equations can arise.

Vector (1+1)-dimensional Systems

In many applications, vector NLS (VNLS) systems are the key governing equations. Physically, the VNLS arises under conditions similar to those described by NLS whenever there are suitable multiple wave trains moving with nearly the same group velocity. Moreover, VNLS also models systems where the electromagnetic field has more than one component. For example, in optical fibers and waveguides, the propagating electric field has two polarized components transverse to the direction of propagation. The dimensionless system

$$iu_z + \frac{1}{2}u_{tt} + (|u|^2 + |v|^2)u = 0$$

$$iv_z + \frac{1}{2}v_{tt} + (|u|^2 + |v|^2)v = 0$$

was considered by Manakov as an asymptotic model governing the propagation of the electric field in a waveguide, where z is the normalized distance along the waveguide, t is a transverse coordinate and $(u,v)^T$ (the superscript T denotes matrix transpose) are the transverse components of the complex electromagnetic field envelope. Manakov was able to integrate the above VNLS system by the IST method.

Subsequently, Menyuk showed that in optical fibers with constant birefringence, the two polarization components $(u,v)^T$ of the complex electromagnetic field envelope orthogonal to direction of propagation along a fiber satisfy asymptotically the following non-dimensional equations

$$i(u_z + \delta u_t) + \frac{d}{2}u_{tt} + (|u|^2 + \alpha|v|^2)u = 0$$

$$i(v_z - \delta v_t) + \frac{d}{2}v_{tt} + (\alpha|u|^2 + |v|^2)v = 0$$

where δ represents the group velocity "mismatch" between the components u and v, d is the group velocity dispersion and α is a constant depending on the polarization properties of the fiber. The physical phenomenon of birefringence implies that the phase and group velocities of the electromagnetic wave are different for each polarization component. It is important to realize, however, that the derivation of the above equations assumes that certain nonlinear (four-wave mixing) terms are neglected. In the general case, i.e. when $\alpha \neq 1$, the vector NLS system is unlikely to be integrable. However, in a communications environment, due to the distances involved, not only does the birefringence evolve, but it does so randomly and on a scale much faster than the distances required for communication transmission. In this case, Menyuk showed, after averaging over the fast birefringence fluctuations, the relevant equation is the above but with $\alpha = 1$ and $\delta = 0$ that is, it reduces to the integrable VNLS derived by Manakov, which therefore attains broader relevance.

As indicated above, the Manakov system is integrable, and it possesses vector soliton solutions. In the focusing case that is, with a plus sign in front of the cubic nonlinear terms these are bright solitons whose shape is the same as that of the bright solitons of the scalar NLS equation, multiplied by a constant polarization vector. Unlike scalar solitons, however, the collision of solitons with internal degrees of freedom (e.g. vector or matrix solitons) can be highly nontrivial: even though the collision is elastic, in the sense that the total energy of each soliton is conserved, there can be a significant redistribution of energy among the components. It has been shown by Soljacic that the parameters controlling the energy switching between components exhibit nontrivial transformation of information. This set forth the experimental foundations of computation with solitons. Despite the vector nature of the problem, one can show that the multisoliton interaction process is nevertheless pair-wise and the net result of the interaction is independent of the order in which such collisions occur. This interaction property can be related to the fact that the map determining the interaction of two solitons satisfies the Yang-Baxter relation.

The defocusing VNLS equation -- namely, Equations with a minus sign in front of the nonlinear terms -- admits “dark-dark soliton” solutions; i.e., solitons which have dark solitonic behavior in both components, as well as “dark-bright” soliton solutions, which contain one dark and one bright component. Although the mathematical properties of VNLS have been investigated for decades, the IST for the vector system under non-vanishing boundary conditions has been developed only recently.

Scalar Multidimensional Systems

The NLS equation in 2 spatial dimensions, i.e.

$$i\psi_t + \Delta\psi + |\psi|^2\psi = 0, \quad \mathrm{x} = (x, y) \in \mathbb{R}^2$$

has been investigated shortly after the early studies on the one-dimensional equation. Note that in optics the transverse Laplacian, here simply indicated by Δ, describes wave diffraction. Remarkable early direct numerical simulations and scaling arguments by Kelley indicated wave collapse could occur. Vlasov *et al* showed that for a purely cubic nonlinearity in a self-focusing nonlinear medium, the phenomenon of wave collapse takes place and the light beam blows up in a finite time. The proof that a finite-time singularity can occur in equation $i\psi_t + \Delta\psi + |\psi|^2\psi = 0, \quad \mathrm{x} = (x, y) \in \mathbb{R}^2$ is remarkably straight forward and it is based on the virial theorem. One can also prove rigorously that, for initial conditions for which the Hamiltonian $H = \int (|\nabla\psi|^2 - (1/2)|\psi|^4)d\mathrm{x}$ is negative, there exists a time t_* such that the quantity.

$$\int |\nabla \psi|^2 \, dx$$

becomes infinite as t approaches t_*, which in turn implies that ψ also becomes infinite as $t \to t_*$ (blowup in finite time). It is worth mentioning that near blowup the solution displays universal scaling properties.

Results are also available for the more general NLS equation in d spatial dimensions and with generic power nonlinearity:

$$i\psi_t + \Delta_d \psi + |\psi|^{2\sigma} \psi = 0, \quad x \in \mathbb{R}^d,$$

where Δ_d is the d-dimensional Laplacian. More precisely, one has the following cases:

- Supercritical ($\sigma d > 2$): the solution blows up.
- Critical ($\sigma d = 2$): blowup can occur or global solution can exist.
- Subcritical ($\sigma d < 2$): global solutions exist.

The first proof of global existence of solutions to the focusing NLS equation in the sub-critical dimension was given by Ginibre and Velo. There are many references to this interesting subject.

Finally, we mention the Zakharov system for Langmuir turbulence in plasmas, where the mean field obeys a dynamical equation.

Benney-Roskes and Davey-Stewartson Models

Yet another example of the wide applicability of NLS-type systems is given by nonlocal nonlinear Schrodinger systems, where the NLS-type field is coupled to an underlying mean term. The underlying governing systems of equations typically have quadratic nonlinearities; this happens for example in water waves and $\chi^{(2)}$ nonlinear-optical media. Such systems -- which we refer to here as *NLS with mean terms*, or NLSM for brevity -- are also referred to as Benney-Roskes or Davey-Stewartson type. The key difference in the derivation of such systems from that of the usual NLS equation is the presence of an additional mean term that corresponds to the zeroth harmonic of the field at the fundamental frequency. This leads to a system of equations that describes the nonlocal-nonlinear coupling between a dynamic field that is associated with the first harmonic (with a "cascaded" effect from the second harmonic), and a static field that is associated with the mean term (i.e., the zero-th harmonic). In water waves, with finite depth, the NLSM equation was first derived by Benney and Roskes in 1969. It was re-derived a few years later by Davey and Stewartson, who put it in a convenient form:

$$iu_z + \frac{1}{2}(\sigma_1 u_{xx} + u_{yy}) + \sigma_2 |u|^2 u - \rho u \phi_x = 0$$

$$\phi_{xx} + \nu \phi_{yy} = (|u|^2)_x$$

where σ_1, σ_2, ν and ρ are real; u is proportional to the amplitude of the first harmonic and υ the mean (zeroth harmonic) in the slowly varying envelope expansion. The system reduces to the classical

NLS equation when ρ=0, because in that case the mean field ϕ does not couple to the harmonic field u. In addition, when ν=0 the first equation reduces to a classical NLS with the cubic term $(\sigma_2-\rho)|u|^2u$. In optics $\rho>0$, whereas in water waves $\rho<0$. In either case, the NLSM system is a nonlocal system of equations; analysis and numerical simulations indicate that when σ_1, σ_2, $\nu>0$ collapse can occur in the NLSM equations in the same spirit as in NLS collapse. Near the collapse region the solution asymptotically attains the form of a special localized wave -which satisfies a stationary system of equations, obtained by setting $u = e^{i\lambda z}F(x, y)$,$\phi=G(x, y)$, λ constant.

Interestingly, when $\sigma_{1,}\sigma_2$ are ±1, the above (2+1)-dimensional equations are integrable by IST. This limit corresponds to the shallow water wave limit of the Benney-Roskes/Davey-Stewartson system.

Vector Multidimensional Systems

Probably the best known multidimensional NLS system is the extension of the 1+1 vector NLS dimensional equation described above in optics, taking into account transverse variations. In the multidimensional case it takes the form

$$i(u_z+\delta u_t)+\frac{d}{2}u_{tt}+\frac{1}{2}\Delta u+(|u|^2+\alpha|v|^2)u=0$$
$$i(v_z+\delta v_t)+\frac{d}{2}v_{tt}+\frac{1}{2}\Delta u+(\alpha|u|^2+|v|^2)v=0$$

where δ represents the group velocity "mismatch" between the transverse components $(u,v)^T$ of the electromagnetic field envelope, d is the dimensionless dispersion coefficient, α is a constant depending on the polarization properties of the medium and Δ is the two-dimensional Laplacian in the directions transverse to the direction of propagation z.

Ablowitz, Biondini and Blair also derived vector extensions of the NLSM equations in electromagnetics from Maxwell's equations in $\chi^{(2)}$ nonlinear-optical media. These equations reduce to the above system when the $\chi^{(2)}$ contributions vanish. The vector NLSM system takes the following form

$$i(u_{j,z}+\delta_j u_{j,t})+\frac{1}{2}(\Delta u_j+d_{1,}ju_{j,tt})+(M_{1,j}\left|u_j\right|^2+M_{2,\bar{j}}\left|u_{\bar{j}}\right|^2+M_{3,j\phi j}+M_{4,\bar{j}}\phi_{\bar{j}})u_j=0$$

$$\phi_{j,xx}+s_{1,j}\phi_{j,yy}+s_{2,j}\phi_{j,tt}+s_{3,\bar{j}}\phi_{\bar{j},xy}+(N_{1,j}\partial_t^2+N_{2,j}\partial_x^2+N_{3,j}\partial_y^2+N_{4,j}\partial_{xy}^2)(\left|u_j\right|^2-\left|u_{\bar{j}}\right|^2)=0$$

here j=1, 2; $\bar{j}$ = 2, 1, δ_j represents the group velocity "mismatch", d_1,j, $s_{k,j}$ are coefficients related to linear dispersion relations and $M_{k,j}$, $N_{k,j}$ depend on the nonlinear coefficients

$\chi^{(2)}$,$\chi^{(3)}$.

Discrete Models

Discrete (1+1)-Dimensional Systems

Both the NLS and the VNLS equations admit integrable and non-integrable discretizations which, besides being used as numerical schemes for the continuous counterparts, also have physical applications as discrete systems.

An important discretization of NLS is the following

$$i\frac{d}{dt}q_n = \frac{1}{h^2}(q_{n+1} - 2q_n + q_{n-1}) \pm |q_n|^2 (q_{n+1} + q_{n-1})$$

which is referred to here as the integrable discrete NLS (IDNLS) or also sometimes in the literature as the Ablowitz-Ladik equation. It is an $O(h^{2})$ finite-difference approximation of NLS which is integrable via the IST and has soliton solutions on the infinite lattice.

If the nonlinear term in the IDNLS is changed to $2|q_n|^2 q_n$, one obtains an equation which is often called the Discrete Nonlinear Schrodinger (DNLS) equation. Since there are regimes where DNLS exhibits chaotic dynamics it is likely to be not integrable.

The DNLS equation describes a simple model for a lattice of coupled anharmonic oscillators. In one spatial dimension, the equation in its simplest form is

$$i\frac{d}{dz}\psi_n + \psi_{n+1} + \psi_{n-1} + \gamma |\psi_n|^2 \psi_n = 0$$

where ψ_n is the complex mode amplitude of the oscillator at site n and γ is an anharmonic parameter, and it is a standard finite difference approximation to the NLS equation. The prototypical application of DNLS is given by coupled nonlinear optical waveguides "etched" into a suitable optical material and well separated from each other in, say, the x-direction (or n-direction), with propagation occurring in the longitudinal, direction z. Starting from Maxwell›s equations, one can model the governing wave equation for the electromagnetic field in the x-direction in Kerr nonlinear materials in terms of the following nonlinear Helmholtz equation

$$\Psi_{zz} + \Psi_{xx} + (f(x) + \delta |\Psi|^2)\Psi = 0$$

where $|\delta| \ll 1$ and $f(x)$ models the linear index of refraction. Then one expands the solution Ψ in terms of a suitable series of functions

$$\Psi = \sum_{m=-\infty}^{\infty} E_m(\delta z)\psi(x - md)e^{-i\lambda o z}$$

where d is the spacing of the waveguide array, and $\psi_m = \psi(x-md)$ has one bound state $\lambda_{0.}$ Assuming the eigenfunctions to be localized corresponding to waveguides that are well separated and imposing the condition of maximal balance, the resulting equation for En turns out to be the DNLS equation.

The DNLS equation was first derived in the context of nonlinear optics by Christodoulides and Joseph. The equation had previously been studied by Davydov in molecular biology and Su in condensed matter physics. The paper by Eilbeck was the first to point out the universal nature of the discrete NLS equation, and reported a number of applications, especially for periodic solutions with small periods. Experimentally, discrete solitons were observed in a nonlinear optical array by Eisenberg. The experimental results involving waveguide arrays are remarkable in how clearly solitons are formed. Figure is a schematic of the waveguide array and indicates that the input is a laser beam strongly focused towards the center ridge. Figure shows the power measured for three experiments at the end of the waveguide array. The experiments correspond to low input power (70W), medium input power (320W) and large input power (500W). One can see that at low power the beam diffracts whereas focusing occurs at high input power. The bottom most portion of the figure indicates strong self focusing and a discrete soliton is formed.

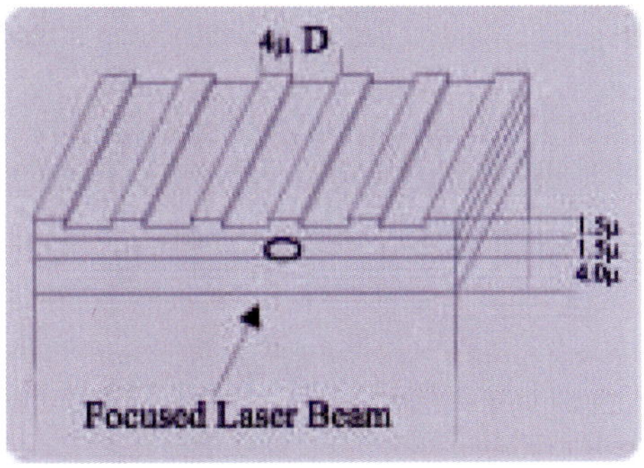

Figure: Optical waveguide structure.

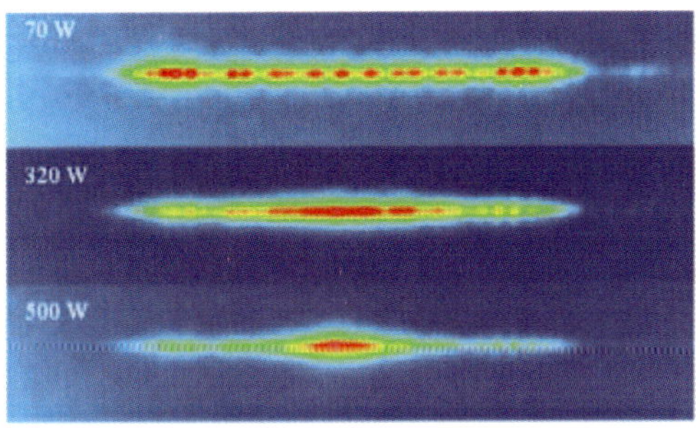

Figure: Output power at the end of the waveguide array corresponding to different input power values.

Considerable interest involving the DNLS equation has appeared recently, following the experimental progress in the fields of nonlinear optical waveguide arrays and Bose-Einstein condensates trapped in periodic potentials arising from optical standing waves.

While DNLS equations are not transformable to the integrable discrete systems presented above, the latter nevertheless provide useful insight into discrete equations and solitary wave phenomena. For example, the use of soliton perturbation theory to elucidate the role of the on-site nonlinearity as a non-integrable perturbation to the IDNLS equation.

One can also consider discretizations of vector NLS equations, both integrable and nonintegrable. The integrable system is given by the following system

$$i\frac{dq_n}{dt} = \frac{1}{h^2}[q_{n+1} - 2q_n + q_{n-1}] \pm || q_n |^2 (q_{n+1} + q_{n-1})$$

where q_n is an N-component vector. Equation $i\frac{dq_n}{dt} = \frac{1}{h^2}[q_{n+1} - 2q_n + q_{n-1}] \pm || q_n |^2 (q_{n+1} + q_{n-1})$ for $q_n = q(nh)$ in the limit $h \to 0$, $nh = x$ gives VNLS. The discrete VNLS system is also integrable. It is also worth mentioning that Eilbeck carried out an extensive study (using path-following) of a coupled DNLS system (i.e., the vector generalization of equation $i\frac{d}{dz}\psi_n + \psi_{n+1} + \psi_{n-1} + \gamma |\psi_n|^2 \psi_n = 0$ arising in crystalline acetanilide.

Scalar Discrete (2+1)-Dimensional Systems

In the 2+1-dimensional case and for a square lattice, the DNLS equation is readily generalized to

$$i\frac{d\psi_{n,m}}{dt}+\left|\psi_{n,m}\right|^{2}\psi_{n,m}+\psi_{n+1,m}+\psi_{n-1,m}+\psi_{n,m+1}+\psi_{n,m-1}-4\psi_{n,m}=0 \quad n,m\in\mathbb{Z}$$

and stationary solutions with frequency ω can be found. Surveys of recent results on the subject can be found in Kevrekidis and Eilbeck and Johansson. Single-site peaked discrete soliton solutions of $i\frac{d\psi_{n,m}}{dt}+\left|\psi_{n,m}\right|^{2}\psi_{n,m}+\psi_{n+1,m}+\psi_{n-1,m}+\psi_{n,m+1}+\psi_{n,m-1}-4\psi_{n,m}=0 \quad n,m\in\mathbb{Z}$ were first studied by Mezentsev. The solution can be smoothly continued from a single-site solution at the anti-continuum (large-amplitude) limit $\omega\to\infty$ to the so called ground state solution of the continuous 2D NLS equation in the small-amplitude limit $\omega\to 0$. There is an instability-threshold at $\omega\sim 1$, so that the solution is stable for larger ω (discrete branch) and unstable for smaller ω (continuous branch). The stability change is characterized by a change of slope in the dependence of $N(\omega)$, where $N(\omega)$ is the L^2 norm of the mode with frequency ω. The value of the excitation number N at the minimum is nonzero, and thus there is an excitation threshold for its creation, in contrast to the 1D case. The effect of this excitation threshold in 2D was proposed by Kalosakas to be experimentally observable in terms of a delocalizing transition of Bose-Einstein condensates in optical lattices. The dynamics resulting from the instability on the unstable branch is, in the initial stage, similar to the collapse of the unstable ground state solution of the continuous 2D NLS equation, with increased localization and blow-up of the central peak. In contrast to the continuum case, however, the peak amplitude must remain finite, and the result is a highly localized 'pulson' state where the peak intensity oscillates between the central site and its four nearest neighbors. This process is referred to as 'quasicollapse' by Laedke.

Finally, it is worth mentioning that two-dimensional lattices admit a new type of localized solutions, 'vortex-breathers', with no counterpart in 1D. For the DNLS model, vortex-breathers for a square 2D lattice were first obtained by Johansson 1998. Typically they become unstable as the coupling is increased.

Dispersion Management

As mentioned above, the NLS equation is an asymptotic approximation (via a quasi-monchromatic wave expansion) of Maxwell's equations with cubic nonlinear polarization terms. Recall that, for the NLS equation in the normalized form given by equation $iq_z+q_{tt}\pm 2|q|^2 q=0$, q is related to the slowly varying complex envelope of the electromagnetic field in Maxwell›s equations. Based on this relationship, it was predicted by Hasegawa and Tappert in 1973 that solitons could propagate in optical fibers.

Soliton propagation in fibers was demonstrated experimentally in 1980 in a seminal paper by Mollenauer, Stolen and Gordon, and work on optical solitons continued in the following years. Then, in a major development at the end of the 1980's, Erbium-doped fiber amplifiers (EDFAs) were developed, and started to be used in communication systems to counteract fiber loss. The use of all-optical amplification eliminated the need for the electronic regeneration of optical signals at various intervals throughout the system, At the same time, however, it resulted in an optically transparent transmission line, which meant that perturbations could now grow unimpeded from the beginning to the end of the line. Moreover, signal amplification by stimulated emission in always accompanied by spontaneous emission which manifests as noise, and which severely limited the transmission distances through which signals could propagate successfully. The use of frequency filters was proposed to reduce this difficulty. In 1990's, system developers also started envisioning multi-channel communication systems, in which signals in numerous carrier frequencies are simultaneusly launched in the same fiber, usually in evenly spaced frequency "windows", a technique called wavelength-division multiplexing (WDM). The interactions between localized pulses in different channels, however, resulted serious penalties associated with WDM transmission. The technology of dispersion management was then developed in order to deal effectively with these difficulties. A dispersion-managed (DM) system consists in a periodic concatenation of fibers with different dispersion characteristics, and turns the large variations in the fiber dispersion to an advantage. For example, in a two-step DM system, between every two optical amplifiers (whose spacing is usually between 40 and 100km) one has a fiber with one dispersion sign that is connected to another fiber with the opposite dispersion sign. Dispersion management is now widely used in modern communication systems, and it significantly reduces the various penalties that arise in multi-channel communications systems, such as four-wave mixing and frequency and timing shifts. Moreover, dispersion management also alleviates noise-induced penalties such as the Gordon-Haus and Gordon-Mollenauer effects. As a result, most long-distance optical fiber submarine systems today are dispersion-managed. A schematic diagram of a DM system is given in figure. The figure shows the multiplexed input, a short pre-compensation fiber, amplifiers (EDFA's: erbium doped fiber amplifiers), two fibers with different dispersion characteristics between each successive EDFA, and a short pre-compensation fiber before de-multiplexing. The pre- and post-compensation fiber sections are used to adjust for the net average dispersion.

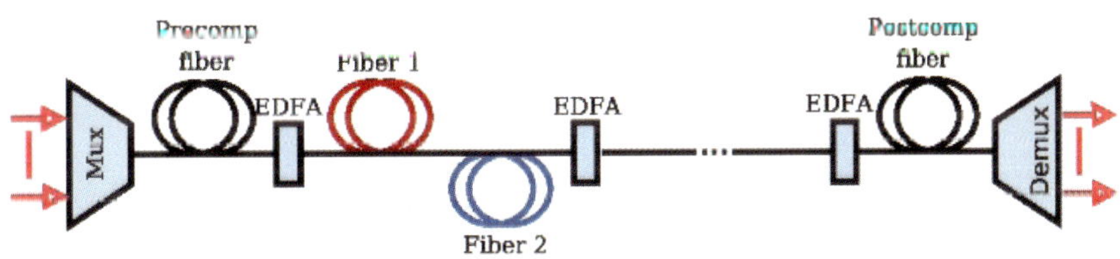

Figure: Schematic diagram of a dispersion-managed communication system.

Mathematically speaking, the normalized equation used to describe DM systems is the NLS equation with rapidly varying coefficients, given by

$$iu_z + \frac{d(z)}{2}u_{tt} + g(z)|u|^2 u = 0$$

where $d(z)$ is the dispersion, which in communications applications is large, and $g(z)$ is the loss-gain coefficient. For example, in a two-step DM system, between every two optical amplifiers (say 40km apart) has one part of the fiber with one dispersion sign; i.e. $d(z)=d_1$ which is fused to the remaining part which has the opposite dispersion sign $d(z)=d_2$ with $d_1d_2 < 0$. The choice of normalizations can be made such that the loss-gain coefficient has unit average in a dispersion map. A good understanding of the dynamical behavior is usually obtained by studying the "lossless" case $g(z)=1$.

As mentioned above, NLS equations are derived by asymptotic analysis from Maxwell's equations. Given the disparate scales involved in Maxwell's equations, i.e. operable wavelengths are near 1.5μm and transmission distances up to 10,000 km yield differing scales on the order of $10^{13;}$ hence asymptotic analysis and reductions to NLS equations are critically important. Researchers do not -- cannot, in fact -- solve Maxwell's equations through the exceedingly long distances that are required when dealing with such transmission problems. Numerical studies of the NLS equation in these DM environments show that localized pulse solutions exist; they are termed DM solitons. Asymptotic analysis (via the method of multiple scales) of such NLS equations with rapidly varying coefficients leads one to a nonlocal equation which governs strongly dispersion-managed communications systems. This asymptotic equation, termed the DMNLS equation (dispersion managed nonlinear Schrodinger equation) is nonlocal and it admits solitary wave solutions. In figure below we show both a classical (sech profile soliton and a DM soliton obtained from the DMNLS equation obtained from equation $iu_z + \frac{d(z)}{2} u_{tt} + g(z)|u|^2 u = 0$.

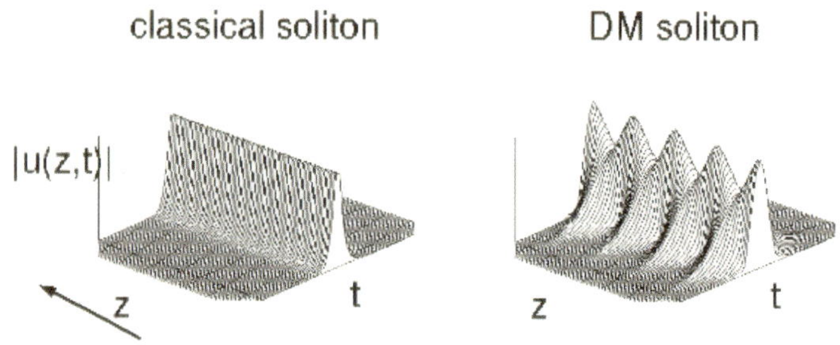

Figure: Classical and DM solitons.

Interestingly DM systems also give rise to other special solutions of the DMNLS equation, in particular quasi-linear pulse solutions, which are more like linear modes, but have a nonlinear phase contribution, and are used more frequently in communication applications than DM solitons. Dispersion management is also used to generate ultra-short soliton pulses on the order of a few femto-seconds in mode-locked lasers such as Ti-sapphire, Sr-fosterite and fiber lasers, amongst others. Here the dispersion compensation is designed to allow the system to be in the net anomalous dispersion regime. In figure a typical mode-locked Ti-sapphire laser system is depicted and in figure an associated pulse train.

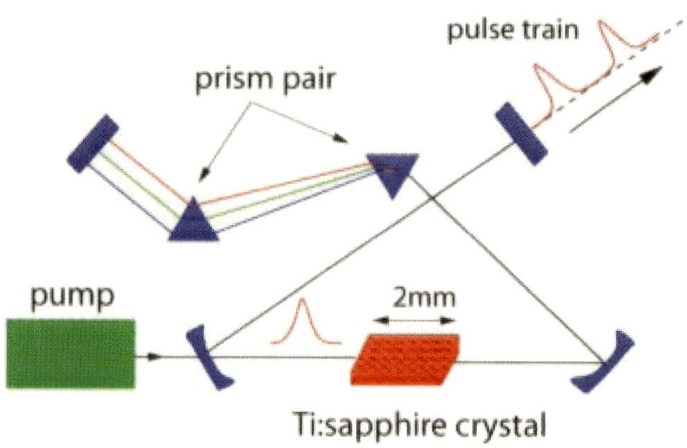

Figure: Dispersion managed Ti: sapphire laser system.

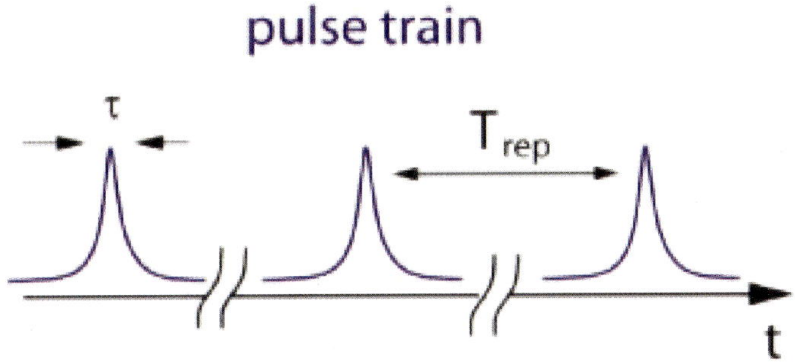

Figure: Dispersion managed Ti: corresponding pulse train.

Typical values are the following. Pulse width

$$\tau = 10$$

fs=10^{-14}sec; repetition time between pulses

$$T_{rep}=10$$

ns=10^{-8} sec and repetition frequency

$$f_{rep} = 1/T_{rep}=100$$

Mathcmatical Framework

Inverse Scattering Transform (IST)

As mentioned above, there are a number of equations of NLS type, both continuous and discrete, which are solvable by IST. More precisely, the following NLS-type equations are solvable by IST: the scalar NLS equation in 1+1 dimensions for both rapidly decaying initial data ($q \to 0$ as $t \to \pm\infty$) and for data which decays tends to a constant background ($|q| \to |q_{0|}$ as $t \to \pm\infty$, with $q_0 \neq 0$); the vector NLS equation in 1+1 dimensions, also both for rapidly decaying initial data, and for data which decays rapidly to a constant background); the scalar and vector integrable discrete NLS equations -- i.e., equation $i\frac{d}{dt}q_n = \frac{1}{h^2}(q_{n+1} - 2q_n + q_{n-1}) \pm |q_n|^2 (q_{n+1} + q_{n-1})$ and $i\frac{dq_n}{dt} = \frac{1}{h^2}[q_{n+1} - 2q_n + q_{n-1}] \pm || q_n|^2 (q_{n+1} + q_{n-1})$ for decaying data, and, in the scalar case, also for nondecaying data. Finally, a special case of the multi-dimensional Benney-Roskes equation, called the Davey-Stewartson equation with $\sigma 1, \sigma 2 = \pm 1$ is also solvable by IST.

The IST is a method that allows one to linearize a class on nonlinear evolution equations. An essential pre-requisite of the IST method is the association of the nonlinear PDE with a pair of linear problems. We say that the operator pair X, T is a Lax pair for the nonlinear equation

$$q_t = F[x, t, q, q_x, q_{xx}, \ldots], \; q = q(x,t),$$

if the compatibility condition of the over determined linear system of differential equations

$$v_x = \mathrm{X}v, \; v_t = \mathrm{T}v$$

[i.e., the equality of the mixed derivatives $v_{xt} = v_{tx}$] is identically satisfied provided q solves the nonlinear PDE. Here X, T are in general matrix functions of q, q_x, q_{xx},... and of a complex parameter k which is assumed to be time-independent.

The solution of the initial-value problem by IST proceeds in three steps, as follows:

1. The direct problem - the transformation of the initial data from the original "physical" variables ($q(x,0)$) to the transformed "scattering" variables ($S(k,0)$);
2. Time dependence - the evolution of the transformed data often according to simple, explicitly solvable evolution equations (i.e., finding $S(k, t)$);
3. The inverse problem - the recovery of the evolved solution ($q(x, t)$) from the evolved solution in the transformed variables ($S(k, t)$).

Both the direct and the inverse problem make use of the first operator in the Lax pair, so-called scattering problem, while the time evolution is determined by the second operator in the Lax pair.

$$q(x,0)\xrightarrow{\text{Forward Integral Equation}} S(k,0)=\left\{\rho(k,0),\{k_j,C_j(0)\}_{j=1}^{J}\right\}$$

$$\downarrow \text{Simple ODE}$$

$$q(x,t)\xleftarrow{\text{Riemann-Hilbert Equation}} S(k,t)=\left\{\rho(k,t),\{k_j,C_j(t)\}_{j=1}^{J}\right\}$$

In the direct problem, the first step is to construct eigenfunction solutions of the associated linear problem. These eigenfunctions depend on both the original spatial variables and the spectral parameter (eigenvalue) k. Second, with these eigenfunctions, one determines *scattering data* that are independent of the original spatial variables. In the inverse problem, the first step is the recovery of the eigenfunctions from the (evolved) scattering data. Finally, one recovers the solution in the original variables from these (evolved) eigenfunctions.

The analyticity properties of the eigenfunctions in terms of the scattering parameter are key to the formulation of the inverse problem. For the NLS systems discussed above, the eigenfunctions are sectionally analytic functions of the scattering parameter and the inverse problem is therefore formulated as a generalized Riemann-Hilbert problem, which in turn is transformed into a system of linear algebraic-integral equations.

In conclusion, we mention that with the IST machinery one can do more than solve the initial-value problem; one can also construct special solutions by positing an elementary ansatz for the transformed variables and then applying the inverse transformation to obtain the corresponding solution in the physical variables. In general, the soliton and multisoliton solutions are constructed in this way. Also, the IST provides an effective way to study the asymptotic (long-term) behavior of the solutions (e.g., via the nonlinear steepest descent technique), as well as a way to study their stability (using the so-called squared eigenfunctions).

Direct Methods

Solutions to certain nonlinear PDEs can be obtained by using direct methods. For instance, one can effectively transform the nonlinear equation into a bilinear differential equation, often called the Hirota bilinear form, and then find exact solutions by a perturbation technique. In the case of the NLS equation (Hirota 1973), substituting $u = G/F$, F real, into $iu_t + u_{xx} + |u|^2 u = 0$, gives the bilinear form

$$\frac{1}{F^2}(iD_t + D_x^2)G \cdot F - \frac{G}{F^3}(D_x^2 F \cdot F - GG^*) = 0$$

where the bilinear operator is defined by

$$D_x^m D_t^n a \cdot b = (\partial_x - \partial_{x'})^m (\partial_t - \partial_{t'})^n a(x,t)b(x',t')|_{x'=x,t'=t}$$

The bilinear equation can be decoupled by choosing

$$(iD_t + D_x^2)G \cdot F = 0, \quad D_x^2 F \cdot F = GG^*,$$

where upon

$$|u|^2 = \frac{GG^*}{F^2} = \frac{D_x^2 F \cdot F}{F^2} = 2(log F)_{xx}$$

A one soliton solution can then be obtained by taking $G = e^{\eta_1}$ and $F = 1 + e^{\eta_1 + \eta_1^* + \phi_{1,1}}$, with $\eta_1 = p_1 x - \Omega_1 t + x_0$, $\Omega_1 = -ip_1^2$ and $e^{\phi_{1,1}} = (p1 + p_1^*)^{-2} / 2$. Multisoliton solutions can be obtained in a similar way.

References

- Dirac, P. A. M. (1958). The Principles of Quantum Mechanics (Fourth ed.). Oxford. ISBN 0-19-851208-2
- Schrodinger-equation, science: britannica.com, Retrieved 12 April 2018
- Derivation-from-wave-mechanics, schrodinger-equation: brilliant.org, Retrieved 22 May 2018
- Nonlinear-Schrodinger-systems:-continuous-and-discrete: scholarpedia.org, Retrieved 30 March 2018
- Baym, Gordon (1969). Lectures on Quantum Mechanics. W. A. Benjamin. ISBN 978-0805306675

3

Prominent Quantum Mechanics Equations

An understanding of quantum mechanics requires an examination of the principal quantum mechanical equations, such as Dirac equation, Klein–Gordon equation, Pauli equation and Rydberg equation. These have been thoroughly discussed in this chapter.

Dirac Equation

Dirac Equation was discovered in the late 1920s by physicist Paul Dirac. It remains highly influential as it brought together two of the most important ideas in science: quantum mechanics and Einstein's special theory of relativity. As a result, Dirac's equation describes how particles like electrons behave when they travel close to the speed of light.

The electron Dirac equation has the form

$$i\hbar\frac{\partial}{\partial t}\psi = \alpha \cdot \mathrm{p}\psi + \beta m\psi,$$

where m is the mass of the electron and we continue to work in units in which the speed of light is unity. The state ψ is a four-component spinor and the operators α and β are the 4 × 4 matrices

$$\alpha = \begin{pmatrix} 0 & \sigma \\ \sigma & 0 \end{pmatrix},$$

$$\beta = \begin{pmatrix} \mathrm{I} & 0 \\ 0 & -\mathrm{I} \end{pmatrix},$$

where σ is a vector, the components of which are the familiar Pauli matrices, 0 is the 2 × 2 zero matrix and I is the unit matrix. The square of each of these matrices is the unit matrix and they mutually anticommute

$$\alpha_i\alpha_j + \alpha_j\alpha_i = 2\delta_{ij},$$

$$\beta\alpha_i + \alpha_i\beta = 0.$$

We can identify the positive quantity $\rho = \psi^{\dagger}\psi$ with the probability density for the electron position by normalizing it so that integration over all space gives unity. This density obeys an associated continuity equation, which may be derived directly from the Dirac equation

$$\frac{\partial \rho}{\partial t} + \nabla \cdot \mathrm{j} = 0,$$

Where j is identified as the probability current

$$\mathrm{j} = \psi^{\dagger} \alpha \psi$$

This leads us to identify α as the velocity operator for the electron. The eigenvalues of the three operators, α_x, α_y and α_z are all ± 1, which each correspond to the velocity of light. We can resolve this apparent conflict with experience by noting that a wave-packet formed by superposing energy eigen states moves with the expected group velocity given by the ratio of the average momentum and the average energy. There is a remnant of the velocity eigenvalues in the rapid oscillations associated with the positive- and negative-energy components of the state. This, Zitterbewegung has a frequency in excess of $2m/\hbar$ or, equivalently, a spatial period comparable to the Compton wavelength of the electron.

Dirac Equation for a Free Particle

The Dirac Hamiltonian takes the form

$$H = c\vec{\alpha} \cdot \mathbf{P} + \beta mc^2$$

where

$$\vec{\alpha} = \begin{pmatrix} 0 & \vec{\sigma} \\ \vec{\sigma} & 0 \end{pmatrix} \qquad \beta = \begin{pmatrix} I & 0 \\ 0 & -I \end{pmatrix}$$

Using $\mathrm{P} = (\hbar / i)\nabla$, in the coordinate basis, the Dirac equation for a free particle reads

$$\left[-i\hbar c\vec{\alpha} \cdot \nabla + \beta mc^2 \right] \Psi(\mathbf{r},) = i\hbar \frac{\partial}{\partial t} \Psi(\mathbf{r},t)$$

Since the operator on the left side is a 4×4 matrix, the wave function $\Psi\text{ü}\mathbf{r}\ t$ is actually a four-component vector of functions of $\mathbf{r}$ and t:

$$\Psi\text{ü}\mathbf{r}\ t = \begin{pmatrix} \Psi_1\text{ü}\mathbf{r}\ t \\ \Psi_2\text{ü}\mathbf{r}\ t \\ \Psi_3\text{ü}\mathbf{r}\ t \\ \Psi_4\text{ü}\mathbf{r}\ t \end{pmatrix}$$

which is called a four-component *Dirac spinor*. In order to generate an eigenvalue problem, we look for a solution of the form

$$\Psi(\mathbf{r},t) = \psi(\mathbf{r}) e^{-iEt/\hbar}$$

which, when substituted into the Dirac equation gives the eigenvalue equation

$$\left[-i\hbar c\vec{\alpha} \cdot \nabla + \beta mc^2 \right] \psi(\mathbf{r}) = E\psi(\mathbf{r})$$

Note that, since H is only a function of P, then [P, H] = 0 so that the eigenvalues **p** of **P** can be used to characterize the states. In particular, we look for free-particle (plane-wave) solutions of the form:

$$\psi_{\mathbf{p}}(\mathbf{r}) = u_{\mathbf{p}} e^{i\mathbf{p}\cdot\mathbf{r}/\hbar}$$

where $u_{\mathbf{p}}$ is a four-component vector which satisfies

$$\left[c\vec{\alpha}\cdot\mathbf{p} + \beta mc^2\right] u_{\mathbf{p}} = E u_{\mathbf{p}}$$

Since the matrix on the left is expressible in terms of 2×2 blocks, we look for $u_{\mathbf{p}}$ in the form of a vector composed of two two-component vectors:

$$u_{\mathbf{p}} = \begin{pmatrix} \phi_{\mathbf{p}} \\ \chi_{\mathbf{p}} \end{pmatrix}$$

Therefore, writing the equation in matrix form, we find

$$\begin{pmatrix} mc^2 & c\vec{\sigma}\cdot\mathrm{p} \\ c\vec{\sigma}\cdot\mathrm{p} & -mc^2 \end{pmatrix}\begin{pmatrix} \phi_{\mathbf{p}} \\ \chi_{\mathbf{p}} \end{pmatrix} = \begin{pmatrix} \mathrm{E} & 0 \\ 0 & \mathrm{E} \end{pmatrix}\begin{pmatrix} \phi_{\mathbf{p}} \\ \chi_{\mathbf{p}} \end{pmatrix}$$

or

$$\begin{pmatrix} E - mc^2 & -c\vec{\sigma}\cdot\mathrm{p} \\ -c\vec{\sigma}\cdot\mathrm{p} & E + mc^2 \end{pmatrix}\begin{pmatrix} \phi_{\mathbf{p}} \\ \chi_{\mathbf{p}} \end{pmatrix} = 0$$

which yields two equations

$$\left(E - mc^2\right)\phi_{\mathbf{p}} - c\vec{\sigma}\cdot\mathrm{p}\,\chi_{\mathbf{p}} = 0$$

$$-c\vec{\sigma}\cdot\mathrm{p}\,\phi_{\mathbf{p}} + \left(E - mc^2\right)\chi_{\mathbf{p}} = 0$$

From the second equation:

$$\chi_{\mathbf{p}} = \frac{c\vec{\sigma}\cdot\mathbf{p}}{E + mc^2}\phi_{\mathbf{p}}$$

Note, one could also solve the first for $\phi_{\mathbf{p}}$ and obtain

$$\phi_{\mathbf{p}} = \frac{c\vec{\sigma}\cdot\mathbf{p}}{E - mc^2}\chi_{\mathbf{p}}$$

Using the first of these, then a single equation for $\phi_{\mathbf{p}}$ can be obtained

$$\left(E-mc^2\right)\left(E+mc^2\right)\phi_{\mathbf{p}}-c^2\left(\vec{\sigma}\cdot\mathbf{p}\right)^2\phi_{\mathbf{p}}=0$$

However,

$$\left(\vec{\sigma}\cdot\mathbf{p}\right)^2=\left(\vec{\sigma}\cdot\mathbf{p}\right)\left(\vec{\sigma}\cdot\mathbf{p}\right)=\mathbf{p}\cdot\mathbf{p}+i\,\vec{\sigma}\cdot(\mathbf{p}\times\mathbf{p})=p^2$$

Hence, we have the condition

$$\left[E^2-\left((mc^2)+c^2p^2\right)\right]\phi_{\mathbf{p}}=0$$

Since $\phi_{\mathbf{p}}\neq 0$ the equation is only satisfied if the quantity in the brackets vanishes, which yields the eigen values

$$E=E_{\mathbf{p}}=\pm\sqrt{p^2c^2+m^2c^4}$$

We see that the eigenvalues can be positive or negative. A plot of the energy levels is shown below:

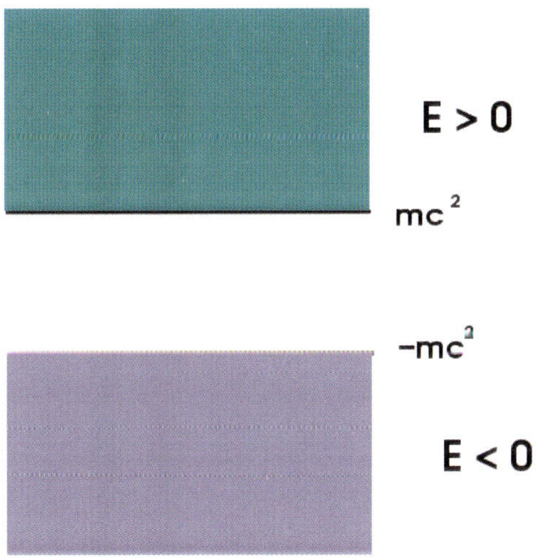

There is a continuum for $E_{\mathbf{p}}>mc^2$ (turquoise) and for $E_{\mathbf{p}}<-mc^2$ (periwinkle). There is also a gap between $-mc^2$ and mc^2.

We will show that for $E>0$, an appropriate solution is to take

$$\phi_{\mathbf{p}}=\begin{pmatrix}1\\0\end{pmatrix}\quad or\quad\begin{pmatrix}0\\1\end{pmatrix}$$

If this is the case, then

$$\chi_{\mathbf{p}}=\frac{c\vec{\sigma}\cdot\mathbf{p}}{E_{\mathbf{p}}+mc^2}\begin{pmatrix}1\\0\end{pmatrix}\quad or\quad\frac{c\vec{\sigma}\cdot\mathbf{p}}{E_{\mathbf{p}}+mc^2}\begin{pmatrix}0\\1\end{pmatrix}$$

However,

$$\vec{\sigma}\cdot\mathbf{p} = \begin{pmatrix} p_z & p_x - ip_y \\ p_x + ip_y & -p_z \end{pmatrix}$$

so that

$$\chi_{\mathbf{p}} = \begin{pmatrix} cp_z/(E_{\mathbf{p}} + mc^2) \\ c\left(p_x + ip_y\right)/(E_{\mathbf{p}} + mc^2) \end{pmatrix} \quad or \quad \begin{pmatrix} c\left(p_x - ip_y\right)/(E_{\mathbf{p}} + mc^2) \\ -cp_z/(E_{\mathbf{p}} + mc^2) \end{pmatrix}$$

so that the full solution $u_{\mathbf{p}}$ is

$$u_{\mathbf{p}} = \begin{pmatrix} 1 \\ 0 \\ cp_z/(E_{\mathbf{p}} + mc^2) \\ c\left(p_x + ip_y\right)/(E_{\mathbf{p}} + mc^2) \end{pmatrix} \quad or \quad \begin{pmatrix} 0 \\ 1 \\ c\left(p_x - ip_y\right)/(E_{\mathbf{p}} + mc^2) \\ -cp_z/(E_{\mathbf{p}} + mc^2) \end{pmatrix}$$

Note that when $\mathbf{p} = 0$ the third and fourth components of $u_{\mathbf{p}}$ vanish. In this case, energy is just $E_0 = mc^2$ and the full time-dependent wave function becomes

$$\Psi(t) \rightarrow \begin{pmatrix} 1 \\ 0 \\ 0 \\ 0 \end{pmatrix} e^{-imc^2 t/\hbar} \quad or \quad \begin{pmatrix} 0 \\ 1 \\ 0 \\ 0 \end{pmatrix} e^{-imc^2 t/\hbar}$$

Which are both forward propagating solutions. These correspond to particle solutions, in particular, a spin-*1/2* particle propagating forward in time with an energy equal to the rest mass energy.

When $E < 0$, we take

$$\chi_{\mathbf{p}} = \begin{pmatrix} 1 \\ 0 \end{pmatrix} \quad or \quad \begin{pmatrix} 0 \\ 1 \end{pmatrix}$$

so that

$$\phi_{\mathbf{p}} = \frac{c\vec{\sigma}\cdot\mathbf{p}}{E_{\mathbf{p}} - mc^2}\chi_{\mathbf{p}} = \frac{-c\vec{\sigma}\cdot\mathbf{p}}{|E_{\mathbf{p}}| + mc^2}\chi_{\mathbf{p}}$$

By the same reasoning, the solution for $u_{\mathbf{p}}$ is

$$u_{\mathbf{p}} = \begin{pmatrix} -cp_z/\left(\left|E_{\mathbf{p}}\right| + mc^2\right) \\ -c\left(p_x + ip_y\right)/\left(\left|E_{\mathbf{p}}\right| + mc^2\right) \\ 1 \\ 0 \end{pmatrix} \quad or \quad \begin{pmatrix} -c\left(p_x - ip_y\right)/\left(\left|E_{\mathbf{p}}\right| + mc^2\right) \\ cp_z/(E_{\mathbf{p}} + mc^2) \\ 0 \\ 1 \end{pmatrix}$$

so that in the limit $\mathbf{p} = 0$, and $E_0 = -mc^2$,

$$\Psi(t)\rightarrow\begin{pmatrix}0\\0\\1\\0\end{pmatrix}e^{imc^2t/\hbar}\quad or\quad\begin{pmatrix}0\\0\\0\\1\end{pmatrix}e^{imc^2t/\hbar}$$

which describes particles moving backward in times. Thus, the interpretation is that the negative energy solutions correspond to anti-particles, the components, ϕ_p and χ_p of u_p corresponds to the particle and anti-particle components, respectively. Thus, the Dirac equation no only describes spin but it also includes particle and the corresponding anti-particle solutions.

In the non-relativistic limit, for $E>o$, we have

$$E_p \approx mc^2 + \frac{\mathbf{p}^2}{2m}$$

so that

$$\chi_p = \frac{c\vec{\sigma}\cdot\mathbf{p}}{2mc^2 + \mathbf{p}^2/2m}\phi_p$$

since $mc^2 \gg \mathbf{p}^2/2m$, it follows that

$$\chi_p \ll \phi_p$$

Neglecting it, and recalling that for $E>0$,

$$\phi_p - \begin{pmatrix}1\\0\end{pmatrix}\quad or\quad\begin{pmatrix}0\\1\end{pmatrix}$$

the eigenfunctions reduce to

$$\psi_p(\mathbf{r}) = \begin{pmatrix}1\\0\\0\\0\end{pmatrix}e^{i\mathbf{p}\cdot\mathbf{r}/\hbar}\quad or\quad\begin{pmatrix}0\\1\\0\\0\end{pmatrix}e^{i\mathbf{p}\cdot\mathbf{r}/\hbar}$$

The lower component has become redundant, and the eigenfunctions just correspond to those of a free particle with an attached spin eigenfunction $\begin{pmatrix}1\\0\end{pmatrix}$ or $\begin{pmatrix}0\\1\end{pmatrix}$ for $m_s = \hbar/2$ or $-\hbar/2$, respectively. For $E>0$, the lower component χ_p is called the minor component and the upper component ϕ_p is called the major component.

Physical Interpretation

Identification of Observables

The critical physical question in a quantum theory is—what are the physically observable quantities defined by the theory? According to the postulates of quantum mechanics, such quantities are defined by Hermitian operators that act on the Hilbert space of possible states of a system. The eigenvalues of these operators are then the possible results of measuring the corresponding physical quantity. In the Schrödinger theory, the simplest such object is the overall Hamiltonian, which represents the total energy of the system. If we wish to maintain this interpretation on passing to the Dirac theory, we must take the Hamiltonian to be

$$H = \gamma^0 \left[mc^2 + c\gamma^k \left(p_k - qA_k \right) \right] + qA^0.$$

where, as always, there is an implied summation over the twice-repeated index $k = 1, 2, 3$. This looks promising, because we see by inspection the rest energy of the particle and, in case $A = 0$, the energy of a charge placed in an electric potential qA^0. What about the term involving the vector potential? In classical electrodynamics, the energy of a charge moving in an applied potential is

$$H = c\sqrt{\left(p - qA\right)^2 + m^2c^2} + qA^0.$$

Thus, the Dirac Hamiltonian is fundamentally distinguished from its classical counterpart, and we must take great care to correctly identify what is observable in this theory. Much of the apparently paradoxical behavior implied by the Dirac equation amounts to a misidentification of these observables.

Hole Theory

The negative E solutions to the equation are problematic, for it was assumed that the particle has a positive energy. Mathematically speaking, however, there seems to be no reason for us to reject the negative-energy solutions. Since they exist, we cannot simply ignore them, for once we include the interaction between the electron and the electromagnetic field, any electron placed in a positive-energy eigenstate would decay into negative-energy eigenstates of successively lower energy. Real electrons obviously do not behave in this way, or they would disappear by emitting energy in the form of photons.

To cope with this problem, Dirac introduced the hypothesis, known as hole theory, that the vacuum is the many-body quantum state in which all the negative-energy electron eigenstates are occupied. This description of the vacuum as a "sea" of electrons is called the Dirac sea. Since the Pauli exclusion principle forbids electrons from occupying the same state, any additional electron would be forced to occupy a positive-energy eigenstate, and positive-energy electrons would be forbidden from decaying into negative-energy eigenstates.

If an electron is forbidden from simultaneously occupying positive-energy and negative-energy eigenstates, then the feature known as Zitterbewegung, which arises from the interference of positive-energy and negative-energy states, would have to be considered to be an unphysical prediction of time-dependent Dirac theory. This conclusion may be inferred from the explanation of hole theory given in the preceding paragraph. Recent results have been published in Nature in which the Zitterbewegung feature was simulated in a trappedion experiment. This experiment impacts the hole interpretation if one infers that the physics-laboratory experiment is not merely a check on the mathematical correctness of a Dirac-equation solution but the measurement of a real effect whose detectability in electron physics is still beyond reach.

Dirac further reasoned that if the negative-energy eigenstates are incompletely filled, each unoccupied eigenstate – called a hole – would behave like a positively charged particle. The hole possesses a *positive* energy since energy is required to create a particle–hole pair from the vacuum. As noted above, Dirac initially thought that the hole might be the proton, but Hermann Weyl pointed out that the hole should behave as if it had the same mass as an electron, whereas the proton is over 1800 times heavier. The hole was eventually identified as the positron, experimentally discovered by Carl Anderson in 1932.

It is not entirely satisfactory to describe the "vacuum" using an infinite sea of negative-energy electrons. The infinitely negative contributions from the sea of negative-energy electrons have to be canceled by an infinite positive "bare" energy and the contribution to the charge density and current coming from the sea of negative-energy electrons is exactly canceled by an infinite positive "jellium" background so that the net electric charge density of the vacuum is zero. In quantum field theory, a Bogoliubov transformation on the creation and annihilation operators (turning an occupied negative-energy electron state into an unoccupied positive energy positron state and an unoccupied negative-energy electron state into an occupied positive energy positron state) allows us to bypass the Dirac sea formalism even though, formally, it is equivalent to it.

In certain applications of condensed matter physics, however, the underlying concepts of "hole theory" are valid. The sea of conduction electrons in an electrical conductor, called a Fermi sea, contains electrons with energies up to the chemical potential of the system. An unfilled state in the Fermi sea behaves like a positively charged electron, though it is referred to as a "hole" rather than a "positron". The negative charge of the Fermi sea is balanced by the positively charged ionic lattice of the material.

In Quantum Field Theory

In quantum field theories such as quantum electrodynamics, the Dirac field is subject to a process of second quantization, which resolves some of the paradoxical features of the equation.

Other Formulations

The Dirac equation can be formulated in a number of other ways.

As a Differential Equation in One Real Component

Generically (if a certain linear function of electromagnetic field does not vanish identically), three out of four components of the spinor function in the Dirac equation can be algebraically eliminated, yielding an equivalent fourth-order partial differential equation for just one component. Furthermore, this remaining component can be made real by a gauge transform.

Curved Spacetime

This article has developed the Dirac equation in flat spacetime according to special relativity. It is possible to formulate the Dirac equation in curved spacetime.

The Algebra of Physical Space

This article developed the Dirac equation using four vectors and Schrödinger operators. The Dirac equation in the algebra of physical space uses Clifford algebra over the real numbers, a type of geometric algebra.

Klein–Gordon Equation

The *Klein-Gordon equation* is the linear partial differential equation which is the equation of motion of a freescalar field of possibly non-vanishing mass *m* on some (possibly curved) spacetime (Lorentzian manifold): it is the relativistic wave equation with inhomogenety the mass $m^{2.}$

The structure of the Klein-Gordon equation appears also in the equations of motion of richer fields than just scalar fields, where now the underlying field bundle may more generally be some vector bundle. Therefore the fundamental solutions of the Klein-Gordon equation, called the *propagators* pervades all of relativistic perturbative quantum field theory.

The Klein-Gordon equation can be expressed as follows

$$\nabla^2\psi - \frac{1}{c^2}\frac{\partial^2\psi}{\partial t^2} = -\left(\frac{E_0}{c\hbar}\right)^2\psi$$

First, show that the classical Schrödinger equation obtains from the Klein-Gordon equation in the classical limit of the kinetic energy being much less than the rest energy of a particle. Having done this focus will then be turned to generalizing the Klein-Gordon equation. This will be done for the case of a four dimensional space then generalized to include higher dimensions.

Connecting the Klein-Gordon Equation to the Schrödinger Equation

The Klein-Gordon equation above becomes the Schrödinger equation for a free particle under the condition that the dynamical energy is approximately equal to the rest energy. For now assume a four-dimensional Minkowski space.

The Klein-Gordon equation can be reworked as follows

$$-\left(\frac{\hbar}{2m}\right)^2\cdot\nabla^2\psi + \left(\frac{\hbar^2}{2E_0}\right)\cdot\frac{\partial^2\psi}{\partial t^2} + \left(\frac{E_0}{2}\right)\psi + i\hbar.\frac{\partial\psi}{\partial t} = i\hbar.\frac{\partial\psi}{\partial t}$$

With $E_o = m_{rest}\cdot c^2$. It can be shown above that a fixed energy is possible under the condition

$$i\hbar.\frac{\partial\psi}{\partial t} = E_D\psi$$

In the classical limit $E_D \approx E_o$ and so

$$i\hbar.\frac{\partial\psi}{\partial t} \approx E_0\psi$$

$$\frac{\partial\psi}{\partial t} \approx \frac{E_0}{i\hbar}\psi$$

Under this approximation the Klein Gordon equation simplifies to

$$-\left(\frac{\hbar}{2m}\right)^2 \cdot \nabla^2\psi + E_0\psi = i\hbar.\frac{\partial\psi}{\partial t}$$

This is the Schrödinger equation with the rest energy included. By comparison with the classical conservation of energy the total energy (kinetic plus rest plus potential energy) E is represented by the portion on the right side. It is therefore postulated that the total energy operator E will always be simply

$$E_{op} = i\hbar.\frac{\partial}{\partial t}$$

so the relativistic quantum mechanical equations go to the Schroedinger equation in the classical limit.

Generalizing the Klein-Gordon Equation

The Laplacian for the space described by the scale factors in *(2b)* is

$$\Box_0^2\phi = \frac{\partial^2\phi}{\partial x^2} + \frac{\partial^2\phi}{\partial y^2} + \frac{\partial^2\phi}{\partial z^2} - \frac{1}{c^2}.\frac{\partial^2\psi}{\partial t^2}$$

Again, the zero subscript denotes a special relativistic coordinate system where no forces are present.

In this light the Klein-Gordon equation above can be rewritten as

$$\Box_0^2\psi = -\left(\frac{E_0}{c\hbar}\right)^2\psi$$

Which suggests a generalization of the Klein-Gordon equation as follows

$$\Box^2\psi = -\left(\frac{E_0}{c\hbar}\right)^2\psi$$

However as we allow for higher dimensions this will not reduce down to the original Klein-Gordon equation for the case of a particle in a free space. To correct for this the generalized Klein-Gordon equation can be modified as follows

$$\Box^2\psi + \left[\nabla_0^2\psi - \frac{1}{c^2}.\frac{\partial^2\psi}{\partial t^2}\right] = -\left(\frac{E_0}{c\hbar}\right)^2\psi + \Box_0^2\psi$$

With the del operator involving only the three spatial coordinates and the L denoting that the geometry of a free space is to be used.

Conserved Current

The conserved current associated to the U(1) symmetry of a complex field $\varphi(x) \in \mathbb{C}$ satisfying the Klein–Gordon equation reads

$$\partial_\mu J^\mu(x) = 0, \qquad J^\mu(x) \equiv \varphi^*(x)\partial^\mu\varphi(x) - \varphi(x)\partial^\mu\varphi^*(x).$$

The form of the conserved current can be derived systematically by applying Noether's theorem to the U(1) symmetry. We will not do so here, but simply give a proof that this conserved current is correct.

Proof using algebraic manipulations from the KG equation

From the Klein–Gordon equation for a complex field $\varphi(x)$ of mass m written in covariant notation

$$(\Box + m^2)\varphi(x) = 0,$$

and its complex conjugate

$$(\Box + m^2)\varphi^*(x) = 0,$$

we have, multiplying by the left respectively by $\varphi^*(x)$ and $\varphi(x)$ (and omitting for brevity the explicit x dependence),

$$\varphi^*(\Box + m^2)\varphi = 0,$$

$$\varphi(\Box + m^2)\varphi^* = 0.$$

Subtracting the former from the latter we obtain

$$\varphi^*\Box\varphi - \varphi\Box\varphi^* = 0$$

From which we obtain the conservation law for the Klein–Gordon field:

$$\partial_\mu J^\mu(x) = 0, \qquad J^\mu(x) \equiv \varphi^*(x)\partial^\mu\varphi(x) - \varphi(x)\partial^\mu\varphi^*(x).$$

Relativistic Free Particle Solution

The Klein–Gordon equation for a free particle can be written as

$$\nabla^2\psi - \frac{1}{c^2}\frac{\partial^2}{\partial t^2}\psi = \frac{m^2c^2}{\hbar^2}\psi$$

We look for plane wave solutions of the form

$$\psi(\mathbf{r}, t) = e^{i(\mathbf{k}\cdot\mathbf{r} - \omega t)}$$

for some constant angular frequency $\omega \in \mathbb{R}$ and wave number $k \in \mathbb{R}^3$. Substitution gives the *dispersion relation*:

$$-|\mathbf{k}|^2 + \frac{\omega^2}{c^2} = \frac{m^2c^2}{\hbar^2}.$$

Energy and momentum are seen to be proportional to ω and k:

$$\langle \mathbf{p} \rangle = \left\langle \psi \middle| -i\hbar\nabla \middle| \psi \right\rangle = \hbar\mathbf{k},$$

$$\langle E \rangle = \left\langle \psi \middle| i\hbar\frac{\partial}{\partial t} \middle| \psi \right\rangle = \hbar\omega.$$

So the dispersion relation is just the classic relativistic equation:

$$\langle E \rangle^2 = m^2c^4 + \langle \mathbf{p} \rangle^2 c^2.$$

For massless particles, we may set m = 0, recovering the relationship between energy and momentum for massless particles:

$$\langle E \rangle = |\langle \mathbf{p} \rangle| c.$$

Action

The Klein Gordon equation can also be derived via a variational method by considering the action:

$$S = \int \left(-\frac{\hbar^2}{m} \eta^{\mu\nu} \partial_\mu \bar{\psi} \partial_\nu \psi - \ddot{u} \; {}^2\bar{\psi}\psi \right) d^4$$

where ψ is the Klein–Gordon field and m is its mass. The complex conjugate of ψ is written ψ. If the scalar field is taken to be real-valued, then ψ = ψ.

Applying the formula for the Hilbert stress–energy tensor to the Lagrangian density (the quantity inside the integral), we can derive the stress–energy tensor of the scalar field. It is

$$T^{\mu\nu} = \frac{\hbar^2}{m}\left(\eta^{\mu\alpha}\eta^{\nu\beta} + \eta^{\mu\beta}\eta^{\nu\alpha} - \eta^{\mu\nu}\eta^{\alpha\beta}\right)\partial_\alpha \bar{\psi} \partial_\beta \psi - \eta^{\mu\nu} mc^2 \bar{\psi}\psi.$$

By integration of the time–time component T^{00} over all space, one may show that both the positive and negative frequency plane wave solutions can be physically associated with particles with *positive* energy. This is not the case for the Dirac equation and its energy–momentum tensor.

Electromagnetic Interaction

There is a simple way to make any field interact with electromagnetism in a gauge invariant way: replace the derivative operators with the gauge covariant derivative operators. This is because to maintain symmetry of the physical equations for the wavefunction φ under a local $U(1)$ gauge transformation $\varphi \to \varphi' = \exp(i\theta)\varphi$ where $\theta(t, \mathbf{x})$ is a locally variable phase angle, which transformation redirects the wavefunction in the complex phase space defined by $\exp(i\theta) = \cos\theta + i\sin\theta$, it is required that ordinary derivatives ∂_μ be replaced by gauge-covariant derivatives $D_\mu = \partial_\mu - ieA_\mu$

while the gauge fields transform as $eA_\mu \to eA'_\mu = eA_\mu + \partial_\mu \theta$. The Klein–Gordon equation therefore becomes:

$$D_\mu D^\mu \varphi = -(\partial_t - ieA_0)^2 \varphi + (\partial_i - ieA_i)^2 \varphi = m^2 \varphi$$

in natural units, where A is the vector potential. While it is possible to add many higher order terms, for example,

$$D_\mu D^\mu \varphi + AF^{\mu\nu} D_\mu \varphi D_\nu (D_\alpha D^\alpha \varphi) = 0$$

these terms are not renormalizable in 3 + 1 dimensions.

The field equation for a charged scalar field multiplies by i, which means the field must be complex. In order for a field to be charged, it must have two components that can rotate into each other, the real and imaginary parts.

The action for a charged scalar is the covariant version of the uncharged action:

$$S = \int_x \left(\partial_\mu \varphi^{ü} + ieA_\mu \varphi\ \right)\left(\partial_\nu \varphi - ieA_\nu \varphi\right)\eta^{\mu\nu} = \int_x /D\varphi|$$

Gravitational Interaction

In general relativity, we include the effect of gravity by replacing partial with covariant derivatives and the Klein–Gordon equation becomes (in the mostly pluses signature)

$$0 = -g^{\mu\nu}\nabla_\mu \nabla_\nu \psi + \frac{m^2c^2}{\hbar^2}\psi = -g^{\mu\nu}\nabla_\mu(\partial_\nu \psi) + \frac{m^2c^2}{\hbar^2}\psi$$

$$= -g^{\mu\nu}\partial_\mu \partial_\nu \psi + g^{\mu\nu}\Gamma^\sigma{}_{\mu\nu}\partial_\sigma \psi + \frac{m^2c^2}{\hbar^2}\psi$$

or equivalently

$$\frac{-1}{\sqrt{-g}}\partial_\mu\left(g^{\mu\nu}\sqrt{-g}\partial_\nu \psi\right) + \frac{m^2c^2}{\hbar^2}\psi = 0$$

where $g^{\alpha\beta}$ is the inverse of the metric tensor that is the gravitational potential field, g is the determinant of the metric tensor, ∇_μ is the covariant derivative and $\Gamma^\sigma{}_{\mu\nu}$ is the Christoffel symbol that is the gravitational force field.

Pauli Equation

The Pauli equation is the equation for the electron with spin. Its charge is $-e$, its mass is m_e, its spin is $\vec{S} = \frac{1}{2}\hbar\vec{\sigma}$, and its magnetic moment is

$$\vec{M} = -2\mu_B \vec{S}$$

where μ_B is the Bohr magneton.

In a magnetic field $\vec{B}$, the magnetic moment gives rise to an energy

$$-\vec{M}\cdot\vec{B} = \mu_B\vec{\sigma}\cdot\vec{B}$$

and we note that this is a 2 × 2 matrix.

The Hamiltonian is now

$$\hat{H} = \frac{1}{2m_e}\left(\hat{p}+e\vec{A}\right)^2 - eV + U + \mu_B\vec{\sigma}\cdot\vec{B}$$

To make sense mathematically, the first three terms are understood to be multiplied into the unit 2 × 2 matrix

The wave function must now be a column matrix: we write

$$\Psi = \begin{pmatrix}\psi_1\\ \psi_2\end{pmatrix}$$

and hence the Pauli equation:

$$i\hbar.\frac{\partial\Psi}{\partial t} = \hat{H}\Psi$$

In evaluating $\left(\hat{p}-e\hat{A}\right)^2$ we must remember the fundamental commutation relations. We have

$$\left(\hat{p}-e\vec{A}\right)^2 = \left(\hat{p}_x - eA_x\right)^2 + \left(\hat{p}_y - eA_y\right)^2 + \left(\hat{p}_z - eA_z\right)^2$$

and

$$\begin{aligned}\left(\hat{p}_x - eA_x\right)^2 &= \left(\hat{p}_x - eA_x\right)\left(\hat{p}_x - eA_x\right)\\ &= \hat{p}_x^2 - e\left(A_x\hat{p}_x + \hat{p}_xA_x\right) + e^2A_x^2\end{aligned}$$

hence

$$\left(\hat{p}-e\vec{A}\right)^2 = \hat{p}^2 - e\vec{A}.\hat{p} + i\hbar e\nabla.\vec{A} + e^2A^2$$

The term proportional to A_2 is negligibly small except in certain stars with gigantic magnetic fields. In applications to atomic, nuclear and solid state physics this term can be safely neglected.

Spin-1/2 Particles

The usual Schrodinger equation for a spin-0 particle of mass M in a potential $V(r)$,

$$-\frac{\hbar^2}{2M}\nabla^2\psi + V(r)\psi = i\hbar\frac{\partial\psi}{\partial t},$$

can be obtained from the classical Hamiltonian $H = p^2/2M + V$, using the fact that the momentum operator in the coordinate representation is given by $-i\hbar\nabla$. In the case of a spin-1/2 particle, the wave function is not a complex-valued function but a two-component spinor

$$\psi(r,t) = \begin{pmatrix} \psi^1(r,t) \\ \psi^2(r,t) \end{pmatrix},$$

which under a rotation through an angle α about the axis defined by a unit vector n transforms into

$$\psi' = \left(cos\frac{1}{2}\alpha I - i\,sin\,\frac{1}{2}\alpha\; n\cdot\sigma \right)\psi,$$

where I is the identity 2 × 2 matrix and $\sigma = (\sigma_1, \sigma_2, \sigma_3)$ is formed by the Pauli matrices

$$\sigma_1 = \begin{pmatrix} 0 & 1 \\ 1 & 0 \end{pmatrix},$$

$$\sigma_2 = \begin{pmatrix} 0 & -i \\ i & 0 \end{pmatrix},$$

$$\sigma_3 = \begin{pmatrix} 1 & 0 \\ 0 & -1 \end{pmatrix},$$

Thus, for an infinitesimal rotation,

$$\psi' \simeq \psi - i\alpha\frac{1}{2}n.\sigma\,\psi,$$

which means that $n \cdot S = (1/2)\hbar n \cdot \sigma$ is the operator corresponding to the component of the spin angular momentum operator along n.

The Pauli matrices satisfy

$$\sigma_i\sigma_j = \delta_{ij}\, I + i\varepsilon_{ijk}\sigma_k,$$

where ε_{ijk} is totally antisymmetric with $\varepsilon_{123} = 1$, $i, j,... = 1, 2, 3$, and there is summation over repeated indices. Since the entries of the Pauli matrices are constant, making use of Eq. $(\sigma_i\sigma_j = \delta_{ij}\, I + i\varepsilon_{ijk}\sigma_k,)$,

$$(\sigma\cdot\nabla)^2 = \sigma_i\frac{\partial}{\partial x_i}\sigma_j\frac{\partial}{\partial x_j} = \sigma_i\sigma_j\frac{\partial}{\partial x_i}\frac{\partial}{\partial x_j} = \begin{pmatrix} \nabla^2 & 0 \\ 0 & \nabla^2 \end{pmatrix},$$

where the x_i are Cartesian coordinates; therefore, the equation

$$-\frac{}{M}(\sigma.\nabla)\;\psi + V(r)\psi = i\;\frac{\partial}{\partial t}$$

for the two-component spinor (2) implies that each component of ψ (ψ^1 and ψ^2) satisfies the Schrodinger equation ($-\frac{\hbar^2}{2M}\nabla^2\psi + V(r)\psi = i\hbar\frac{\partial\psi}{\partial t}$,) and conversely. However, when there is a magnetic field present this equivalence disappears and Eq. ($-\frac{\hbar^2}{2M}(\sigma.\nabla)^2\psi + V(r)\psi = i\hbar\frac{\partial\psi}{\partial t}$) leads to a coupling between the two components of the spinor ψ.

The standard procedure to take into account the interaction of a particle of electric charge q with an electromagnetic field consists in replacing the partial derivatives $\partial/\partial x_i$ and $\partial/\partial t$ by $\partial/\partial x_i - iqA_i/(\hbar c)$ and $\partial/\partial t + iq\varphi/\hbar$, respectively, where $A = (A_1, A_2, A_3)$, and ϕ are potentials of the electromagnetic field. In this manner, for a spin-0 charged particle, from Eq. ($-\frac{\hbar^2}{2M}\nabla^2\psi + V(r)\psi = i\hbar\frac{\partial\psi}{\partial t}$,) one obtains

$$-\frac{\hbar^2}{2M}\left[\nabla^2\psi - \frac{2iq}{\hbar c}A.\nabla\psi - \frac{iq}{\hbar c}(\nabla\cdot A)\psi - \left(\frac{q}{\hbar c}\right)^2 A^2\psi\right] + V(r)\psi + q\phi\psi = i\hbar\frac{\partial\psi}{\partial t}.$$

and, similarly, making use of Eqs. ($\sigma_i\sigma_j = \delta_{ij}I + i\varepsilon_{ijk}\sigma_k$,), Eq. ($-\frac{\hbar^2}{2M}(\sigma.\nabla)^2\psi + V(r)\psi = i\hbar\frac{\partial\psi}{\partial t}$) yields

$$-\frac{\hbar^2}{2M}\left[\nabla^2\psi - \frac{2iq}{\hbar c}A.\nabla\psi - \frac{iq}{\hbar c}(\nabla\cdot A)\psi - \left(\frac{q}{\hbar c}\right)^2 A^2\psi + \frac{q}{\hbar c}B\cdot\sigma\psi\right] + V(r)\psi + q\phi\psi = i\hbar\frac{\partial\psi}{\partial t}.$$

When $B = 0$, Eq. above reduces to two independent equations of the form (7), one for each component of the spinor ψ. However, when $B \neq 0$, the components of ψ are coupled through the term

$$-B\cdot\frac{q}{Mc}\frac{1}{2}\hbar\sigma\psi.$$

Recalling that the energy of a magnetic dipole moment μ in a magnetic field B is equal to $-\mu\cdot B$, it follows that a charged spin-1/2 particle obeying Eq. ($-\frac{\hbar^2}{2M}\left[\nabla^2\psi - \frac{2iq}{\hbar c}A.\nabla\psi - \frac{iq}{\hbar c}(\nabla\cdot A)\psi - \left(\frac{q}{\hbar c}\right)^2 A^2\psi + \frac{q}{\hbar c}B\cdot\sigma\psi\right] + V(r)\psi + q\phi\psi = i\hbar\frac{\partial\psi}{\partial t}$.) behaves as if it had a magnetic dipole moment represented by the operator

$$\mu = \frac{q}{Mc}\frac{1}{2}\hbar\sigma = \frac{q}{Mc}S.$$

(By contrast with the electric charge, which is a "c-number", the magnetic dipole moment associated with the particle is an operator.) Equation ($\mu = \frac{q}{Mc}\frac{1}{2}\hbar\sigma = \frac{q}{Mc}S.$) shows that the ratio of the magnetic dipole moment to the intrinsic angular momentum is equal to

$$\frac{q}{Mc}$$

Before considering an analog of Eq. ($-\frac{\hbar^2}{2M}(\sigma.\nabla)^2\psi + V(r)\psi = i\hbar\frac{\partial\psi}{\partial t}$) applicable to spin-3/2 particles, it will be convenient to write Eq. ($-\frac{\hbar^2}{2M}(\sigma.\nabla)^2\psi + V(r)\psi = i\hbar\frac{\partial\psi}{\partial t}$) making use of the two-component spinor notation that will be employed in the treatment of spin-3/2 particles.

The entries of the Pauli matrices (4) will be denoted by $B(A, B, \quad 1,2)$, so that $\sigma_i^A B$ stands for the entry in the A-th row and B-th column of the matrix σ_i. The spinor indices, such as those of the spinor (2), and of the Pauli matrices, will be lowered or raised following the convention

$$\varphi_A = \varepsilon_{AB}\varphi^B, \qquad \varphi^A = \varphi_B\varepsilon^{BA},$$

where

$$(\varepsilon_{AB}) \equiv \begin{pmatrix} 0 & 1 \\ -1 & 0 \end{pmatrix} \equiv (\varepsilon^{AB}).$$

$(Thus,\ \phi_1 = \phi^2,\ \phi_2 = -\phi^1.)$ *Hence,* $\varepsilon^A B = \delta_B^A$ *and*

$$\phi_A\psi^A = \varepsilon_{AB}\phi^B\psi^A = -\phi^B\varepsilon_{BA}\psi^A = -\phi^B\psi_B$$
$$= -\phi^A\psi_A.$$

Any tensor with Cartesian components $t_{ij\cdots k}$ has a spinor equivalent defined by

$$t_{ABCD\cdots MN} \equiv \frac{1}{\sqrt{2}}\sigma^i\ AB\frac{1}{\sqrt{2}}\sigma^j\mathrm{CD}\cdots$$
$$\times\frac{1}{\sqrt{2}}\sigma^k{}_{\mathrm{MN}}\mathrm{t}_{\mathrm{ij}\cdots\mathrm{k}},$$

where, following the conventions stated above, $\sigma_{iAB} = \varepsilon_{AC}\,\sigma_i^C\,B$. (Since we are considering here Cartesian coordinates only, the tensor indices are lowered or raised by means of the metric tensor δ_{ij} and its inverse δ^{ij}; hence, $\sigma_{iAB} = \sigma^i{}_{AB}$.) An explicit computation shows that

$$\sigma_{iAB} = \sigma_{iBA}.$$

Furthermore, since the Pauli matrices have a vanishing trace, from Eq. ($\sigma_i\sigma_j = \delta_{ij}\ I + i\varepsilon_{ijk}\sigma_k$,) we obtain $tr\ (\sigma_i\,\sigma_j) = 2\delta_{ij}$, i.e., $\sigma_i AB\sigma_j BA = 2\delta_{ij}$ or, equivalently

$$\sigma_i^{AB}\sigma_{jAB} = -2\delta_{ij}.$$

Hence, from Eqs. ($t_{ABCD\cdots MN} \equiv \frac{1}{\sqrt{2}}\sigma^i\ AB\frac{1}{\sqrt{2}}\sigma^j\mathrm{CD}\cdots \times\frac{1}{\sqrt{2}}\sigma^k{}_{\mathrm{MN}}\mathrm{t}_{\mathrm{ij}\cdots\mathrm{k}}$) and ($\sigma_i^{AB}\sigma_{jAB} = -2\delta_{ij}$) we find that, if t_{AB} and s_{AB} are the spinor sequivalents of t_i and s_i, respectively

$$t^{AB}\, s_{AB} = -t^i\, s_i.$$

According to the definition, we shall write

$$\partial_{AB} = \frac{1}{\sqrt{2}}\sigma^i{}_{AB}\partial_i,$$

where $\partial_i \equiv \partial/\partial x^i$. Thus, the Schrodinger–Pauli equation ($-\frac{\hbar^2}{2M}(\sigma.\nabla)^2\psi + V(r)\psi = i\hbar\frac{\partial\psi}{\partial t}$) ¨ can be expressed as

$$-\frac{\hbar^2}{M}\partial^A{}_C\partial^C{}_B\psi^B + V(r)\psi^A = i\hbar\frac{\partial\psi^A}{\partial t}.$$

We can see that Eq. ($-\frac{\hbar^2}{M}\partial^A{}_C\partial^C{}_B\psi^B + V(r)\psi^A = i\hbar\frac{\partial\psi^A}{\partial t}$.) is equivalent to two decoupled Schrodinger equations using the fact that if ¨ $\phi_{AB} = -\phi_{BA}$, then

$$\phi_{AB} = \frac{1}{2}\phi^R{}_R\varepsilon_{AB}.$$

Indeed, any 2 × 2 antisymmetric matrix must be proportional to (ε_{AB}) and, as can be readily seen,

$$\begin{aligned}\phi_{AB} &= \phi_{12}\,\varepsilon_{AB} = \frac{1}{2}(\phi_{12} - \phi_{21})\varepsilon_{AB} \\ &= \frac{1}{2}(\phi^2{}_2 + \phi^1{}_1)\,\varepsilon_{AB} = \frac{1}{2}\phi^R{}_R\,\varepsilon_{AB}.\end{aligned}$$

Owing to Eq. ($-\phi^A\psi_A$), and the fact that $\partial_{AB} = \partial_{BA}$

$$\partial_{AC}\partial^C{}_B = -\partial_A{}^C\partial_{CB} = -\partial_{CB}\partial_A{}^C = -\partial_{BC}\partial^C{}_A,$$

hence,

$$\begin{aligned}\partial_{AC}\partial^C{}_B &= \frac{1}{2}\varepsilon_{AB}\partial^R{}_C\partial^C{}_R = -\frac{1}{2}\varepsilon_{AB}\,\partial^{RC}\partial_{RC} \\ &= \frac{1}{2}\varepsilon_{AB}\nabla^2,\end{aligned}$$

which is equivalent to

$$\partial^A{}_C\partial^C{}_B = \frac{1}{2}\delta^A{}_B\nabla^2,$$

so that, in effect, Eq. ($-\frac{\hbar^2}{M}\partial^A{}_C\partial^C{}_B\psi^B + V(r)\psi^A = i\hbar\frac{\partial\psi^A}{\partial t}$.) amounts to

$$-\frac{\hbar^2}{2M}\nabla^2\psi^A + V(r)\psi^A = i\hbar\frac{\partial\psi^A}{\partial t}.$$

When there is an electromagnetic field present, we replace $\partial^A{}_B$ by $\partial^A{}_B - (iq/\hbar c)A^A{}_B$ and $\partial/\partial t$ by $\partial/\partial t + (iq/\hbar)\phi$ in Eq. ($-\frac{\hbar^2}{M}\partial^A{}_C\partial^C{}_B\psi^B + V(r)\psi^A = i\hbar\frac{\partial\psi^A}{\partial t}$.), and we obtain

$$-\frac{\hbar^2}{M}\left(\partial^A{}_C - \frac{iq}{\hbar c}A^A{}_C\right)\left(\partial^C{}_B - \frac{iq}{\hbar c}A^C{}_B\right)\psi^B$$
$$+V(r)\psi^A + q\phi\psi^A = i\hbar\frac{\partial\psi^A}{\partial t},$$

which is equivalent to

$$-\frac{\hbar^2}{M}\Big[\partial^A{}_C\partial^C{}_B\psi^B - \frac{iq}{\hbar c}\left(\partial^A{}_C A^C{}_B\right)\psi^B$$
$$-\frac{iq}{\hbar c}A^C{}_B\partial^A{}_C\psi^B - \frac{iq}{\hbar c}A^A{}_C\partial^C{}_B\psi^B$$
$$-\left(\frac{q}{\hbar c}\right)^2 A^A{}_C A^C{}_B\psi^B\Big] + V(r)\psi^A + q\phi\psi^A$$
$$= i\hbar\frac{\partial\psi^A}{\partial t}.$$

In order to reduce this last expression we begin by noticing that

$$\partial_{AC}A^C{}_B = \frac{1}{2}\left(\partial_{AC}A^C{}_B + \partial_{BC}A^C{}_A\right)$$
$$+\frac{1}{2}\left(\partial_{AC}A^C{}_B - \partial_{BC}A^C{}_A\right)$$
$$= \partial_{(A|C|}A^C{}_{B)} + \varepsilon_{AB}\partial^R{}_C A^C{}_R,$$

where the parenthesis denotes symmetrization on the indices enclosed $\left(\text{e.g.}, \mathrm{M}_{(AB)} = (1/2)(\mathrm{M}_{AB} + \mathrm{M}_{BA})\right)$ and the indices between bars are excluded from the symmetrization. The first term in the right-hand side of the last equality is the spinor equivalent of $\left(i/\sqrt{2}\right)\nabla\times A$, which follows from the fact that the spinor equivalent of the Levi-Civita symbol ε_{ijk} or $\varepsilon_{ABCDEG} = \left(i/\sqrt{2}\right)\left(\varepsilon_{AC}\varepsilon_{BE}\varepsilon_{DG} + \varepsilon_{BD}\varepsilon_{AG}\varepsilon_{CE}\right)$, while the last term is equal to $(1/2)\varepsilon_{AB}\nabla\cdot A$. Making use of Eqs. ($-\phi^A\psi_A$.) and ($\phi_{AB} = \frac{1}{2}\phi^R{}_R\varepsilon_{AB}$.) we find that

$$A^C{}_B\partial_{AC} + A_{AC}\partial^C{}_B = A^C{}_B\partial_{AC} - A^C{}_A\partial_{BC}$$
$$= \varepsilon_{BA}A^{CR}\partial_{RC} = \varepsilon_{AB}A.\nabla.$$

Finally, by analogy with Eq. ($\partial^A{}_C\partial^C{}_B = \frac{1}{2}\delta^A{}_B\nabla^2$,) $A^A{}_C A^C{}_B = (1/2)\delta^A_B A^2$. Thus, equation can be also be written as

$$-\frac{\hbar^2}{2M}\left[\nabla^2\psi^A + \frac{\sqrt{2}q}{\hbar c}B^A{}_B\psi^B - \frac{iq}{\hbar c}(\nabla.A)\psi^A\right.$$

$$-\frac{2iq}{\hbar c} A.\nabla\psi^{A} - \left(\frac{q}{\hbar c}\right)^{2} A^{2}\psi^{A}\Bigg] + V(r)\psi^{A}$$

$$+q\phi\psi^{A} = i\hbar\frac{\partial\psi^{A}}{\partial t},$$

where B_{AB} denotes the spinor equivalent of B, and one can verify that this expression coincides with Eq. $-\frac{\hbar^{2}}{2M}\left[\nabla^{2}\psi - \frac{2iq}{\hbar c} A.\nabla\psi - \frac{iq}{\hbar c}(\nabla\cdot A)\psi - \left(\frac{q}{\hbar c}\right)^{2} A^{2}\psi + \frac{q}{\hbar c} B\cdot\sigma\psi\right] + V(r)\psi + q\phi\psi = i\hbar\frac{\partial\psi}{\partial t}$.

Spin-3/2 Particles

A spin-3/2 particle is described by a totally symmetric three index spinor field, ψ^{ABC}, which under rotations transforms according to

$$\psi'^{ABC} = U^{A}{}_{R}U^{B}{}_{S}U^{C}{}_{T}\psi^{RST},$$

where $(U^{A}{}_{B})$ is the SU(2) matrix appearing in Eq. ($\psi' = \left(cos\frac{1}{2}\alpha I - i\,sin\,\frac{1}{2}\alpha\; n{\cdot}\sigma\right)\psi,$), namely

$$U^{A}{}_{B} = cos\frac{1}{2}\alpha\delta^{A}_{B} - i\,\sqrt{2}\,sin\frac{1}{2}\alpha\; n^{A}B$$

and n_{AB} is the spinor equivalent of the unit vector n. Hence, for an infinitesimal rotation,

$$\psi'^{ABC} \simeq \left(\delta^{A}_{R} - \frac{i\alpha}{\sqrt{2}}n^{A}{}_{R}\right)\left(\delta^{B}_{S} - \frac{i\alpha}{\sqrt{2}}n^{B}{}_{S}\right)$$

$$\times\left(\delta^{C}_{T} - \frac{i\alpha}{\sqrt{2}}n^{C}{}_{T}\right)\psi^{RST}$$

$$\simeq \psi^{ABC} - \frac{3i}{\sqrt{2}}\alpha n^{(A}{}_{R}\psi^{BC)R},$$

which implies that the operator $n\cdot S$ given by

$$(n\,.\,S\psi)^{ABC} = \frac{3\hbar}{\sqrt{2}}n^{(A}{}_{R}\psi^{BC)R}$$

corresponds to the component of the spin along n.

By analogy with the Schrodinger–Pauli equation ($-\frac{\hbar^{2}}{M}\partial^{A}{}_{C}\partial^{C}{}_{B}\psi^{B} + V(r)\psi^{A} = i\hbar\frac{\partial\psi^{A}}{\partial t}$.), for a spin-3/2 particle we propose the equation

$$-\frac{\hbar^{2}}{M}\partial^{(A}{}_{R}\partial^{|R}{}_{S}\psi^{S|BC)} + V(r)\psi^{ABC} = i\hbar\frac{\partial\psi^{ABC}}{\partial t}$$

which, owing to Eq. ($\partial^{A}{}_{C}\partial^{C}{}_{B} = \frac{1}{2}\delta^{A}{}_{B}\nabla^{2}$), means that, in the absence of an electromagnetic field, each component of ψ^{ABC} satisfies the Schrodinger equation $-\frac{\hbar^{2}}{2M}\nabla^{2}\psi + V(r)\psi = i\hbar\frac{\partial\psi}{\partial t}$. (It should be noticed that, even in the absence of an electromagnetic field, the components of the spinor field

ψ^{ABC} may be coupled among themselves if a non-Cartesian basis is employed, in which case the partial derivatives appearing in Eq. ($i\hbar\frac{\partial\psi^{ABC}}{\partial t}$) have to be replaced by covariant derivatives. Instead of the symmetrized second derivatives $\partial^{(A}{}_{R}\partial^{|R}{}_{S}\psi^{S|BC)}$ appearing in Eq. ($i\hbar\frac{\partial\psi^{ABC}}{\partial t}$), one could also consider $\partial^{(A}{}_{R}\chi^{|R|BC)}$, with $\chi^{RBC} \equiv \partial^{(R}{}_{S}\psi^{|S|BC)}$; however, in the latter case, each component ψ^{ABC} would not satisfy the Schrodinger equation.

By combining Eq. ($i\hbar\frac{\partial\psi^{ABC}}{\partial t}$) and its complex conjugate one obtains the continuity equation

$$\frac{\partial}{\partial t}\left(\hat{\psi}_{ABC}\psi^{ABC}\right)+\frac{\hbar}{iM}\partial_{RS}\left(\psi^{SBC}\partial^{AR}\hat{\psi}_{ABC}\right.$$
$$\left.+\hat{\psi}^{SBC}\partial^{AR}\psi_{ABC}\right)=0,$$

where $\hat{\psi}_{ABC} \equiv \overline{\psi^{ABC}}$; hence, $\hat{\psi}_{ABC}\psi^{ABC}$ is real and positive.

As in the case of the relativistic description of spin-3/2 particles, instead of three-index spinors, one can employ wave functions with one spinor index and one tensor index. In the present case, we can define

$$\psi_{iA} \equiv -\frac{1}{\sqrt{2}}\sigma i^{BC}\psi_{ABC},$$

then the symmetry of ψ_{ABC} in its three indices is expressed by the condition $\sigma^{i\,AB}\psi_{iA}=0$.

When there is an electromagnetic field present, we replace $\partial^{A}{}_{B}$ by $\partial^{A}{}_{B}-(iq/\hbar c)A^{A}{}_{B}$ and $\partial/\partial t$ by $\partial/\partial t + (iq/\hbar)\phi$, in Eq. ($i\hbar\frac{\partial\psi^{ABC}}{\partial t}$), which yields

$$-\frac{\hbar^2}{M}\left(\partial^{(A}{}_{R}-\frac{iq}{\hbar c}A^{(A}{}_{R}\right)\left(\partial^{|R}{}_{S}-\frac{iq}{\hbar c}A^{|R}{}_{S}\right)\psi^{S|BC)}$$
$$+V(r)\psi^{ABC}+q\phi\psi^{ABC}=i\hbar\frac{\partial\psi^{ABC}}{\partial t}.$$

Following the same steps as in Eq. ($i\hbar\frac{\partial\psi^{A}}{\partial t}$) we obtain

$$-\frac{\hbar^2}{2M}\left[\nabla^2\psi^{ABC}+\frac{\sqrt{2}q}{\hbar c}B^{(A}{}_{S}\psi^{BC)S}\right.$$
$$-\frac{iq}{\hbar c}(\nabla.\mathrm{A})\psi^{ABC}-\frac{2iq}{\hbar c}\mathrm{A}.\nabla\psi^{ABC}$$
$$\left.-\left(\frac{q}{\hbar c}\right)^2\mathrm{A}^2\psi^{ABC}\right]+V(r)\psi^{ABC}$$
$$+q\phi\psi^{ABC}=i\hbar\frac{\partial\psi^{ABC}}{\partial t}.$$

Thus, taking into account Eq. ($(n\,.\,S\psi)^{ABC} = \frac{3\hbar}{\sqrt{2}} n^{(A}{}_{R}\psi^{BC)R}$), one concludes that in the present case there is an interaction with the magnetic field of the form $-\mu \cdot B$, with

$$\mu = \frac{q}{3Mc} S$$

that corresponds to the gyromagnetic ratio

$$\frac{q}{3Mc}.$$

Owing to the difference between the gyromagnetic ratios (10) and (30), in both cases, the greatest eigenvalue of the operator μ is given by

$$\mu_{\max} = \frac{|q|}{Mc}\frac{\hbar}{2}.$$

Equations ($-\frac{\hbar^2}{M}\partial^A{}_C\partial^C{}_B\psi^B + V(r)\psi^A = i\hbar\frac{\partial\psi^A}{\partial t}.$) and ($i\hbar\frac{\partial\psi^{ABC}}{\partial t}$) can be readily generalized for any value of the spin. A spin-s particle would be represented by a totally symmetric 2s-index spinor field, $\psi^{AB\cdots L}$, satisfying

$$-\frac{\hbar^2}{M}\partial^{(A}{}_R\partial^{|R}{}_S\psi^{S|B\cdots L)} + V(r)\psi^{AB\cdots L} = i\hbar\frac{\partial\psi^{AB\cdots L}}{\partial t}$$

and if the particle has electric charge q, the gyromagnetic ratio would be given by

$$\frac{q}{3sMc}.$$

As pointed out above, under the rotation corresponding to the SU(2) matrix ($U^A{}_B$), the Cartesian components of a spinor $\psi^{AB\ldots L}$ transform according to

$$\psi'^{AB\cdots L} = \ddot{u}^A{}_P{}^B{}_Q\cdots{}^L{}_R\psi^{PQ\cdots R}.$$

Since each term in Eq. ($i\hbar\frac{\partial\psi^{AB\cdots L}}{\partial t}$) transforms in the same manner as $\psi^{AB\ldots L}$, the validity of Eq. ($i\hbar\frac{\partial\psi^{AB\cdots L}}{\partial t}$) in a given Cartesian frame implies its validity in any Cartesian frame obtained from the original one by means of a rotation. This can be seen as a consequence of the fact that a contraction of the form $\partial^A{}_R\psi^{RB\ldots L}$ transforms as a 2s-index spinor since

$$U_{RM}U^R{}_N = \det(U^A{}_B)\varepsilon_{NM} = \varepsilon_{NM}$$

and therefore

$$\begin{aligned}
&\partial'^{A}{}_{R}\,\psi'^{RB\ldots L} \\
&= \left(U^{A}{}_{P}U_{RM}\partial^{PM}\right)U^{R}{}_{N}U^{B}{}_{Q}\cdots U^{L}{}_{S}\psi^{NQ\ldots S} \\
&= U_{RM}U^{R}{}_{N}U^{A}{}_{P}U^{B}{}_{Q}\cdots U^{L}{}_{S}\partial^{PM}\psi^{NQ\ldots S} \\
&= \varepsilon_{NM}U^{A}{}_{P}U^{B}{}_{Q}\cdots U^{L}{}_{S}\partial^{PM}\psi^{NQ\ldots S} \\
&= U^{A}{}_{P}U^{B}{}_{Q}\cdots U^{L}{}_{S}\partial^{P}{}_{N}\psi^{NQ\ldots S}
\end{aligned}$$

On the other hand, under the Galilean transformation

$$r' = r - vt, \qquad t' = t,$$

where v is constant, using the chain rule, one finds that

$$\frac{\partial}{\partial x'^{i}} = \frac{\partial}{\partial x^{i}}, \qquad \frac{\partial}{\partial t'} = v^{i}\frac{\partial}{\partial x^{i}} + \frac{\partial}{\partial t'}$$

then, assuming that

$$\psi'^{\mathrm{AB}\ldots\mathrm{L}} = \psi^{AB\ldots L}\,exp\left(-i\frac{m\,\mathrm{v.r}}{\hbar} + i\frac{mv^{2}t}{2\hbar}\right),$$

a straightforward computation, making use of Eq. ($\partial^{A}{}_{C}\partial^{C}{}_{B} = \frac{1}{2}\delta^{A}{}_{B}\nabla^{2}$,), shows that Eq. ($i\hbar\frac{\partial\psi^{AB\cdots L}}{\partial t}$) is form-invariant under Galilean transformations.

Rydberg Equation

The Rydberg formula is the mathematical formula to determine the wavelength of light emitted by an electron moving between the energy levels of an atom. When an electron transfers from one atomic orbital to another, its energy changes. When an electron shifts from an orbital with high energy to a lower energy state, a photon of light is generated. A photon of light gets absorbed by the atom when the electron moves from low energy to a higher energy state. The Rydberg Formula applies to the spectra of various elements and is given by

$$\bar{\upsilon} = \frac{1}{\lambda} = R\left(\frac{1}{n_1^2} - \frac{1}{n_2^2}\right)$$

Where,

n_1 and n_2 are integers and n_2 is always greater than n_1.

R is constant, called Rydberg constant and formula is usually written as

$$\bar{\upsilon} = R_H\left(\frac{1}{n_1^2} - \frac{1}{n_2^2}\right)$$

The modern value of Rydberg constant is 109677.57 cm-1 and the most precise known physical constant.

$$\bar{\upsilon} = \frac{1}{\lambda} = 109680\left(\frac{1}{n_1^2} - \frac{1}{n_2^2}\right)cm^{-1}\left(n_2 > n_1\right)$$

Example

Determine the v¯ for the transition wherein n_1=6 to $_{n2}$ = 3 in a hydrogen atom.

Solution:

Since, the mass of nucleus is much greater than mass of an electron,

Therefore, R = 109737.32 cm^{-1}

v¯ = 109737 x (1/132 – 1/162)

v¯ = 9144.78 cm^{-1}

Determine the wavelength of the second line in the Paschen series and prove that this line lies near to the visible, infrared region.

Example

According to Paschen series, n_1 = 3 and n_2 = 4, 5...

Therefore, the second line in the Paschen series is set by putting n_1 = 3 and n_2 = 5

$$\bar{\upsilon} = \frac{1}{\lambda} = 109680\left(\frac{1}{n_1^2} - \frac{1}{n_2^2}\right)cm^{-1}\left(n_2 > n_1\right)$$

Substitute the above values in the equation

v¯ = 7.799 x 103 cm^{-1}

λ = 1.282 x 10^{-4} cm = 1282 nm which is in near infrared region.

References

- Greiner, W. (2000). Relativistic Quantum Mechanics. Wave Equations (3rd ed.). Springer Verlag. ISBN 3-5406-74578
- Klein-Gordon-equation: ncatlab.org, Retrieved 12 April 2018
- Jump up^ Akhmeteli, Andrey (2011). "One real function instead of the Dirac spinor function" (PDF). Journal of Mathematical Physics. 52 (8): 082303. arXiv:1008.4828. Bibcode:2011JMP....52h2303A. doi:10.1063/1.3624336
- Tommy Ohlsson (22 September 2011). Relativistic Quantum Physics: From Advanced Quantum Mechanics to Introductory Quantum Field Theory. Cambridge University Press. p. 86. ISBN 978-1-139-50432-4
- Feshbach, H.; Villars, F. (1958). "Elementary relativistic wave mechanics of spin 0 and spin 1/2 particles". Rev. Mod. Phys. 30 (1). Bibcode:1958RvMP...30...24F. doi:10.1103/RevModPhys.30.24

Quantum Number

In a quantum system, quantum numbers describe the values of conserved quantities. They can have integral or half-integral values. This chapter closely examines the different quantum numbers, such as principal, azimuthal, magnetic, spin and total angular momentum quantum numbers, for a complete understanding of quantum numbers.

The quantum numbers are basically represents the address of an electrons in atom. These quantum numbers represent the location, energy level and spin of electron in atom. These quantum numbers are useful for representing the electron configuration. Quantum numbers are of four types –

- Principal quantum number (n)
- Orbital or Azimuthal quantum number (l)
- Magnetic quantum number (m or m_l)
- Spin magnetic quantum number (m_s)

Rules Governing the Allowed Combinations of Quantum Numbers

- The three quantum numbers (*n*, *l*, and *m*) that describe an orbital are integers: 0, 1, 2, 3, and so on.
- The principal quantum number (*n*) cannot be zero. The allowed values of *n* are therefore 1, 2, 3, 4, and so on.
- The angular quantum number (*l*) can be any integer between 0 and $n - 1$. If $n = 3$, for example, *l* can be either 0, 1, or 2.
- The magnetic quantum number (*m*) can be any integer between -*l* and +*l*. If $l = 2$, *m* can be either -2, -1, 0, +1, or +2.

Possible Combinations of Quantum Numbers

There is only one orbital in the $n = 1$ shell because there is only one way in which a sphere can be oriented in space. The only allowed combination of quantum numbers for which $n = 1$ is the following.

n	*l*	*m*	
1	0	0	1s

There are four orbitals in the $n = 2$ shell.

n	l	m	
2	0	0	2s
2	1	−1	
2	1	0	2p
2	1	1	

There is only one orbital in the 2s subshell. But, there are three orbitals in the 2p subshell because there are three directions in which a p orbital can point. One of these orbitals is oriented along the X axis, another along the Y axis, and the third along the Z axis of a coordinate system, as shown in the figure below. These orbitals are therefore known as the $2p_x$, $2p_y$, and $2p_z$ orbitals.

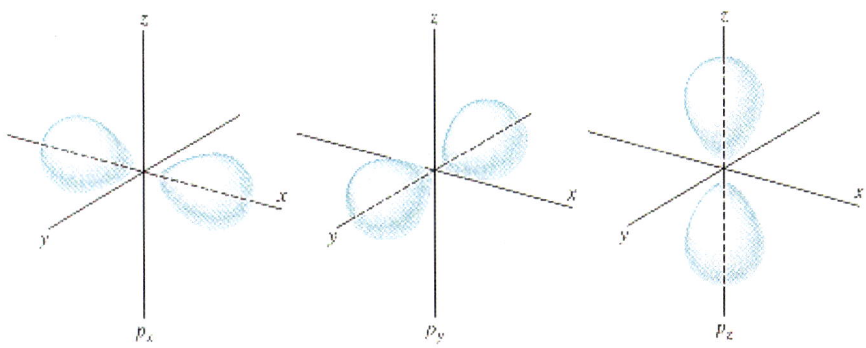

There are nine orbitals in the n = 3 shell.

n	l	m	
3	0	0	3s
3	1	−1	
3	1	0	3p
3	1	1	
3	2	−2	
3	2	−1	
3	2	0	3d
3	2	1	
3	2	2	

There is one orbital in the 3s subshell and three orbitals in the 3p subshell. The n = 3 shell, however, also includes 3d orbitals.

The five different orientations of orbitals in the 3d subshell are shown in the figure below. One of these orbitals lies in the XY plane of an XYZ coordinate system and is called the $3d_{xy}$ orbital. The $3d_{xz}$ and $3d_{yz}$ orbitals have the same shape, but they lie between the axes of the coordinate system in the XZ and YZ planes. The fourth orbital in this subshell lies along the X and Y axes and is called the $3d_{x^2-y^2}$ orbital. Most of the space occupied by the fifth orbital lies along the Z axis and this orbital is called the $3d_{z^2}$ orbital.

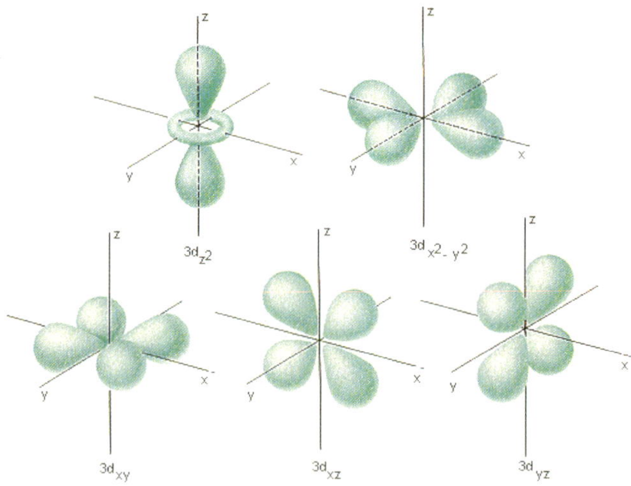

The number of orbitals in a shell is the square of the principal quantum number: $1^2 = 1$, $2^2 = 4$, $3^2 = 9$. There is one orbital in an s subshell ($l = 0$), three orbitals in a p subshell ($l = 1$), and five orbitals in a d subshell ($l = 2$). The number of orbitals in a subshell is therefore $2(l) + 1$.

Before we can use these orbitals we need to know the number of electrons that can occupy an orbital and how they can be distinguished from one another. Experimental evidence suggests that an orbital can hold no more than two electrons.

Total Angular Momenta Numbers

Total Momentum of a Particle

When one takes the spin–orbit interaction into consideration, the L and S operators no longer commute with the Hamiltonian, and their eigenvalues therefore change over time. Thus another set of quantum numbers should be used. This set includes

1. The total angular momentum quantum number:

   $j = |\ell \pm s|,$

   which gives the total angular momentum through the relation

   $J^2 = \hbar^2 j(j + 1).$

2. The projection of the total angular momentum along a specified axis:

   $m_j = -j, -j + 1, -j + 2, \ldots, j - 2, j - 1, j$

   analogous to the above and satisfies

   $m_j = m_\ell + m_s$ and $|m_\ell + m_s| \leq j.$

3. Parity

This is the eigenvalue under reflection: positive (+1) for states which came from even ℓ and negative (−1) for states which came from odd ℓ. The former is also known as even parity and the latter as odd parity, and is given by

$$P = (-1)^{\ell}.$$

For example, consider the following 8 states, defined by their quantum numbers:

#	n	ℓ	m_ℓ	m_s
1.	2	1	1	+1/2
2.	2	1	1	−1/2
3.	2	1	0	+1/2
4.	2	1	0	−1/2
5.	2	1	−1	+1/2
6.	2	1	−1	−1/2
7.	2	0	0	+1/2
8.	2	0	0	−1/2

$\ell + s$	$\ell - s$	$m_\ell + m_s$
3/2	~~1/2~~	3/2
3/2	1/2	1/2
3/2	1/2	1/2
3/2	1/2	−1/2
3/2	1/2	−1/2
3/2	~~1/2~~	−3/2
1/2	−1/2	1/2
1/2	−1/2	−1/2

The quantum states in the system can be described as linear combination of these 8 states. However, in the presence of spin–orbit interaction, if one wants to describe the same system by 8 states that are eigenvectors of the Hamiltonian (i.e. each represents a state that does not mix with others over time), we should consider the following 8 states:

j	m_j	parity	
3/2	3/2	odd	(coming from state (1) above)
3/2	1/2	odd	(coming from states (2) and (3) above)
3/2	−1/2	odd	(coming from states (4) and (5) above)
3/2	−3/2	odd	(coming from state (6) above)
1/2	1/2	odd	(coming from states (2) and (3) above)
1/2	−1/2	odd	(coming from states (4) and (5) above)
1/2	1/2	even	(coming from state (7) above)
1/2	−1/2	even	(coming from state (8) above)

Nuclear Angular Momentum Quantum Numbers

In nuclei, the entire assembly of protons and neutrons (nucleons) has a resultant angular momentum due to the angular momenta of each nucleon, usually denoted 'I'. If the total angular momentum of a neutron is $j_n = \ell + s$ and for a proton is $j_p = \ell + s$ (where s for protons and neutrons happens to be ½ again) then the nuclear angular momentum quantum numbers I are given by:

$$I = |j_n - j_p|,\ |j_n - j_p| + 1,\ |j_n - j_p| + 2, \ldots, (j_n + j_p) - 2,\ (j_n + j_p) - 1,\ (j_n + j_p)$$

Parity with the number I is used to label nuclear angular momentum states, examples for some isotopes of Hydrogen (H), Carbon (C), and Sodium (Na) are;

${}^{1}_{1}\mathrm{H}\, I = (1/2)^{+}$	${}^{9}_{6}C\, I = (3/2)^{-}$	${}^{20}_{11}\mathrm{Na}\, I = 2^{+}$
${}^{2}_{1}\mathrm{H}\, I = 1^{+}$	${}^{10}_{6}C\, I = 0^{+}$	${}^{21}_{11}\mathrm{Na}\, I = (3/2)^{+}$
${}^{3}_{1}\mathrm{H}\, I = (1/2)^{+}$	${}^{11}_{6}C\, I = (3/2)^{-}$	${}^{22}_{11}\mathrm{Na}\, I = 3^{+}$
	${}^{12}_{6}C\, I = 0^{+}$	${}^{23}_{11}\mathrm{Na}\, I = (3/2)^{+}$
	${}^{13}_{6}C\, I = (1/2)^{-}$	${}^{24}_{11}\mathrm{Na}\, I = 4^{+}$
	${}^{14}_{6}C\, I = 0^{+}$	${}^{25}_{11}\mathrm{Na}\, I = (5/2)^{+}$
	${}^{15}_{6}C\, I = (1/2)^{+}$	${}^{26}_{11}\mathrm{Na}\, I = 3^{+}$

The reason for the unusual fluctuations in I, even by differences of just one nucleon, are due to the odd/even numbers of protons and neutrons - pairs of nucleons have a total angular momentum of zero (just like electrons in orbitals), leaving an odd/even numbers of unpaired nucleons. The property of nuclear spin is an important factor for the operation of NMR spectroscopy in organic chemistry, and MRI in nuclear medicine, due to the nuclear magnetic moment interacting with an external magnetic field.

Elementary Particles

Elementary particles contain many quantum numbers which are usually said to be intrinsic to them. However, it should be understood that the elementary particles are quantum states of the standard model of particle physics, and hence the quantum numbers of these particles bear the same relation to the Hamiltonian of this model as the quantum numbers of the Bohr atom does to its Hamiltonian. In other words, each quantum number denotes a symmetry of the problem. It is more useful in quantum field theory to distinguish between space time and internal symmetries.

Typical quantum numbers related to spacetime symmetries are spin (related to rotational symmetry), the parity, C-parity and T-parity (related to the Poincaré symmetry of spacetime). Typical internal symmetries are lepton number and baryon number or the electric charge.

Multiplicative Quantum Numbers

A minor but often confusing point is as follows: most conserved quantum numbers are additive, so in an elementary particle reaction, the *sum* of the quantum numbers should be the same before and after the reaction. However, some, usually called a *parity*, are multiplicative; i.e., their *product* is conserved. All multiplicative quantum numbers belong to a symmetry (like parity) in which applying the symmetry transformation twice is equivalent to doing nothing (involution).

Principal Quantum Number

This represents the main shell to which to electron belongs. These energy levels are denoted by the symbols K, L, M, N... for n = 1, n= 2, n = 3... respectively. The values of n are 1, 2, 3, ... Here, n=1 refers to the K shell, n=2 represents L shell, n=3 represents M shell etc.

Thus the value of n also represents the order of the shell from the nucleus. 'n' refers to the ionization energy of the electrons in that energy level.

How can we say that Principal quantum number indicates the size of the Orbital?

We could observe that K shell (n=1) has least energy whereas L shell (n=2) has more energy compared to K shell. Hence, the value of 'n' also represents energy of the electron.

As the value of 'n' increases, size of the shell increases. Maximum number of electrons in any shell is $2n^2$. Hence the the size of the orbital can be determined using this quantum number.

Principal quantum number corresponds to energy levels in the Bohr model of the atom. It describes how far the electron is from the nucleus. The value of n is always an integer from 1 i.e. $n \geq 1$, such as n = 1, n= 2, n = 3... 7 and so on.

As the value of n increases, size of orbital also increases. Maximum number of electrons that is possible in a shell is $2n^2$ and maximum number of electrons in an energy level is $4n^2$.

For example:

For energy level 1, n = 1, maximum no. of electrons possible = $2(1)^2 = 2$.

For energy level 2, n = 2, maximum no. of electrons possible = $2(2)^2 = 8$.

For energy level 3, n = 3, maximum no. of electrons possible = $2(3)^2 = 18$.

For energy level 4, n = 4 maximum no. of electrons possible = $2(4)^2 = 32$.

Principal Quantum Number Chart

For any principal quantum number n, value of angular momentum quantum number u is 0 to u - 1, value of magnetic quantum number is $-l$ to $+l$.

Following chart shows the principal quantum number chart.

Quantum numbers			No of Quantum states	
n	l	m_l	In the sub shell	In the Main level
1	0	0	2	2
2	0	0	2	8
2	1	-1, 0, +1	6	8
3	0	0	2	18
3	1	-1, 0, +1	6	18
3	2	-2,-1,0,+1,+2	10	18

4	0	0	2	32
4	1	-1, 0, +1	6	32
4	2	-2,-1,0,+1,+2	10	32
4	3	-3,-2,-1, 0,+1,+2,+3	14	32

As shown in above chart,

For n = 1, value of ι = 0, the Magnetic moment is $m_ι = 0$

For n = 2, value of ι = 0, 1, the Magnetic moment is $m_ι$ = 0, (- 1, 0, + 1).

For n = 3, value of ι = 0, 1, 2, the Magnetic moment is $m_ι$ = 0, (- 1, 0, + 1), (- 2, - 1, 0, + 1, + 2).

Thus we can calculate Azimuthal quantum number (ι) and magnetic quantum number ($m_ι$) using the principal quantum number n.

Derivation

There are a set of quantum numbers associated with the energy states of the atom. The four quantum numbers *n*, ℓ, *m*, and *s* specify the complete and unique quantum state of a single electron in an atom, called its wave function or orbital. Two electrons belonging to the same atom cannot have the same no of four quantum numbers, due to the Pauli exclusion principle. The wavefunction of the Schrödinger wave equation reduces to the three equations that when solved lead to the first three quantum numbers. Therefore, the equations for the first three quantum numbers are all interrelated. The principal quantum number arose in the solution of the radial part of the wave equation as shown below.

The Schrödinger wave equation describes energy eigenstates with corresponding real numbers E_n and a definite total energy, the value of E_n. The bound state energies of the electron in the hydrogen atom are given by:

$$E_n = \frac{E_1}{n^2} = \frac{-13.6 \text{ eV}}{n^2}, \quad n = 1, 2, 3, \ldots$$

The parameter *n* can take only positive integer values. The concept of energy levels and notation were taken from the earlier Bohr model of the atom. Schrödinger's equation developed the idea from a flat two-dimensional Bohr atom to the three-dimensional wave function model.

In the Bohr model, the allowed orbits were derived from quantized (discrete) values of orbital angular momentum, *L* according to the equation

$$\mathrm{L} = n \cdot \hbar = n \cdot \frac{h}{2\pi}$$

where n = 1, 2, 3, ... and is called the principal quantum number, and *h* is Planck's constant. This formula is not correct in quantum mechanics as the angular momentum magnitude is described by the azimuthal quantum number, but the energy levels are accurate and classically they correspond to the sum of potential and kinetic energy of the electron.

The principal quantum number n represents the relative overall energy of each orbital. The energy level of each orbital increases as its distance from the nucleus increases. The sets of orbitals with the same n value are often referred to as electron shells or energy levels.

The minimum energy exchanged during any wave-matter interaction is the product of the wave frequency multiplied by Planck's constant. This causes the wave to display particle-like packets of energy called quanta. The difference between energy levels that have different n determine the Emission spectrum of the element.

In the notation of the periodic table, the main shells of electrons are labeled:

K (n = 1), L (n = 2), M (n = 3), etc.

based on the principal quantum number.

The principal quantum number is related to the radial quantum number, n_r, by:

$$n = n_r + \ell + 1$$

where ℓ is the azimuthal quantum number and n_r is equal to the number of nodes in the radial wave function.

The definite total energy for a particle motion in a common Coulomb field and with a discrete spectrum, is given by:

$$E_n = -\frac{Z^2\hbar^2}{2m_0 a_B^2 n^2} = -\frac{Z^2 e^4 m_0}{2\hbar^2 n^2},$$

where:

- a_B is the Bohr radius,
- n is the principal quantum number.

This discrete energy spectrum resulted from the solution of the quantum mechanical problem on the electron motion in the Coulomb field, coincides with the spectrum that was obtained with the help application of the Bohr-Sommerfeld quantization rules to the classical equations. The radial quantum number determines the number of nodes of the radial wave function $R(r)$.

Azimuthal Quantum Number

A quantum number for an atomic orbital is called azimuthal quantum number. This number describes the shape of the orbital and also determines the orbital angular momentum.

Other than principal quantum number (n), spectroscopic notation, magnetic quantum number (m) and the spin quantum number (s) – the azimuthal quantum number is another set of quantum numbers which describe the unique quantum state of an electron.

It is also termed as the orbital angular momentum quantum number, orbital quantum number or second quantum number, and is symbolized as ℓ.

Azimuthal quantum number describes the shape of orbital. It is denoted by. Values of are from zero to n-1.

For s-orbital, ℓ = 0

For p-orbita, ℓ = 1

For d-orbital, ℓ = 2

For f-orbital, ℓ = 3

With the help of the value of azimuthal quantum number we can determine the total number of energy sub-levels in a given energy level.

Angular Momentum Quantum Numbers

- Intrinsic (or spin) angular momentum quantum number, or simply spin quantum number.
- Orbital angular momentum quantum number.
- Magnetic quantum number, related to the orbital momentum quantum number.
- Total angular momentum quantum number.

The relation between angular momentum of the electron and ι is

Angular Momentum = $\sqrt{\iota(\iota \quad)}$—

where h = Plancks constant.

The various sub-levels or sub-shells are also designated by letters s, p, d, f depending upon the values of ι as shown below:

Value of ι	0	1	2	3
Symbol of Sublevel	s	p	d	f

Each value of ιι represents a particular sub-shell or sub-level. For e.g. for n = 3, ι can have three values, i.e, ι = 0,1,2. This means that there are three subshells in the third energy level.

Each of the different azimuthal quantum number states can take 2(2 ι + 1) electrons.

Principal Quantum number n	Azimuthal Quantum number (ι)	Orbital	Shape of the Orbital.
1	0	s	Spherical
2	0	s	Spherical
2	1	p	Dumble
3	0	s	Spherical
3	1	p	Polar dumble
3	2	d	Clover Leaf

4	0	s	Spherical
4	1	p	Polar dumble
4	2	d	Clover leaf
4	3	f	Complex

These orbital can take even more complex shapes according to the values of angular quantum number. A sphere ($\iota = 0$) can be oriented in space in only one way. However orbital that has polar or cloverleaf shapes can point in different directions. To describe the orientation in space of a particular orbital, we therefore need a third quantum number which is known as the magnetic quantum number (m_ι).

There is only one type of s orbital, three types of p orbital (p_x, p_y and p_z), five types of d orbital (d_{z^2}, d_{xy}, d_{zy}, d_{xz} and $d_{x^2-y^2}$) and 7 types of f orbital. These subtypes of orbital are due to magnetic quantum number which describes the orientation of orbital in space or position of electron in the orbital.

Following diagram shows the shape of orbital depending on the value of ι.

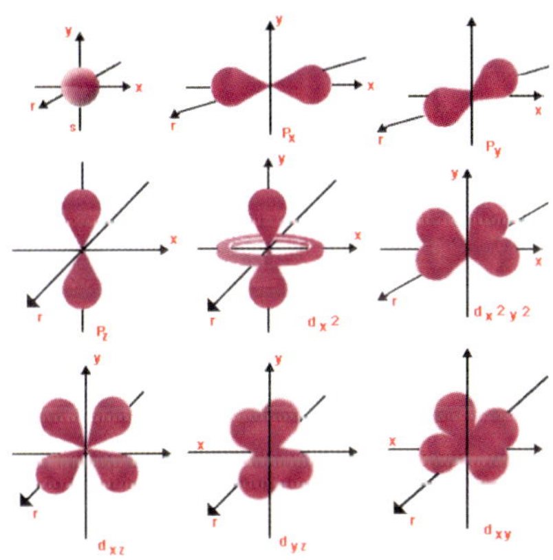

Azimuthal quantum number also determines the orbital angular momentum of an electron. The magnitude of orbital angular momentum (L) is given through the following relation:

$$L^2 = \bar{h}^2 \iota(\iota+1)$$

Here ι is azimuthal quantum number, $\bar{h}$ is reduced planck's constant and is equal to $\frac{h}{2\pi}$.

Addition of Quantized Angular Momenta

Given a quantized total angular momentum $\vec{j}$ which is the sum of two individual quantized angular momenta $\vec{\ell_1}$ and $\vec{\ell_2}$,

$$\vec{j} = \vec{\ell_1} + \vec{\ell_2}$$

the quantum number j associated with its magnitude can range from $|\ell_1 - \ell_2|$ to $\ell_1 + \ell_2$ in integer steps where ℓ_1 and ℓ_2 are quantum numbers corresponding to the magnitudes of the individual angular momenta.

Total Angular Momentum of an Electron in the Atom

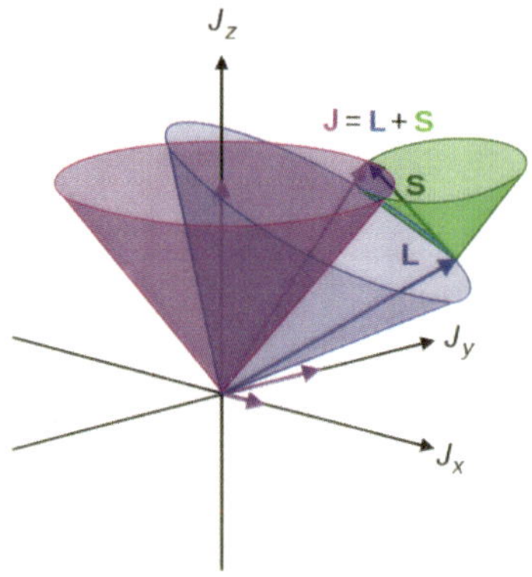

"Vector cones" of total angular momentum J (purple), orbital L (blue), and spin S (green). The cones arise due to quantum uncertainty between measuring angular momentum components

Due to the spin-orbit interaction in the atom, the orbital angular momentum no longer commutes with the Hamiltonian, nor does the spin. These therefore change over time. However the total angular momentum J does commute with the Hamiltonian and so is constant. J is defined through

$$\vec{J} = \vec{L} + \vec{S}$$

L being the orbital angular momentum and S the spin. The total angular momentum satisfies the same commutation relations as orbital angular momentum, namely

$$[J_i, J_j] = i\hbar\epsilon_{ijk}J_k$$

from which follows

$$\left[J_i, J^2\right] = 0$$

where J_i stand for J_x, J_y, and J_z.

The quantum numbers describing the system, which are constant over time, are now j and m_j, defined through the action of **J** on the wave function ψ

$$J^2\psi = \hbar^2 j(j+1)\psi$$
$$J_z\psi = \hbar m_j\psi$$

So that j is related to the norm of the total angular momentum and m_j to its projection along a specified axis.

As with any angular momentum in quantum mechanics, the projection of J along other axes cannot be co-defined with J_z, because they do not commute.

Relation between New and old Quantum Numbers

j and m_j, together with the parity of the quantum state, replace the three quantum numbers ℓ, m_ℓ and m_s (the projection of the spin along the specified axis). The former quantum numbers can be related to the latter.

Furthermore, the eigenvectors of j, s, m_j and parity, which are also eigenvectors of the Hamiltonian, are linear combinations of the eigenvectors of ℓ, s, m_ℓ and m_s.

Magnetic Quantum Number

The magnetic quantum number of an electron is one of the four quantum numbers that state the position of the electron with respect to the nucleus. The other three are-

- Principal quantum number
- Azimuthal quantum number
- Spin quantum number

Magnetic quantum number is the third on the list between azimuthal and spin quantum numbers. It splits the subshells (such as s, p, d,f) into individual orbitals and places the electron in one of them. It defines the orientation in space of a given orbital of a particular energy (n) and shape (I).

In each subshell, the number of orbitals is given as 2+1, where is the azimuthal quantum number. By that process, we can find out the orbital in every subshell.

Since magnetic quantum number is dependant on azimuthal quantum number, its value ranges from – to +, including 0.

For example, s subshell has only one orbital, the p subshell has 3 and d has 5 and so on.

Subshells	Orbitals	Maximum Number of Electrons
S	1	2
P	3	6
D	5	10
F	7	14

Find Magnetic Quantum Number

Quantum numbers in general, and magnetic quantum number in particular can be derived while solving the Schrödinger equation. However, boundaries or constraints need to be put in place to match the equations with the physical world. Magnetic quantum number can be derived from solving the azimuthal equation of the hydrogen Schrodinger equation.

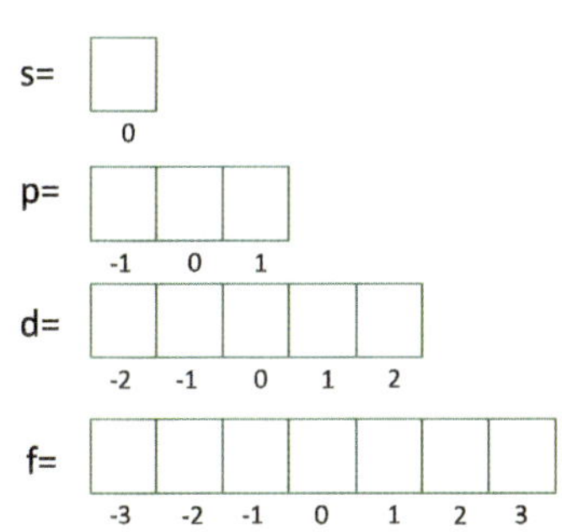

Determine Magnetic Quantum Number

To determine Magnetic quantum number of an electron, we should first find the type of sub shell (s, p, d, f) or angular momentum quantum number. Angular momentum quantum number can be known from principal quantum number. Hence from principal quantum number, we can deduce magnetic quantum number as follows.

For principal quantum number n, angular momentum quantum number $\iota = n - 1$. Further for each value of possible angular momentum quantum number ι, value of magnetic quantum number is: $-\iota$ to $+\iota$.

For example: For vanadium (V) if the principal quantum number n is 3.

$\therefore \iota = n - 1 = 3 - 1 = 2.$

$m_\iota = (2\iota + 1) = (2(2) + 1) = 5$ possible values

Thus the possible values are - 2, - 1, 0, + 1, + 2.

For vanadium configuration of last sub shell is $3d^3$.

Here value of magnetic quantum number of last filled electron is 0 hence value of $m_\iota = 0$ for the electron of highest energy (last filled electron) in Vanadium.

Magnetic Spin Quantum Number

Magnetic quantum number $m\iota$ shows the orientation of orbital spin in three dimensional space around the nucleus or the position of electron in the orbital. Its values lies from $-\iota$ to $+\iota$. The maximum possible number of electrons in any orbital (position) is two (2).

Hence there are two possible electrons in an orbital with two spins i.e. up and down.

As shown s sublevel has 1 position therefore it contains maximum 2 electrons.

$m_\iota = 0.$

p sublevel has 3 possible orbital therefore it can have maximum of $2 \times 3 = 6$ electrons as shown:

$m_\iota = -1, 0, +1.$

d sublevel has 5 possible orbital therefore it can have maximum of $2 \times 5 = 10$ electrons as shown:

$m_\iota = -2, -1, 0, +1, +2.$

f sublevel has 7 possible orbital therefore it can have maximum of $2 \times 7 = 14$ electrons as shown:

$m_\iota = -3, -2, -1, 0, +1, +2, +3.$

Following table gives idea of maximum number of electron accumulating the orbital:

Sub shell	Number of Orbitals	Maximum number of electrons
s (ı = 0)	1	2
p (ı = 1)	3	6
d (ı = 2)	5	10
f (ı = 3)	7	14

Consider an example of oxygen atom which has 8 electrons: 2 electrons in inner shell (K) and 6 electrons in outer shell (L).

For innermost shell n = 1,

∴ ı = 1 − 1 = 0 (s sub shell). mı = 0. Therefore possible orbital is 1s.

Outer shell has n = 2,

∴ ı = 2 -1 = 1.

This shell is made up of single s orbital and three p orbital (p_x, p_y, p_z).

To fill these orbital we have some rules:

Fill the electrons with lower energy first. Hence 2s orbital will be filled first and then 2p orbital. 2 electrons out of the 6 electrons in outer shell will be filled in 2s orbital. Now the 3rd, 4th and 5th electron will be filled in p_x, p_y, p_z orbital respectively. 6th electron will be filled again in p_x orbital.

As a Component of Angular Momentum

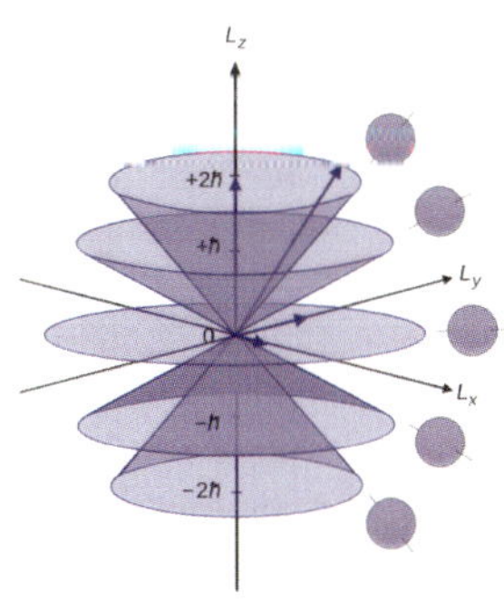

Illustration of quantum mechanical orbital angular momentum. The cones and plane represent possible orientations of the angular momentum vector for $\ell = 2$ and $m = -2, -1, 0, 1, 2$. Even for the extreme values of m, the z component of this vector is less than its total magnitude.

The axis used for the polar coordinates in this analysis is chosen arbitrarily. The quantum number m refers to the projection of the angular momentum in this arbitrarily-chosen direction, conventionally called the z direction or quantization axis. L_z, the magnitude of the angular momentum in the z direction, is given by the formula:

$$L_z = m\hbar.$$

This is a component of the atomic electron's total angular momentum, $\mathbf{L}$, whose magnitude is related to the azimuthal quantum number of its subshell ℓ by the equation:

$$L = \hbar\sqrt{\ell(\ell+1)}$$

where $\hbar$ is the reduced Planck constant. Note this $L=0$ for $\ell=0$ and approximates $L=(\ell+0.5)\hbar$ for high ℓ. It is not possible to measure the angular momentum of the electron in all three axes simultaneously. These properties were first demonstrated in the Stern-Gerlach experiment, by Otto Stern and Walther Gerlach.

The energy of any wave is its frequency multiplied by Planck's constant. The wave displays particle-like packets of energy called quanta. The formula for the quantum number of each quantum state uses Planck's reduced constant, which only allows particular or discrete or quantized energy levels.

Effect in Magnetic Fields

The quantum number m refers, loosely, to the direction of the angular momentum vector. The magnetic quantum number m only affects the electron›s energy if it is in a magnetic field because in the absence of one, all spherical harmonics corresponding to the different arbitrary values of m are equivalent. The magnetic quantum number determines the energy shift of an atomic orbital due to an external magnetic field (the Zeeman effect) — hence the name *magnetic* quantum number. However, the actual magnetic dipole moment of an electron in an atomic orbital arrives not only from the electron angular momentum, but also from the electron spin, expressed in the spin quantum number.

Since each electron has a magnetic moment in a magnetic field, it will be subject to a torque which tends to make the vector $\mathbf{L}$ parallel to the field, a phenomenon known as Larmor precession.

Spin Quantum Number

It is observed that the electron is not only revolving around the nucleus but it is also spinning on its own axis. The spin of the electron produces a small magnetic field as a result of which the electron behaves as a 'tiny magnet'.

Spin quantum number describes the spin orientation of the electron. It is designated by s. Since the electron can spin only in two ways:

1. Clockwise
2. Anti-Clockwise

Therefore, the spin quantum number can take only two values:

$$+\frac{1}{2} \text{ or} -\frac{1}{2}.$$

$+\frac{1}{2}$ spin is denoted by a vertical arrow pointing upwards and

$-\frac{1}{2}$ spin is denoted by a vertical arrow pointing downwards.

Number of Electrons that Accumulate an Orbital

Spin quantum number describes the direction in which an electron spins in an orbital. There are only two directions possible, clockwise and anticlockwise. Two electrons in an orbital can never have same spin i.e. spin of electrons in same orbital is opposite.

∴ Maximum number of electrons is 2 in an orbital.

Experiment

The beam of hydrogen atoms passes through a heterogeneous magnetic field; it divides itself into two parts. The atoms of these two parts have motions in opposite directions. The H atoms traveling through heterogeneous magnetic field is the following diagram.

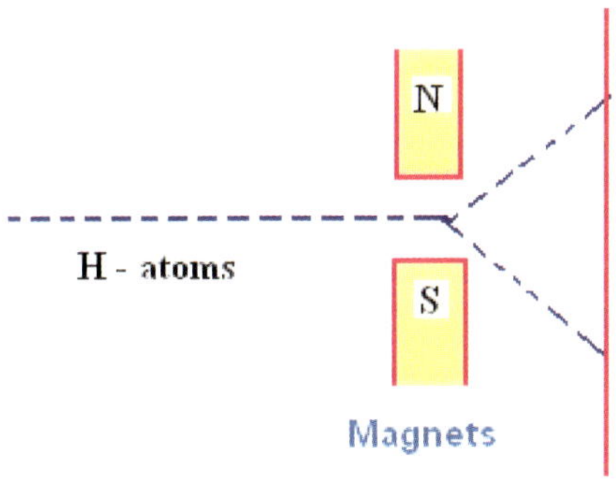

This observation suggests that all atoms do not have the same magnetic property. This kind of dissimilarity is possible only if some electrons have spin motion in single direction and others have spin motion in opposite direction.

To account for this observation, Goudsmit Uhlenbeck stated in 1925, that the electrons have quantized spin quantum number of angular momentum.

Magnetic Spin Quantum Number

Magnetic spin quantum number is the secondary spin quantum number which ranges from –s to **+s** in steps of one. This generates total (2s + 1) values of m_s.

The allowed values for s are non-negative integers or half-integers. All known particles are composed of fermions (for example the electron, proton or neutron) possess half integral values. All fermions have spin quantum number = $\frac{1}{2}$, Delta baryons, which decay into protons and neutrons have spin equal to $\frac{3}{2}$ while bosons (e.g., photon, mesons) have integer spin values (1 till now but theories predicts s = 0 for Higgs boson and s= 2 for graviton which is hypothetical carrier of gravity).

The Spin of a particle in any projected direction say **z** is:

$$s_z = m_s\bar{h}$$

Here s_z is spin in z (arbitrary direction).

m_s is magnetic spin quantum number or secondary spin quantum number.

$\hbar$ is reduced Planck's Constant.

$\therefore$ The allowed values of spin quantum number are 0, $\frac{1}{2}$, 1, $\frac{3}{2}$, 2, etc.

Derivation

As a solution for a certain partial differential equation, the quantized angular momentum can be written as:

$$\|\mathbf{s}\| = \sqrt{s(s+1)}\,\hbar$$

where

$\mathbf{s}$ is the quantized spin vector

$\|\mathbf{s}\|$ is the norm of the spin vector

s is the spin quantum number associated with the spin angular momentum

$\hbar$ is the reduced Planck constant.

Algebra

The algebraic theory of spin is a carbon copy of the Angular momentum in quantum mechanics theory. First of all, spin satisfies the fundamental commutation relation:

$$[S_i, S_j] = i\hbar\epsilon_{ijk}S_k,$$
$$\left[S_i, S^2\right] = 0$$

where ε_{lmn} is the (antisymmetric) Levi-Civita symbol. This means that it is impossible to know two coordinates of the spin at the same time because of the restriction of the uncertainty principle.

Next, the eigenvectors of S^2 and S_z satisfy:

$$S^2\,|s,m_s\rangle = \hbar^2 s(s+1)\,|s,m_s\rangle$$
$$S_z\,|s,m_s\rangle = \hbar m_s\,|s,m_s\rangle$$
$$S_\pm\,|s,m_s\rangle = \hbar\sqrt{s(s+1) - m_s(m_s \pm 1)}\,|s,m_s \pm 1\rangle$$

where $S_\pm = S_x \pm iS_y$ are the creation and annihilation (or "raising" and "lowering" or "up" and "down") operators.

Electron Spin

The spin angular momentum is characterized by a quantum number; s = 1/2 specifically for electrons. In a way analogous to other quantized angular momenta, L, it is possible to obtain an expression for the total spin angular momentum:

$$S = \hbar\sqrt{\frac{1}{2}\left(\frac{1}{2}+1\right)} = \frac{\sqrt{3}}{2}\hbar$$

where

$\hbar$ is the reduced Planck constant.

The hydrogen spectra fine structure is observed as a doublet corresponding to two possibilities for the z-component of the angular momentum, where for any given direction z:

$$\mathbf{S}_z = \pm\frac{1}{2}\hbar$$

whose solution has only two possible z-components for the electron. In the electron, the two different spin orientations are sometimes called "spin-up" or "spin-down".

The spin property of an electron would give rise to magnetic moment, which was a requisite for the fourth quantum number. The electron spin magnetic moment is given by the formula:

$$\mu_s = -\frac{e}{2m}gS$$

where

e is the charge of the electron

g is the Landé g-factor

and by the equation:

$$\mu_z = \pm\frac{1}{2}g\mu_B$$

where μ_B is the Bohr magneton.

When atoms have even numbers of electrons the spin of each electron in each orbital has opposing orientation to that of its immediate neighbor(s). However, many atoms have an odd number of electrons or an arrangement of electrons in which there is an unequal number of "spin-up" and "spin-down" orientations. These atoms or electrons are said to have unpaired spins that are detected in electron spin resonance.

Detection of Spin

When lines of the hydrogen spectrum are examined at very high resolution, they are found to be closely spaced doublets. This splitting is called fine structure, and was one of the first experimental evidences for electron spin. The direct observation of the electron's intrinsic angular momentum was achieved in the Stern–Gerlach experiment.

The Stern–Gerlach Experiment

The theory of spatial quantization of the spin moment of the momentum of electrons of atoms situated in the magnetic field needed to be proved experimentally. In 1920 (two years before the theoretical description of the spin was created) Otto Stern and Walter Gerlach observed it in the experiment they conducted.

Silver atoms were evaporated using an electric furnace in a vacuum. Using thin slits, the atoms were guided into a flat beam and the beam sent through an in-homogeneous magnetic field before colliding with a metallic plate. The laws of classical physics predict that the collection of condensed silver atoms on the plate should form a thin solid line in the same shape as the original beam. However, the in-homogeneous magnetic field caused the beam to split in two separate directions, creating two lines on the metallic plate.

The phenomenon can be explained with the spatial quantization of the spin moment of momentum. In atoms the electrons are paired such that one spins upward and one downward, neutralizing the effect of their spin on the action of the atom as a whole. But in the valence shell of silver atoms, there is a single electron whose spin remains unbalanced.

The unbalanced spin creates spin magnetic moment, making the electron act like a very small magnet. As the atoms pass through the in-homogeneous magnetic field, the force moment in the magnetic field influences the electron›s dipole until its position matches the direction of the stronger field. The atom would then be pulled toward or away from the stronger magnetic field a specific amount, depending on the value of the valence electron›s spin. When the spin of the electron is $+1/2$ the atom moves away from the stronger field, and when the spin is $-1/2$ the atom moves toward it. Thus the beam of silver atoms is split while traveling through the in-homogeneous magnetic field, according to the spin of each atom›s valence electron.

In 1927 Phipps and Taylor conducted a similar experiment, using atoms of hydrogen with similar results. Later scientists conducted experiments using other atoms that have only one electron in their valence shell: (copper, gold, sodium, potassium). Every time there two lines formed on the metallic plate.

The atomic nucleus also may have spin, but protons and neutrons are much heavier than electrons (about 1836 times), and the magnetic dipole moment is inversely proportional to the mass. So the nuclear magnetic dipole momentum is much smaller than that of the whole atom. This small magnetic dipole was later measured by Stern, Frisch and Easterman.

Spin quantum number (for electrons) can be determined easily by filling up the electrons in orbital. while filling keep in mind to fill the electrons till last orbital in one direction, after reaching last orbital of a sub shell again start filling the orbital from first orbital but now in opposite direction of spin. Normally the electrons are filled with anti clockwise spin or up arrow first. Second electron in the same orbital will then have a down arrow or clockwise spin.

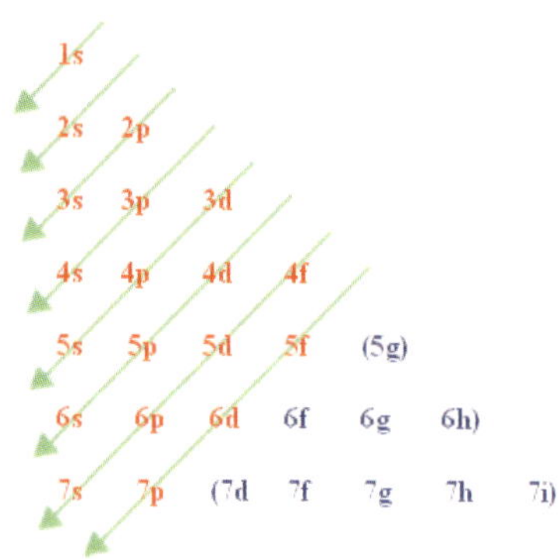

For example for an s sub shell which has only one orbital fill electrons as: up, down $(+\frac{1}{2}, -\frac{1}{2})$.

For p sub shell which has 3 orbital (p_x, p_y, p_z), fill electrons as: up, up, up; then from the first orbital again as down, down, down.

$$\left(\left|+\frac{1}{2}\right|+\frac{1}{2}\left|+\frac{1}{2}\right|+\frac{1}{2}\left|+\frac{1}{2}\right|+\frac{1}{2}\left|+\frac{1}{2}\right|\right)$$

$$\left(\left|+\frac{1}{2}\right|+\frac{1}{2}\left|+\frac{1}{2}\right|-\frac{1}{2}\left|-\frac{1}{2}\right|-\frac{1}{2}\left|-\frac{1}{2}\right|\right).$$

$$\left(\left|+\frac{1}{2}\right|\left|+\frac{1}{2}\right|\left|+\frac{1}{2}\right|\left|+\frac{1}{2}\right|\left|+\frac{1}{2}\right|\left|+\frac{1}{2}\right|\left|+\frac{1}{2}\right|\right)$$

$$\left(\left|+\frac{1}{2}\right|+\frac{1}{2}\left|+\frac{1}{2}\right|-\frac{1}{2}\left|-\frac{1}{2}\right|-\frac{1}{2}\left|-\frac{1}{2}\right|\right)$$

For *d* sub shell which has 5 orbital, fill electrons as: |up| up| up| up| up|, then fill from the first orbital again as down, down, down.

$$\left(\left|+\frac{1}{2}\right|+\frac{1}{2}\left|+\frac{1}{2}\right|+\frac{1}{2}\left|\frac{1}{2}\right|+\frac{1}{2}\left|+\frac{1}{2}\right|+\frac{1}{2}\left|+\frac{1}{2}\right|+\frac{1}{2}\left|+\frac{1}{2}\right|\right)$$

$$\left(\left|+\frac{1}{2}\right|+\frac{1}{2}\left|+\frac{1}{2}\right|+\frac{1}{2}\left|+\frac{1}{2}\right|+\frac{1}{2}\left\|+\frac{1}{2}\right|\frac{1}{2}\left|\frac{1}{2}\right|\frac{1}{2}\left|\frac{1}{2}\right)$$

$$\left(\left|+\frac{1}{2}\right|+\frac{1}{2}\left|+\frac{1}{2}\right|+\frac{1}{2}\left|+\frac{1}{2}\right|-\frac{1}{2}\left|-\frac{1}{2}\right|-\frac{1}{2}\left|-\frac{1}{2}\right|-\frac{1}{2}\left|-\frac{1}{2}\right|\right)$$

For f sub shell which has 7 orbital, fill electrons as: |up| up| up| up| up| up| up|; then from the first orbital again as down, down, down, down, down, down, down

$$\left(\left|+\frac{1}{2}\right|+\frac{1}{2}\left|+\frac{1}{2}\right|+\frac{1}{2}\left|+\frac{1}{2}\right|+\frac{1}{2}\left|+\frac{1}{2}\right|-\frac{1}{2}\left|-\frac{1}{2}\right|-\frac{1}{2}\left|-\frac{1}{2}\right|-\frac{1}{2}\left|-\frac{1}{2}\right|-\frac{1}{2}\left|-\frac{1}{2}\right)$$

Examples,

Consider Selenium (Se) with atomic number 34

Its electron configuration is written as:

$1s^2\ 2s^2\ 2p^6\ 3s^2\ 3p^6\ 4s^2\ 3d^{10}\ 4p^4$

Consider last sub shell i.e. $4p^4$

It is a p sub shell, therefore 3 orbital (p_x, p_y, p_z) are possible.

This sub shell contains 4 electrons as given. Now fill the first three electrons as up, up, up.

Now the last (fourth) electron is again filled in first orbital i.e. p_x with down spin. This is so because two electrons in same orbital cannot have same spins ad since it is already filled with an up electron, new entry will have down spin. Up and down denote anti clockwise and clockwise direction of spin with values $+\frac{1}{2}$ and $-\frac{1}{2}$ for electron.

|up, down| up | up |

So the spin of last electron is down or $-\frac{1}{2}$.

Consider another example of an atom say with last sub shell $3d^7$.

As we know a d sub shell has 5 possible orbital. So fill the first five electrons in the orbital starting from first to fifth with up spin ($+\frac{1}{2}$) as shown:

| up| up | up | up |up |

Now we are left with two more electrons and all the orbital are filled once, so start filling orbital again from first orbital but with down spin ($-\frac{1}{2}$) this time. As we are left with only two electrons we will fill till second orbital only as shown:|down |down|.

Hence electrons are filled in orbital as: (up, down), (up, down), up, up, up.

Energy Levels from the Dirac Equation

In 1928, Paul Dirac developed a relativistic wave equation, now termed the Dirac equation, which predicted the spin magnetic moment correctly, and at the same time treated the electron as a point-like particle. Solving the Dirac equation for the energy levels of an electron in the hydrogen atom, all four quantum numbers including *s* occurred naturally and agreed well with experiment.

Total Spin of an Atom or Molecule

For some atoms the spins of several unpaired electrons (s_1, s_2,...) are coupled to form a *total spin* quantum number S. This occurs especially in light atoms (or in molecules formed only of light atoms) when spin-orbit coupling is weak compared to the coupling between spins or the coupling between orbital angular momenta, a situation known as LS coupling because L and S are constants of motion. Here L is the total orbital angular momentum quantum number.

For atoms with a well-defined S, the multiplicity of a state is defined as (2S+1). This is equal to the number of different possible values of the total (orbital plus spin) angular momentum J for a given (L, S) combination, provided that $S \leq L$ (the typical case). For example, if $S = 1$, there are three states which form a triplet. The eigenvalues of S_z for these three states are $+1\hbar$, 0 and $-1\hbar$. The term symbol of an atomic state indicates its values of L, S, and J.

Total Angular Momentum Quantum Number

The total angular momentum is simply $\underline{J} = \underline{J}_1 + \underline{J}_2$. The maximum and minimum values that $\underline{J}$ can take correspond to the case where either $\underline{J}_1$ and $\underline{J}_2$ are parallel, so that the magnitude of $\underline{J}$ is $|\underline{J}_1| + |\underline{J}_2|$ or antiparallel when it has magnitude $||\underline{J}_1| - |\underline{J}_2||$. This topic discusses the addition of angular momenta for a quantum system.

Total Angular Momentum Operator

In the quantum case, the total angular momentum is represented by the operator

$$\underline{\hat{J}} \equiv \underline{\hat{J}}_1 + \underline{\hat{J}}_2 .$$

We assume that $\underline{\hat{J}}_1$ and $\underline{\hat{J}}_2$ are independent angular momenta, meaning each satisfies the usual angular momentum commutation relations

$$\left[\hat{J}_{nx}, \hat{J}_{ny}\right] = i\hbar\hat{J}_{nz}, \quad \text{etc.,} \quad \left[\hat{J}_n^{\,2}, \hat{J}_{ni}\right] \quad \text{etc.}$$

where $n = 1, 2$ labels the individual angular momenta, $i = x, y, z$ and etc. stands for cyclic permutations. Furthermore, any component of $\underline{\hat{J}}_1$ commutes with any component of $\underline{\hat{J}}_2$:

$$\left[\hat{J}_{1i}, \hat{J}_{2k}\right] = 0, \quad i, k = x, y, z,$$

so that the two angular momenta are compatible. It follows that the four operators $\hat{J}_1^2, \hat{J}_{1z}, \hat{J}_2^2, \hat{J}_{2z}$ are mutually commuting and so must possess a common eigenbasis. This common eigenbasis is known as the uncoupled basis and is denoted $\{|j_1, m_1, j_2, m_2\rangle\}$ in Dirac notation. It has the following properties:

$$\begin{aligned}
\hat{J}_1^2|j_1, m_1, j_2, m_2\rangle &= j_1(j_1+1)\hbar^2|j_1, m_1, j_2, m_2\rangle \\
\hat{J}_{1z}|j_1, m_1, j_2, m_2\rangle &= m_1\hbar|j_1, m_1, j_2, m_2\rangle \\
\hat{J}_2^2|j_1, m_1, j_2, m_2\rangle &= j_2(j_2+1)\hbar^2|j_1, m_1, j_2, m_2\rangle \\
\hat{J}_{2z}|j_1, m_1, j_2, m_2\rangle &= m_2\hbar|j_1, m_1, j_2, m_2\rangle
\end{aligned}$$

It is easy to establish that the total angular momentum operators satisfy the usual commutation relations

$$\left[\hat{J}_x, \hat{J}_y\right] = i\hbar\,\hat{J}_z, \quad \text{etc.,} \quad \left[\hat{J}^2, \hat{J}_i\right] = 0$$

As for any angular momentum operator then, $\hat{J}^2$ has eigenvalues $j(j+1)\hbar^2$ whilst the operator corresponding to the z-component of the total angular momentum has eigenvalues $m\hbar$ with m running between j and $-j$ in integer steps for a given j.

Addition Theorem

The question which then arises is, given two angular momenta, corresponding to angular momentum quantum numbers j_1 and j_2 respectively, what are the allowed values of the total angular momentum quantum number, j? The answer is provided by the Angular Momentum Addition Theorem:

The allowed values of the total angular momentum quantum number j, given two angular momenta corresponding to quantum numbers j_1 and j_2 are:

$$j = j1 + j2,\ j1 + j2 - 1, \ldots, |j1 - j2|$$

and for each of these values of j, m takes on the $(2j+1)$ values

$$m = j,\ j-1, \ldots, -j.$$

It is easy to show that $\hat{J}^2$ commutes with $\hat{J}_1^2$ and $\hat{J}_2^2$ but not with $\hat{J}_{1z}$ or with $\hat{J}_{2z}$ by writing

$$\hat{J}^2 \equiv \left(\underline{\hat{J}}_1 + \underline{\hat{J}}_2\right)^2 = \left\{\hat{J}_1^2 + \hat{J}_2^2 + 2\underline{\hat{J}}_1 \cdot \underline{\hat{J}}_2\right\}$$

The dot product contains the x and y components of the two angular momenta which do not commute with the respective z components.

The operator $\hat{J}_z$ commutes with $\hat{J}_1^2$ and $\hat{J}_2^2$ and so the set of four operators $\hat{J}^2, \hat{J}_z, \hat{J}_1^2, \hat{J}_2^2$ are also a mutually commuting set of operators with a common eigenbasis known as the coupled basis, denoted $\{|j,m,j1,j2|\}$ and satisfying

$$\hat{J}^2|j,m,j1,j2\rangle = j(j+1)\hbar^2|j,m,j1,j2\rangle$$
$$\hat{J}_z|j,m,j1,j2\rangle = m\hbar|j,m,j1,j2\rangle$$
$$\hat{J}_1^2|j,m,j1,j2\rangle = j1(j1+1)\hbar^2|j,m,j1,j2\rangle$$
$$\hat{J}_2^2|j,m,j1,j2\rangle = j2(j2+1)\hbar^2|j,m,j1,j2\rangle$$

These are states of definite total angular momentum and definite z component of total angular momentum but not in general states with definite $\hat{J}_{1z}$ or $\hat{J}_{2z}$. In fact, they are expressible as linear combinations of the states of the uncoupled basis, with coefficients known as Clebsch-Gordan coefficients.

Example: We consider the important case of two spin $-$ particles for which the spin quantum numbers are $s_1 = \frac{1}{2}$ and $s_2 = \frac{1}{2}$ respectively. According to the theorem, the total spin quantum number s takes on the values $s_1 + s_2 \equiv 1$ and $|s_1 + s_2| \equiv 0$ only.

Thus two electrons can have a total spin of 1 or 0 only: these states of definite total spin are referred to as triplet and singlet states respectively, because in the former, there are three possible values of the spin magnetic quantum number, $m_s = 1, 0, -1$, whereas in the latter there is only one such value, $m_s = 0$.

The states of the uncoupled basis are

$$\alpha 1\alpha 2, \quad \alpha 1\beta 2, \quad ,\beta 1\alpha 2, \quad \beta 1\beta 2,$$

where the subscripts 1 and 2 refer to electrons 1 and 2 respectively. The operators $\hat{S}_1^2$ and $\hat{S}_{1z}$ act only on the parts labelled 1, whilst $\hat{S}_2^2$ and $\hat{S}_{2z}$ act only on the parts labelled 2.

It should be clear that that since $\alpha 1\alpha 2$ has $m_{s_1} = \frac{1}{2}$ and $m_{s_2} = \frac{1}{2}$ it must have $m_s = 1$, that is, total z-component of spin $\hbar$, and can therefore only be s = 1 and not s = 0. This is an example of what is known as a stretched state: it has the maximum possible value of the z component of total angular momentum (spin) and must therefore be a member of both the coupled and uncoupled basis: $\alpha 1\alpha 2, \equiv \left| s=1, m_s=1, s1=\frac{1}{2}, s2=\frac{1}{2} \right\rangle$.

A similar argument shows that $\beta 1\beta 2$ has $m_s = -1$ and thus also can only be s = 1, so that $\beta 1\beta 2 \equiv \left| s = 1, m_s = -1, s_1 = \frac{1}{2}, s_2 = \frac{1}{2} \right\rangle$. The remaining two states of the coupled basis are, however, non-trivial linear combinations of the two remaining states of the uncoupled basis:

$$\left| s=1,\ m_s=0,\ s1=\frac{1}{2},\ s2=\frac{1}{2} \right\rangle = \frac{1}{\sqrt{2}}\left[\alpha 1\beta 2+\beta 1\alpha 2\right]$$

$$\left| s=0,\ m_s=0,\ s1=\frac{1}{2},\ s2=\frac{1}{2} \right\rangle - \frac{1}{\sqrt{2}}\left[\alpha 1\beta 2 - \beta 1\alpha 2\right].$$

Proof

We apply the lowering operator, $\hat{S}_-$ for the z component of total spin to the stretched state:

$$\hat{S}_-\alpha 1\,\alpha 2 = \left(\hat{S}_{1-} + \hat{S}_{2-}\right)\alpha 1\,\alpha 2.$$

The left-hand side we write as

$$\hat{S}_-\left|s=1, m_s=1,\ s1=\frac{1}{2},\ s2=\frac{1}{2}\right\rangle = \sqrt{1(1+1)-1(1-1)}\hbar\left|s=1,\ m_s=0,\ s1=\frac{1}{2},\ s2=\frac{1}{2}\right\rangle$$

$$= \sqrt{2}\hbar\left|s=1, m_s=0,\ s1=\frac{1}{2},\ s2=\frac{1}{2}\right\rangle$$

using the usual properties of the lowering operator, whilst we write the right-hand side as

$$\left(\hat{S}_{1-}+\hat{S}_{2-}\right)s1=\frac{1}{2}, m_{s_1}=\frac{1}{2}, s2=\frac{1}{2}, m_{s_2}=\frac{1}{2}\Big\rangle$$

and note that

$$\hat{S}_{1-}\left|s1=\frac{1}{2}, m_{s_1}=\frac{1}{2}, s2=\frac{1}{2}, m_{s_2}=\frac{1}{2}\right\rangle$$

$$= \sqrt{\frac{1}{2}\left(\frac{1}{2}+1\right)-\frac{1}{2}\left(\frac{1}{2}-1\right)}\hbar \left| s_1 = \frac{1}{2}, m_{s_1} = -\frac{1}{2}, s_2 = \frac{1}{2}, m_{s_2} = \frac{1}{2} \right\rangle$$

$$= \hbar \left| s_1 = \frac{1}{2}, m_{s_1} = -\frac{1}{2}, s_2 = \frac{1}{2}, m_{s_2} = \frac{1}{2} \right\rangle \equiv \hbar \beta_1 \alpha_2$$

whilst

$$\hat{S}_{2-} \left| s_1 = \frac{1}{2}, m_{s_1} = \frac{1}{2}, s_2 = \frac{1}{2}, m_{s_2} = \frac{1}{2} \right\rangle =$$

$$\sqrt{\frac{1}{2}\left(\frac{1}{2}+1\right)-\frac{1}{2}\left(\frac{1}{2}-1\right)}\hbar \left| s_1 = \frac{1}{2}, m_{s_1} = \frac{1}{2}, s_2 = \frac{1}{2}, m_{s_2} = -\frac{1}{2} \right\rangle$$

$$= \hbar \left| s_1 = \frac{1}{2}, m_{s_1} = \frac{1}{2}, s_2 = \frac{1}{2}, m_{s_2} = -\frac{1}{2} \right\rangle \equiv \hbar \alpha_1 \beta_2 .$$

Equating the two sides then yields the stated result:

$$\left| s = 1, m_s = 0, s_1 = \frac{1}{2}, s_2 = \frac{1}{2} \right\rangle = \frac{1}{\sqrt{2}} \left[\alpha_1 \beta_2 + \beta_1 \alpha_2 \right].$$

The remaining member of the coupled basis must be a linear combination of $\alpha_1 \beta_2$ and $\beta_1 \alpha_2$ orthogonal to this, which we can take to be

$$\left| s = 0, m_s = 0, s_1 = \frac{1}{2}, s_2 = \frac{1}{2} \right\rangle = \frac{1}{\sqrt{2}} \left[\alpha_1 \beta_2 + \beta_1 \alpha_2 \right].$$

References

- Andrew, A. V. (2006). “2. Schrödinger equation”. Atomic spectroscopy. Introduction of theory to Hyperfine Structure. p. 274. ISBN 978-0-387-25573-6
- Azimuthal-quantum-number, modern-physics: tutorvista.com, Retrieved 30 June 2018
- R.B. Lindsay (1927). “Note on “pendulum” orbits in atomic models”. Proc. Natl. Acad. Sci. 13: 413–419. Bibcode:1927PNAS...13..413L. doi:10.1073/pnas.13.6.413. PMC 1085028. PMID 16587189
- Magnetic-quantum-number, physics: byjus.com, Retrieved 10 April 2018
- Atkins P. and de Paula J. Physical Chemistry (8th ed., W.H.Freeman 2006), p.352 ISBN 0-7167-8759-8
- Magnetic-quantum-number, modern-physics: tutorvista.com, Retrieved 28 May 2018
- Spin-quantum-number, modern-physics: tutorvista.com, Retrieved 14 July 2018

5

Wave Mechanics

A wave is a disturbance in matter or space which is characterized by a transfer of energy. It is described by a wave equation which describes the change of the disturbance with time. This chapter has been structured in a way to provide a comprehensive understanding of wave mechanics.

Uncertainty Principle

Uncertainty principle is also called Heisenberg uncertainty principle or indeterminacy principle a type of statement which articulated (1927) by the German physicist Werner Heisenberg, that the position and the velocity of an object cannot both be measured exactly, at the same time, even in theory. The very concepts of exact position and exact velocity together, in fact, have no meaning in nature.

Ordinary experience provides no clue of this principle. It is easy to measure both the position and the velocity of, say, an automobile, because the uncertainties implied by this principle for ordinary objects are too small to be observed. The complete rule stipulates that the product of the uncertainties in position and velocity is equal to or greater than a tiny physical quantity, or constant $(h/(4\pi)$, where h is Planck's constant, or about 6.6×10^{-34} joule-second). Only for the exceedingly small masses of atoms and subatomic particles does the product of the uncertainties become significant.

Any attempt to measure precisely the velocity of a subatomic particle, such as an electron, will knock it about in an unpredictable way, so that a simultaneous measurement of its position has no validity. This result has nothing to do with inadequacies in the measuring instruments, the technique, or the observer; it arises out of the intimate connection in nature between particles and waves in the realm of subatomic dimensions.

Every particle has a wave associated with it; each particle actually exhibits wavelike behaviour. The particle is most likely to be found in those places where the undulations of the wave are greatest, or most intense. The more intense the undulations of the associated wave become, however, the more ill defined becomes the wavelength, which in turn determines the momentum of the particle. So a strictly localized wave has an indeterminate wavelength; its associated particle, while having a definite position, has no certain velocity. A particle wave having a well-defined wavelength, on the other hand, is spread out; the associated particle, while having a rather precise velocity, may be almost anywhere. A quite accurate measurement of one observable involves a relatively large uncertainty in the measurement of the other.

The uncertainty principle is alternatively expressed in terms of a particle's momentum and position. The momentum of a particle is equal to the product of its mass times its velocity. Thus, the product of the uncertainties in the momentum and the position of a particle equals $h/(4\pi)$ or more. The principle applies to other related (conjugate) pairs of observables, such as energy and time: the product of the uncertainty in an energy measurement and the uncertainty in the time interval during which the measurement is made also equals $h/(4\pi)$ or more. The same relation holds, for an unstable atom or nucleus, between the uncertainty in the quantity of energy radiated and the uncertainty in the lifetime of the unstable system as it makes a transition to a more stable state.

Wave Mechanics Interpretation

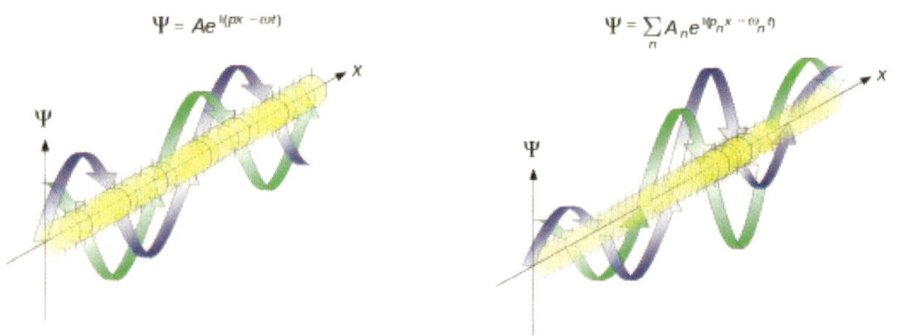

Plane wave Wave packet

Propagation of de Broglie waves in 1d—real part of the complex amplitude is blue, imaginary part is green. The probability (shown as the colour opacity) of finding the particle at a given point x is spread out like a waveform, there is no definite position of the particle. As the amplitude increases above zero the curvature reverses sign, so the amplitude begins to decrease again, and vice versa—the result is an alternating amplitude: a wave.

According to the de Broglie hypothesis, every object in the universe is a wave, i.e., a situation which gives rise to this phenomenon. The position of the particle is described by a wave function $\Psi(x,t)$. The time-independent wave function of a single-moded plane wave of wavenumber k_0 or momentum p_0 is

$$\psi(x) \propto e^{ik_0 x} = e^{ip_0 x/\hbar}.$$

The Born rule states that this should be interpreted as a probability density amplitude function in the sense that the probability of finding the particle between a and b is

$$\mathrm{P}[a \le X \le b] = \int_a^b |\psi(x)|^2 \, \mathrm{d}x.$$

In the case of the single-moded plane wave, $|\psi(x)|^2$ is a uniform distribution. In other words, the particle position is extremely uncertain in the sense that it could be essentially anywhere along the wave packet. Consider a wave function that is a sum of many waves, however, we may write this as

$$\psi(x) \propto \sum_n A_n e^{ip_n x/\hbar},$$

where A_n represents the relative contribution of the mode p_n to the overall total. The figures to the right show how with the addition of many plane waves, the wave packet can become more localized. We may take this a step further to the continuum limit, where the wave function is an integral over all possible modes

$$\psi(x) = \frac{1}{\sqrt{2\pi\hbar}} \int_{-\infty}^{\infty} \varphi(p) \cdot e^{ipx/\hbar}\, dp \ ,$$

with $\varphi(p)$ representing the amplitude of these modes and is called the wave function in momentum space. In mathematical terms, we say that $\varphi(p)$ is the *Fourier transform* of $\psi(x)$ and that x and p are conjugate variables. Adding together all of these plane waves comes at a cost, namely the momentum has become less precise, having become a mixture of waves of many different momenta.

One way to quantify the precision of the position and momentum is the standard deviation σ. Since $|\psi(x)|^2$ is a probability density function for position, we calculate its standard deviation.

The precision of the position is improved, i.e. reduced σ_x, by using many plane waves, thereby weakening the precision of the momentum, i.e. increased σ_p. Another way of stating this is that σ_x and σ_p have an inverse relationship or are at least bounded from below. This is the uncertainty principle, the exact limit of which is the Kennard bound. Click the *show* button below to see a semi-formal derivation of the Kennard inequality using wave mechanics.

Proof of the Kennard Inequality using Wave Mechanics

We are interested in the variances of position and momentum, defined as

$$\sigma_x^2 = \int_{-\infty}^{\infty} x^2 \cdot |\psi(x)|^2\, dx - \left(\int_{-\infty}^{\infty} x \cdot |\psi(x)|^2\, dx \right)^2$$

$$\sigma_p^2 = \int_{-\infty}^{\infty} p^2 \cdot |\varphi(p)|^2\, dp - \left(\int_{-\infty}^{\infty} p \cdot |\varphi(p)|^2\, dp \right)^2 .$$

Without loss of generality, we will assume that the means vanish, which just amounts to a shift of the origin of our coordinates. (A more general proof that does not make this assumption is given below.) This gives us the simpler form

$$\sigma_x^2 = \int_{-\infty}^{\infty} x^2 \cdot |\psi(x)|^2\, dx$$

$$\sigma_p^2 = \int_{-\infty}^{\infty} p^2 \cdot |\varphi(\mathrm{p})|^2\, dp.$$

The function $f(x) = x \cdot \psi(x)$ can be interpreted as a vector in a function space. We can define an inner product for a pair of functions $u(x)$ and $v(x)$ in this vector space:

$$\langle u \mid v \rangle = \int_{-\infty}^{\infty} u^*(x) \cdot v(x) dx,$$

where the asterisk denotes the complex conjugate.

With this inner product defined, we note that the variance for position can be written as

$$\sigma_x^2 = \int_{-\infty}^{\infty} |f(x)|^2\, dx = \langle f \mid f \rangle.$$

We can repeat this for momentum by interpreting the function $\tilde{g}(p) = p \cdot \varphi(p)$ as a vector, but we can also take advantage of the fact that $\psi(x)$ and $\varphi(p)$ are Fourier transforms of each other. We evaluate the inverse Fourier transform through integration by parts:

$$\begin{aligned}
g(x) &= \frac{1}{\sqrt{2\pi\hbar}} \cdot \int_{-\infty}^{\infty} \tilde{g}(p) \cdot e^{ipx/\hbar}\, dp \\
&= \frac{1}{\sqrt{2\pi\hbar}} \int_{-\infty}^{\infty} p \cdot \varphi(p) \cdot e^{ipx/\hbar}\, dp \\
&= \frac{1}{2\pi\hbar} \int_{-\infty}^{\infty} \left[p \cdot \int_{-\infty}^{\infty} \psi(\chi) e^{-ip\chi/\hbar}\, d\chi \right] \cdot e^{ipx/\hbar}\, dp \\
&= \frac{i}{2\pi} \int_{-\infty}^{\infty} \left[\cancel{\psi(\chi) e^{-ip\chi/\hbar} \Big|_{-\infty}^{\infty}} - \int_{-\infty}^{\infty} \frac{d\psi(\chi)}{d\chi} e^{-ip\chi/\hbar}\, d\chi \right] \cdot e^{ipx/\hbar}\, dp \\
&= \frac{-i}{2\pi} \int_{-\infty}^{\infty} \int_{-\infty}^{\infty} \frac{d\psi(\chi)}{d\chi} e^{-ip\chi/\hbar}\, d\chi\, e^{ipx/\hbar}\, dp \\
&= \left(-i\hbar \frac{d}{dx} \right) \cdot \psi(x),
\end{aligned}$$

where the cancelled term vanishes because the wave function vanishes at infinity. Often the term $-i\hbar \frac{d}{dx}$ is called the momentum operator in position space. Applying Parseval's theorem, we see that the variance for momentum can be written as

$$\sigma_p^2 = \int_{-\infty}^{\infty} |\tilde{g}(p)|^2\, dp = \int_{-\infty}^{\infty} |g(x)|^2\, dx = \langle g \mid g \rangle.$$

The Cauchy–Schwarz inequality asserts that

$$\sigma_x^2 \sigma_p^2 = \langle f \mid f \rangle \cdot \langle g \mid g \rangle \geq |\langle f \mid g \rangle|^2 .$$

The modulus squared of any complex number z can be expressed as

$$|z|^2 = \left(\mathrm{Re}(z)\right)^2 + \left(\mathrm{Im}(z)\right)^2 \geq \left(\mathrm{Im}(z)\right)^2 = \left(\frac{z - z^*}{2i} \right)^2 .$$

we let $z = \langle f \mid g \rangle$ and $z^* = \langle g \mid f \rangle$ and substitute these into the equation above to get

$$|\langle f \mid g \rangle|^2 \geq \left(\frac{\langle f \mid g \rangle - \langle g \mid f \rangle}{2i} \right)^2 .$$

All that remains is to evaluate these inner products.

$$\langle f \mid g \rangle - \langle g \mid f \rangle = \int_{-\infty}^{\infty} \psi^*(x) x \cdot \left(-i\hbar \frac{d}{dx} \right) \psi(x) dx$$

$$-\int_{-\infty}^{\infty}\psi^*(x)\left(-i\hbar\frac{d}{dx}\right)\cdot x\psi(x)dx$$

$$=i\hbar\cdot\int_{-\infty}^{\infty}\psi^*(x)\left[\left(-x\cdot\frac{d\psi(x)}{dx}\right)+\frac{d(x\psi(x))}{dx}\right]dx$$

$$=i\hbar\cdot\int_{-\infty}^{\infty}\psi^*(x)\left[\left(-x\cdot\frac{d\psi(x)}{dx}\right)+\psi(x)+\left(x\cdot\frac{d\psi(x)}{dx}\right)\right]dx=i\hbar\cdot\int_{-\infty}^{\infty}\psi^*(x)\psi(x)dx$$

$$=i\hbar\cdot\int_{-\infty}^{\infty}|\psi(x)|^2dx$$

$$=i\hbar$$

Plugging this into the above inequalities, we get

$$\sigma_x^2\sigma_p^2\geq|\langle f\,|\,g\rangle|^2\geq\left(\frac{\langle f\,|\,g\rangle-\langle g\,|\,f\rangle}{2i}\right)^2=\left(\frac{i\hbar}{2i}\right)^2=\frac{\hbar^2}{4}$$

or taking the square root

$$\sigma_x\sigma_p\geq\frac{\hbar}{2}.$$

Note that the only *physics* involved in this proof was that $\psi(x)$ and $\varphi(p)$ are wave functions for position and momentum, which are Fourier transforms of each other. A similar result would hold for *any* pair of conjugate variables

Matrix Mechanics Interpretation

In matrix mechanics, observables such as position and momentum are represented by self-adjoint operators. When considering pairs of observables, an important quantity is the *commutator*. For a pair of operators $\hat{A}$ and $B\hat{}$, one defines their commutator as

$$[\hat{A},\hat{B}]=\hat{A}\hat{B}-\hat{B}\hat{A}.$$

In the case of position and momentum, the commutator is the canonical commutation relation

$$[\tilde{\hat{u}},\ \hat{}\ ]=\ \hbar.$$

The physical meaning of the non-commutativity can be understood by considering the effect of the commutator on position and momentum eigenstates. Let $|\psi\rangle$ be a right eigenstate of position with a constant eigenvalue x_0. By definition, this means that $\hat{x}\,|\,\psi\rangle=x_0\,|\,\psi\rangle$. Applying the commutator to $|\,\psi\rangle$ yields

$$[\hat{x},\hat{p}]\,|\,\psi\rangle=(\hat{x}\hat{p}-\hat{p}\hat{x})\,|\,\psi\rangle=(\hat{x}-x_0\hat{I})\,\hat{p}|\,\psi\rangle=i\hbar\,|\,\psi\rangle,$$

where $\hat{I}$ is the identity operator.

Suppose, for the sake of proof by contradiction, that $|\psi\rangle$ is also a right eigenstate of momentum, with constant eigenvalue p_o. If this were true, then one could write

$$(\hat{x}-x_0\hat{I})\hat{p}|\psi\rangle=(\hat{x}-x_0\hat{I})p_0|\psi\rangle=(x_0\hat{I}-x_0\hat{I})p_0|\psi\rangle=0.$$

On the other hand, the above canonical commutation relation requires that

$$[\hat{x},\hat{p}]|\psi\rangle=i\hbar|\psi\rangle\neq 0.$$

This implies that no quantum state can simultaneously be both a position and a momentum eigenstate.

When a state is measured, it is projected onto an eigenstate in the basis of the relevant observable. For example, if a particle's position is measured, then the state amounts to a position eigenstate. This means that the state is *not* a momentum eigenstate, however, but rather it can be represented as a sum of multiple momentum basis eigenstates. In other words, the momentum must be less precise. This precision may be quantified by the standard deviations,

$$\sigma_x=\sqrt{\langle\hat{x}^2\rangle-\langle\hat{x}\rangle^2}$$

$$\sigma_p=\sqrt{\langle\hat{p}^2\rangle-\langle\hat{p}\rangle^2}.$$

As in the wave mechanics interpretation above, one sees a tradeoff between the respective precisions of the two, quantified by the uncertainty principle.

Robertson–Schrödinger Uncertainty Relations

The most common general form of the uncertainty principle is the *Robertson uncertainty relation*.

For an arbitrary Hermitian operator $\widehat{\mathcal{O}}$ we can associate a standard deviation

$$\sigma_{\mathcal{O}}=\sqrt{\langle\widehat{\mathcal{O}}^2\rangle-\langle\widehat{\mathcal{O}}\rangle^2},$$

where the brackets $\langle\ \rangle$ indicate an expectation value. For a pair of operators $\hat{A}$ and $\hat{B}$, we may define their *commutator* as

$$[\hat{A},\hat{B}]=\hat{A}\hat{B}-\hat{B}\hat{A},$$

In this notation, the Robertson uncertainty relation is given by

$$\sigma_A\sigma_B\geq\left|\frac{1}{2i}\langle[\hat{A},\hat{B}]\rangle\right|=\frac{1}{2}\left|\langle[\hat{A},\hat{B}]\rangle\right|,$$

The Robertson uncertainty relation immediately follows from a slightly stronger inequality, the *Schrödinger uncertainty relation*,

$$\sigma_A^2\sigma_B^2\geq\left|\frac{1}{2}\langle\{\hat{A},\hat{B}\}\rangle-\langle\hat{A}\rangle\langle\hat{B}\rangle\right|^2+\left|\frac{1}{2i}\langle[\hat{A},\hat{B}]\rangle\right|^2,$$

where we have introduced the *anticommutator*,

$$\{\hat{A}, \hat{B}\} = \hat{A}\hat{B} + \hat{B}\hat{A}.$$

Proof of the Schrödinger Uncertainty Relation

The derivation shown here incorporates and builds off of those shown in Robertson, Schrödinger and standard textbooks such as Griffiths. For any Hermitian operator $\hat{A}$, based upon the definition of variance, we have

$$\sigma_A^2 = \langle(\hat{A} - \langle\hat{A}\rangle)\Psi \mid (\hat{A} - \langle\hat{A}\rangle)\Psi\rangle.$$

we let $|f\rangle = |(\hat{A} - \langle\hat{A}\rangle)\Psi\rangle$ and thus

$$\sigma_A^2 = \langle f \mid f\rangle.$$

Similarly, for any other Hermitian operator $\hat{B}$ in the same state

$$\sigma_B^2 = \langle(\hat{B} - \langle\hat{B}\rangle)\Psi \mid (\hat{B} - \langle\hat{B}\rangle)\Psi\rangle = \langle g \mid g\rangle$$

for $|g\rangle = |(\hat{B} - \langle\hat{B}\rangle)\Psi\rangle$.

The product of the two deviations can thus be expressed as

$$\sigma_A^2 \sigma_B^2 = \langle f \mid f\rangle\langle g \mid g\rangle.$$

In order to relate the two vectors $|f\rangle$ and $|g\rangle$, we use the Cauchy–Schwarz inequality which is defined as

$$\langle f \mid f\rangle\langle g \mid g\rangle \geq |\langle f \mid g\rangle|^2,$$

and thus equation $\sigma_A^2 \sigma_B^2 = \langle f \mid f\rangle\langle g \mid g\rangle$ can be written as

$$\sigma_A^2 \sigma_B^2 \geq |\langle f \mid g\rangle|^2.$$

Since $\langle f \mid g\rangle$ is in general a complex number, we use the fact that the modulus squared of any complex number z is defined as $|z|^2 = zz^*$, where z^* is the complex conjugate of z. The modulus squared can also be expressed as

$$|z|^2 = (\mathrm{Re}(z))^2 + (\mathrm{Im}(z))^2 = \left(\frac{z + z^*}{2}\right)^2 + \left(\frac{z - z^*}{2i}\right)^2.$$

we let $z = \langle f \mid g\rangle$ and $z^* = \langle g \mid f\rangle$ and substitute these into the equation above to get

$$|\langle f \mid g\rangle|^2 = \left(\frac{\langle f \mid g\rangle + \langle g \mid f\rangle}{2}\right)^2 + \left(\frac{\langle f \mid g\rangle - \langle g \mid f\rangle}{2i}\right)^2$$

The inner product $\langle f \mid g\rangle$ is written out explicitly as

$$\langle f \mid g\rangle = \langle(\hat{A}-\langle\hat{A}\rangle)\Psi \mid (\hat{B}-\langle\hat{B}\rangle)\Psi\rangle,$$

and using the fact that $\hat{A}$ and $\hat{B}$ are Hermitian operators, we find

$$\begin{aligned}\langle f \mid g\rangle &= \langle\Psi \mid (\hat{A}-\langle\hat{A}\rangle)(\hat{B}-\langle\hat{B}\rangle)\Psi\rangle \\ &= \langle\Psi \mid (\hat{A}\hat{B}-\hat{A}\langle\hat{B}\rangle-\hat{B}\langle\hat{A}\rangle+\langle\hat{A}\rangle\langle\hat{B}\rangle)\Psi\rangle \\ &= \langle\Psi \mid \hat{A}\hat{B}\Psi\rangle-\langle\Psi \mid \hat{A}\langle\hat{B}\rangle\Psi\rangle-\langle\Psi \mid \hat{B}\langle\hat{A}\rangle\Psi\rangle+\langle\Psi \mid \langle\hat{A}\rangle\langle\hat{B}\rangle\Psi\rangle \\ &= \langle\hat{A}\hat{B}\rangle-\langle\hat{A}\rangle\langle\hat{B}\rangle-\langle\hat{A}\rangle\langle\hat{B}\rangle+\langle\hat{A}\rangle\langle\hat{B}\rangle \\ &= \langle\hat{A}\hat{B}\rangle-\langle\hat{A}\rangle\langle\hat{B}\rangle.\end{aligned}$$

Similarly it can be shown that $\langle g \mid f\rangle = \langle\hat{B}\hat{A}\rangle-\langle\hat{A}\rangle\langle\hat{B}\rangle$.

Thus we have

$$\langle f \mid g\rangle-\langle g \mid f\rangle = \langle\hat{A}\hat{B}\rangle-\langle\hat{A}\rangle\langle\hat{B}\rangle-\langle\hat{B}\hat{A}\rangle+\langle\hat{A}\rangle\langle\hat{B}\rangle = \langle[\hat{A},\hat{B}]\rangle$$

and

$$\langle f \mid g\rangle+\langle g \mid f\rangle = \langle\hat{A}\hat{B}\rangle-\langle\hat{A}\rangle\langle\hat{B}\rangle+\langle\hat{B}\hat{A}\rangle-\langle\hat{A}\rangle\langle\hat{B}\rangle = \langle\{\hat{A},\hat{B}\}\rangle-2\langle\hat{A}\rangle\langle\hat{B}\rangle.$$

We now substitute the above two equations above back into equation $|\langle f \mid g\rangle|^2 = \left(\frac{\langle f \mid g\rangle+\langle g \mid f\rangle}{2}\right)^2+\left(\frac{\langle f \mid g\rangle-\langle g \mid f\rangle}{2i}\right)^2$ and get

$$|\langle f \mid g\rangle|^2 = \left(\frac{1}{2}\langle\{\hat{A},\hat{B}\}\rangle-\langle\hat{A}\rangle\langle\hat{B}\rangle\right)^2+\left(\frac{1}{2i}\langle[\hat{A},\hat{B}]\rangle\right)^2.$$

Substituting the above into equation $\sigma_A^2\sigma_B^2 \geq |\langle f \mid g\rangle|^2$ we get the Schrödinger uncertainty relation

$$\sigma_A\sigma_B \geq \sqrt{\left(\frac{1}{2}\langle\{\hat{A},\hat{B}\}\rangle-\langle\hat{A}\rangle\langle\hat{B}\rangle\right)^2+\left(\frac{1}{2i}\langle[\hat{A},\hat{B}]\rangle\right)^2}.$$

This proof has an issue related to the domains of the operators involved. For the proof to make sense, the vector $\hat{B}|\Psi\rangle$ has to be in the domain of the unbounded operator $\hat{A}$, which is not always the case. In fact, the Robertson uncertainty relation is false if $\hat{A}$ is an angle variable and $\hat{B}$ is the derivative with respect to this variable. In this example, the commutator is a nonzero constant—just as in the Heisenberg uncertainty relation—and yet there are states where the product of the uncertainties is zero. This issue can be overcome by using a variational method for the proof, or by working with an exponentiated version of the canonical commutation relations.

Note that in the general form of the Robertson–Schrödinger uncertainty relation, there is no need to assume that the operators $\hat{A}$ and $\hat{B}$ are self-adjoint operators. It suffices to assume that they are merely symmetric operators. (The distinction between these two notions is generally glossed over in the physics literature, where the term *Hermitian* is used for either or both classes of operators.

Examples

Since the Robertson and Schrödinger relations are for general operators, the relations can be applied to any two observables to obtain specific uncertainty relations.

- For position and linear momentum, the canonical commutation relation $[\hat{x},\hat{p}]=i\hbar$ implies the Kennard inequality from above:

$$\sigma_x\sigma_p \geq \frac{\hbar}{2}$$

- For two orthogonal components of the total angular momentum operator of an object:

$$\sigma_{J_i}\sigma_{J_j} \geq \frac{\hbar}{2}|\langle J_k\rangle|,$$

  where i, j, k are distinct and J_i denotes angular momentum along the x_i axis. This relation implies that unless all three components vanish together, only a single component of a system's angular momentum can be defined with arbitrary precision, normally the component parallel to an external (magnetic or electric) field. Moreover, for $[J_x,J_y]=i\hbar\varepsilon_{xyz}J_z$, a choice $\hat{A}=J_x$, $\hat{B}=J_y$, in angular momentum multiplets, $\psi=|j,m\rangle$, bounds the Casimir invariant (angular momentum squared, $\langle J_x^2+J_y^2+J_z^2\rangle$) from below and thus yields useful constraints such as $j(j+1)\geq m(m+1)$, and hence $j\geq m$, among others.

- In non-relativistic mechanics, time is privileged as an independent variable. Nevertheless, in 1945, L. I. Mandelshtam and I. E. Tamm derived a non-relativistic *time–energy uncertainty relation*, as follows. For a quantum system in a non-stationary state ψ and an observable B represented by a self adjoint operator $\hat{B}$, the following formula holds:

$$\sigma_E \frac{\sigma_B}{\left|\frac{\mathrm{d}\langle\hat{B}\rangle}{\mathrm{d}t}\right|} \geq \frac{\hbar}{2}.$$

where σ_E is the standard deviation of the energy operator (Hamiltonian) in the state ψ, σ_B stands for the standard deviation of B. Although the second factor in the left-hand side has dimension of time, it is different from the time parameter that enters the Schrödinger equation. It is a *lifetime* of the state ψ with respect to the observable B: In other words, this is the *time interval* (Δt) after which the expectation value $\langle\hat{B}\rangle$ changes appreciably.

An informal, heuristic meaning of the principle is the following: A state that only exists for a short time cannot have a definite energy. To have a definite energy, the frequency of the state must be defined accurately, and this requires the state to hang around for many cycles, the reciprocal of the required accuracy. For example, in spectroscopy, excited states have a finite lifetime. By the time–energy uncertainty principle, they do not have a definite energy, and, each time they decay, the energy they release is slightly different. The average energy of the outgoing photon has a peak at the theoretical energy of the state, but the distribution has a finite width called the *natural linewidth*. Fast-decaying states have a broad linewidth, while slow decaying states have a narrow linewidth.

The same linewidth effect also makes it difficult to specify the rest mass of unstable, fast-decaying particles in particle physics. The faster the particle decays(the shorter its lifetime), the less certain is its mass (the larger the particle's width).

- For the number of electrons in a superconductor and the phase of its Ginzburg–Landau order parameter

  $\Delta N \Delta\varphi \geq 1.$

A Counter Example

Suppose we consider a quantum particle on a ring, where the wave function depends on an angular variable θ, which we may take to lie in the interval $[0, 2\pi]$. Define "position" and "momentum" operators $\hat{A}$ and $\hat{B}$ by

$$\hat{A}\psi(\theta) = \theta\psi(\theta), \quad \theta \in [0, 2\pi],$$

and

$$\hat{B}\psi = -i\hbar\frac{d\psi}{d\theta},$$

where we impose periodic boundary conditions on $\hat{B}$. Note that the definition of $\hat{A}$ depends on our choice to have θ range from 0 to 2π. These operators satisfy the usual commutation relations for position and momentum operators, $[\hat{A}, \hat{B}] = i\hbar$.

Now let ψ be any of the eigenstates of $\hat{B}$, which are given by $\psi(\theta) = e^{2\pi in\theta}$. Note that these states are normalizable, unlike the eigenstates of the momentum operator on the line. Note also that the operator $\hat{A}$ is bounded, since θ ranges over a bounded interval. Thus, in the state ψ, the uncertainty of B is zero and the uncertainty of is A finite, so that

$$\sigma_A \sigma_B = 0.$$

Although this result appears to violate the Robertson uncertainty principle, the paradox is resolved when we note that ψ is not in the domain of the operator $\hat{B}\hat{A}$, since multiplication by θ disrupts the periodic boundary conditions imposed on $\hat{B}$. Thus, the derivation of the Robertson relation, which requires $\hat{A}\hat{B}\psi$ and $\hat{B}\hat{A}\psi$ to be defined, does not apply. (These also furnish an example of operators satisfying the canonical commutation relations but not the Weyl relations.)

For the usual position and momentum operators $\hat{X}$ and $\hat{P}$ on the real line, no such counterexamples can occur. As long as σ_x and σ_p are defined in the state ψ, the Heisenberg uncertainty principle holds, even if ψ fails to be in the domain of $\hat{X}\hat{P}$ or of $\hat{P}\hat{X}$.

Examples

Quantum Harmonic Oscillator Stationary States

Consider a one-dimensional quantum harmonic oscillator (QHO). It is possible to express the position and momentum operators in terms of the creation and annihilation operators:

$$\hat{x} = \sqrt{\frac{\hbar}{2m\omega}}(a + a^\dagger)$$

$$\hat{p} = i\sqrt{\frac{m\omega\hbar}{2}}(a^\dagger - a).$$

Using the standard rules for creation and annihilation operators on the eigenstates of the QHO,

$$a^\dagger |n\rangle = \sqrt{n+1}\,|n+1\rangle$$

$$a|n\rangle = \sqrt{n}\,|n-1\rangle,$$

the variances may be computed directly,

$$\sigma_x^2 = \frac{\hbar}{m\omega}\left(n + \frac{1}{2}\right)$$

$$\sigma_p^2 = \hbar m\omega\left(n + \frac{1}{2}\right).$$

The product of these standard deviations is then

$$\sigma_x \sigma_p = \hbar\left(n + \frac{1}{2}\right) \ge \frac{\hbar}{2}.$$

In particular, the above Kennard bound is saturated for the ground state $n=0$, for which the probability density is just the normal distribution.

Quantum Harmonic Oscillator with Gaussian Initial Condition

In a quantum harmonic oscillator of characteristic angular frequency ω, place a state that is offset from the bottom of the potential by some displacement x_0 as

$$\psi(x) = \left(\frac{m\Omega}{\pi\hbar}\right)^{1/4} \exp\left(-\frac{m\Omega(x - x_0)^2}{2\hbar}\right),$$

where Ω describes the width of the initial state but need not be the same as ω. Through integration over the propagator, we can solve for the full time-dependent solution. After many cancelations, the probability densities reduce to

$$|\Psi(x,t)|^2 \sim \mathcal{N}\left(x_0 \cos(\omega t), \frac{\hbar}{2m\Omega}\left(\cos^2(\omega t) + \frac{\Omega^2}{\omega^2}\sin^2(\omega t)\right)\right)$$

$$|\Phi(p,t)|^2 \sim \mathcal{N}\left(-m x_0 \omega \sin(\omega t), \frac{\hbar m\Omega}{2}\left(\cos^2(\omega t) + \frac{\omega^2}{\Omega^2}\sin^2(\omega t)\right)\right),$$

where we have used the notation $\mathcal{N}(\mu, \sigma^2)$ to denote a normal distribution of mean μ and variance σ^2. Copying the variances above and applying trigonometric identities, we can write the product of the standard deviations as

$$\sigma_x\sigma_p = \frac{\hbar}{2}\sqrt{\left(\cos^2(\omega t)+\frac{\Omega^2}{\omega^2}\sin^2(\omega t)\right)\left(\cos^2(\omega t)+\frac{\omega^2}{\Omega^2}\sin^2(\omega t)\right)}$$

$$= \frac{\hbar}{4}\sqrt{3+\frac{1}{2}\left(\frac{\Omega^2}{\omega^2}+\frac{\omega^2}{\Omega^2}\right)-\left(\frac{1}{2}\left(\frac{\Omega^2}{\omega^2}+\frac{\omega^2}{\Omega^2}\right)-1\right)\cos(4\omega t)}$$

From the relations

$$\frac{\Omega^2}{\omega^2}+\frac{\omega^2}{\Omega^2} \geq 2, \quad |\cos(4\omega t)| \leq 1,$$

we can conclude the following: (the right most equality holds only when $\Omega=\omega$).

$$\sigma_x\sigma_p \geq \frac{\hbar}{4}\sqrt{3+\frac{1}{2}\left(\frac{\Omega^2}{\omega^2}+\frac{\omega^2}{\Omega^2}\right)-\left(\frac{1}{2}\left(\frac{\Omega^2}{\omega^2}+\frac{\omega^2}{\Omega^2}\right)-1\right)} = \frac{\hbar}{2}.$$

Coherent States

A coherent state is a right eigenstate of the annihilation operator,

$$\hat{a}|\alpha\rangle = \alpha|\alpha\rangle,$$

which may be represented in terms of Fock states as

$$|\alpha\rangle = e^{-\frac{|\alpha|^2}{2}}\sum_{n=0}^{\infty}\frac{\alpha^n}{\sqrt{n!}}|n\rangle$$

In the picture where the coherent state is a massive particle in a QHO, the position and momentum operators may be expressed in terms of the annihilation operators in the same formulas above and used to calculate the variances,

$$\sigma_x^2 = \frac{\hbar}{2m\omega},$$

$$\sigma_p^2 = \frac{\hbar m\omega}{2}.$$

Therefore, every coherent state saturates the Kennard bound

$$\sigma_x\sigma_p = \sqrt{\frac{\hbar}{2m\omega}}\sqrt{\frac{\hbar m\omega}{2}} = \frac{\hbar}{2}.$$

with position and momentum each contributing an amount $\sqrt{\hbar/2}$ in a "balanced" way. Moreover, every squeezed coherent state also saturates the Kennard bound although the individual contributions of position and momentum need not be balanced in general.

Particle in a Box

Consider a particle in a one-dimensional box of length L. The eigenfunctions in position and momentum space are

$$\psi_n(x,t)=\begin{cases}A\sin(k_n x)e^{-i\omega_n t}, & 0<x<L,\\ 0, & \text{otherwise,}\end{cases}$$

and

$$\varphi_n(p,t)=\sqrt{\frac{\pi L}{\hbar}}\frac{n\left(1-(-1)^n e^{-ikL}\right)e^{-i\omega_n t}}{\pi^2 n^2-k^2L^2},$$

where $\omega_n=\frac{\pi^2\hbar n^2}{8L^2 m}$ and we have used the de Broglie relation $p=\hbar k$. The variances of x and p can be calculated explicitly:

$$\sigma_x^2=\frac{L^2}{12}\left(1-\frac{6}{n^2\pi^2}\right)$$

$$\sigma_p^2=\left(\frac{\hbar n\pi}{L}\right)^2.$$

The product of the standard deviations is therefore

$$\sigma_x\sigma_p=\frac{\hbar}{2}\sqrt{\frac{n^2\pi^2}{3}-2}.$$

For all $n=1,2,3,\ldots$, the quantity $\sqrt{\frac{n^2\pi^2}{3}-2}$ is greater than 1, so the uncertainty principle is never violated. For numerical concreteness, the smallest value occurs when $n=1$, in which case

$$\sigma_x\sigma_p=\frac{\hbar}{2}\sqrt{\frac{\pi^2}{3}-2}\approx 0.568\hbar>\frac{\hbar}{2}.$$

Constant Momentum

Assume a particle initially has a momentum space wave function described by a normal distribution around some constant momentum p_0 according to

$$\varphi(p)=\left(\frac{x_0}{\hbar\sqrt{\pi}}\right)^{1/2}\cdot\exp\left(\frac{-x_0^2(p-p_0)^2}{2\hbar^2}\right),$$

where we have introduced a reference scale $x_0 = \sqrt{\hbar / m\omega_0}$, *with* $\omega_0 > 0$ describing the width of the distribution. If the state is allowed to evolve in free space, then the time-dependent momentum and position space wave functions are

$$\Phi(p,t) = \left(\frac{x_0}{\hbar\sqrt{\pi}}\right)^{1/2} \cdot \exp\left(\frac{-x_0^2(p-p_0)^2}{2\hbar^2} - \frac{ip^2t}{2m\hbar}\right),$$

$$\Psi(x,t) = \left(\frac{1}{x_0\sqrt{\pi}}\right)^{1/2} \cdot \frac{e^{-x_0^2p_0^2/2\hbar^2}}{\sqrt{1+i\omega_0 t}} \cdot \exp\left(-\frac{(x-ix_0^2p_0/\hbar)^2}{2x_0^2(1+i\omega_0 t)}\right).$$

Since $\langle p(t)\rangle = p_0$ and $\sigma_p(t) = \hbar / x_0\sqrt{2}$, this can be interpreted as a particle moving along with constant momentum at arbitrarily high precision. On the other hand, the standard deviation of the position is

$$\sigma_x = \frac{x_0}{\sqrt{2}}\sqrt{1+\omega_0^2 t^2}$$

such that the uncertainty product can only increase with time as

$$\sigma_x(t)\sigma_p(t) = \frac{\hbar}{2}\sqrt{1+\omega_0^2 t^2}$$

Additional Uncertainty Relations

Mixed States

The Robertson–Schrödinger uncertainty relation may be generalized in a straightforward way to describe mixed states.

$$\sigma_A^2\sigma_B^2 \geq \left(\frac{1}{2}\mathrm{tr}(\rho\{A,B\}) - \mathrm{tr}(\rho A)\mathrm{tr}(\rho B)\right)^2 + \left(\frac{1}{2i}\mathrm{tr}(\rho[A,B])\right)^2$$

The Maccone–Pati Uncertainty Relations

The Robertson–Schrödinger uncertainty relation can be trivial if the state of the system is chosen to be eigenstate of one of the observable. The stronger uncertainty relations proved by Maccone and Pati give non-trivial bounds on the sum of the variances for two incompatible observables. For two non-commuting observables A and B the first stronger uncertainty relation is given by

$$\sigma_A^2 + \sigma_B^2 \geq \pm i\langle\Psi|[A,B]|\Psi\rangle + |\langle\Psi|(A \pm iB)|\bar{\Psi}\rangle|^2,$$

where $\sigma_A^2 = \langle\Psi|A^2|\Psi\rangle - \langle\Psi|A|\Psi\rangle^2$, $\sigma_B^2 = \langle\Psi|B^2|\Psi\rangle - \langle\Psi|B|\Psi\rangle^2$, $|\bar{\Psi}\rangle$ is a normalized vector that is orthogonal to the state of the system $|\bar{\Psi}\rangle$ and one should choose the sign of $\pm i\langle\Psi|[A,B]|\Psi\rangle$ to make this real quantity a positive number.

The second stronger uncertainty relation is given by

$$\sigma_A^2 + \sigma_B^2 \geq \frac{1}{2} |\langle \bar{\Psi}_{A+B} | (A+B) | \Psi \rangle|^2$$

where $|\bar{\Psi}_{A+B}\rangle$ is a state orthogonal to $|\Psi\rangle$. The form of $|\bar{\Psi}_{A+B}\rangle$ implies that the right-hand side of the new uncertainty relation is nonzero unless $|\Psi\rangle$ is an eigenstate of $(A+B)$. One may note that $|\Psi\rangle$ can be an eigenstate of $(A+B)$ without being an eigenstate of either A or B. However, when $|\Psi\rangle$ is an eigenstate of one of the two observables the Heisenberg–Schrödinger uncertainty relation becomes trivial. But the lower bound in the new relation is nonzero unless $|\Psi\rangle$ is an eigenstate of both.

Phase Space

In the phase space formulation of quantum mechanics, the Robertson–Schrödinger relation follows from a positivity condition on a real star-square function. Given a Wigner function $W(x,p)$ with star product ★ and a function *f*, the following is generally true:

$$\langle f^* \star f \rangle = \int (f^* \star f)\, W(x,p)\, dx\, dp \geq 0.$$

Choosing $f = a + bx + cp,$, we arrive at

$$\langle f^* \star f \rangle = \begin{bmatrix} a^* & b^* & c^* \end{bmatrix} \begin{bmatrix} 1 & \langle x \rangle & \langle p \rangle \\ \langle x \rangle & \langle x \star x \rangle & \langle x \star p \rangle \\ \langle p \rangle & \langle p \star x \rangle & \langle p \star p \rangle \end{bmatrix} \begin{bmatrix} a \\ b \\ c \end{bmatrix} \geq 0.$$

Since this positivity condition is true for *all a*, *b*, and *c*, it follows that all the eigenvalues of the matrix are positive. The positive eigenvalues then imply a corresponding positivity condition on the determinant:

$$\det \begin{bmatrix} 1 & \langle x \rangle & \langle p \rangle \\ \langle x \rangle & \langle x \star x \rangle & \langle x \star p \rangle \\ \langle p \rangle & \langle p \star x \rangle & \langle p \star p \rangle \end{bmatrix} = \det \begin{bmatrix} 1 & \langle x \rangle & \langle p \rangle \\ \langle x \rangle & \langle x^2 \rangle & \left\langle xp + \frac{i\hbar}{2} \right\rangle \\ \langle p \rangle & \left\langle xp - \frac{i\hbar}{2} \right\rangle & \langle p^2 \rangle \end{bmatrix} \geq 0,$$

or, explicitly, after algebraic manipulation,

$$\sigma_x^2 \sigma_p^2 = \left(\langle x^2 \rangle - \langle x \rangle^2 \right)\left(\langle p^2 \rangle - \langle p \rangle^2 \right) \geq \left(\langle xp \rangle - \langle x \rangle \langle p \rangle \right)^2 + \frac{\hbar^2}{4}.$$

Systematic and Statistical Errors

The inequalities above focus on the *statistical imprecision* of observables as quantified by the standard deviation σ. Heisenberg's original version, however, was dealing with the *systematic error*, a disturbance of the quantum system produced by the measuring apparatus, i.e., an observer effect.

If we let ε_A represent the error (i.e., inaccuracy) of a measurement of an observable *A* and η_B the disturbance produced on a subsequent measurement of the conjugate variable *B* by the former measurement of *A*, then the inequality proposed by Ozawa — encompassing both systematic and statistical errors — holds:

$$\varepsilon_A \eta_B + \varepsilon_A \sigma_B + \sigma_A \eta_B \geq \frac{1}{2}\left|\langle[\hat{A}, \hat{B}]\rangle\right|$$

Heisenberg's uncertainty principle, as originally described in the 1927 formulation, mentions only the first term of Ozawa inequality, regarding the *systematic error*. Using the notation above to describe the *error/disturbance* effect of *sequential measurements* (first *A*, then *B*), it could be written as

$$\varepsilon_A \eta_B \geq \frac{1}{2}\left|\langle[\hat{A}, \hat{B}]\rangle\right|$$

The formal derivation of the Heisenberg relation is possible but far from intuitive. It was *not* proposed by Heisenberg, but formulated in a mathematically consistent way only in recent years. Also, it must be stressed that the Heisenberg formulation is not taking into account the intrinsic statistical errors σ_A and σ_B. There is increasing experimental evidence that the total quantum uncertainty cannot be described by the Heisenberg term alone, but requires the presence of all the three terms of the Ozawa inequality.

Using the same formalism, it is also possible to introduce the other kind of physical situation, often confused with the previous one, namely the case of *simultaneous measurements* (*A* and *B* at the same time):

$$\varepsilon_A \varepsilon_B \geq \frac{1}{2}\left|\langle[\hat{A}, \hat{B}]\rangle\right|$$

The two simultaneous measurements on *A* and *B* are necessarily *unsharp* or *weak*.

It is also possible to derive an uncertainty relation that, as the Ozawa's one, combines both the statistical and systematic error components, but keeps a form very close to the Heisenberg original inequality. By adding Robertson

$$\sigma_A \sigma_B \geq \frac{1}{2}\left|\langle[\hat{A}, \hat{B}]\rangle\right|$$

and Ozawa relations we obtain

$$\varepsilon_A \eta_B + \varepsilon_A \sigma_B + \sigma_A \eta_B + \sigma_A \sigma_B \geq \left|\langle[\hat{A}, \hat{B}]\rangle\right|.$$

The four terms can be written as:

$$(\varepsilon_A + \sigma_A)(\eta_B + \sigma_B) \geq \left|\langle[\hat{A}, \hat{B}]\rangle\right|.$$

Defining:

$$\overline{\varepsilon}_A \equiv (\varepsilon_A + \sigma_A)$$

as the *inaccuracy* in the measured values of the variable A and

$$\overline{\eta}_B \equiv (\eta_B + \sigma_B)$$

as the *resulting fluctuation* in the conjugate variable B, Fujikawa established an uncertainty relation similar to the Heisenberg original one, but valid both for *systematic and statistical errors*:

$$\overline{\varepsilon}_A \overline{\eta}_B \geq \left| \langle [\hat{A}, \hat{B}] \rangle \right|$$

Quantum Entropic Uncertainty Principle

For many distributions, the standard deviation is not a particularly natural way of quantifying the structure. For example, uncertainty relations in which one of the observables is an angle has little physical meaning for fluctuations larger than one period. Other examples include highly bimodal distributions, or unimodal distributions with divergent variance.

A solution that overcomes these issues is an uncertainty based on entropic uncertainty instead of the product of variances. While formulating the many-worlds interpretation of quantum mechanics in 1957, Hugh Everett III conjectured a stronger extension of the uncertainty principle based on entropic certainty. This conjecture, also studied by Hirschman and proven in 1975 by Beckner and by Iwo Bialynicki-Birula and Jerzy Mycielski is that, for two normalized, dimensionless Fourier transform pairs $f(a)$ and $g(b)$ where

$$f(a) = \int_{-\infty}^{\infty} g(b)\, e^{2\pi i a b}\, db \text{ and } g(b) = \int_{-\infty}^{\infty} f(a)\, e^{-2\pi i a b}\, da$$

the Shannon information entropies

$$H_a = \int_{-\infty}^{\infty} f(a) \log(f(a)) da,$$

and

$$H_b = \int_{-\infty}^{\infty} g(b) \log(g(b)) db$$

are subject to the following constraint,

$$H_a + H_b \geq \log(e/2)$$

where the logarithms may be in any base.

The probability distribution functions associated with the position wave function $\psi(x)$ and the momentum wave function $\varphi(x)$ have dimensions of inverse length and momentum respectively, but the entropies may be rendered dimensionless by

$$H_x = -\int |\psi(x)|^2 \ln(x_0 |\psi(x)|^2) dx = -\left\langle \ln(x_0 |\psi(x)|^2) \right\rangle$$

$$H_p = -\int |\varphi(p)|^2 \ln(p_0 |\varphi(p)|^2) dp = -\left\langle \ln(p_0 |\varphi(p)|^2) \right\rangle$$

where x_0 and p_0 are some arbitrarily chosen length and momentum respectively, which render the arguments of the logarithms dimensionless. Note that the entropies will be functions of these chosen parameters. Due to the Fourier transform relation between the position wave function $\psi(x)$ and the momentum wavefunction $\varphi(p)$, the above constraint can be written for the corresponding entropies as

$$H_x + H_p \geq \log\left(\frac{eh}{2x_0 p_0}\right)$$

where h is Planck's constant.

Depending on one's choice of the $x_0 p_0$ product, the expression may be written in many ways. If $x_0 p_0$ is chosen to be h, then

$$H_x + H_p \geq \log\left(\frac{e}{2}\right)$$

If, instead, $x_0 p_0$ is chosen to be $\hbar$, then

$$H_x + H_p \geq \log(e\pi)$$

If x_0 and p_0 are chosen to be unity in whatever system of units are being used, then

$$H_x + H_p \geq \log\left(\frac{eh}{2}\right)$$

where h is interpreted as a dimensionless number equal to the value of Planck's constant in the chosen system of units.

The quantum entropic uncertainty principle is more restrictive than the Heisenberg uncertainty principle. From the inverse logarithmic Sobolev inequalities

$$H_x \leq \frac{1}{2}\log(2e\pi\sigma_x^2 / x_0^2),$$

$$H_p \leq \frac{1}{2}\log(2e\pi\sigma_p^2 / p_0^2),$$

(equivalently, from the fact that normal distributions maximize the entropy of all such with a given variance), it readily follows that this entropic uncertainty principle is *stronger than the one based on standard deviations*, because

$$\sigma_x \sigma_p \geq \frac{\hbar}{2}\exp\left(H_x + H_p - \log\left(\frac{eh}{2x_0 p_0}\right)\right) \geq \frac{\hbar}{2}.$$

In other words, the Heisenberg uncertainty principle, is a consequence of the quantum entropic uncertainty principle, but not vice versa. A few remarks on these inequalities. First, the choice of base e is a matter of popular convention in physics. The logarithm can alternatively be in any base, provided that it be consistent on both sides of the inequality. Second, recall the Shannon entropy has been used, *not* the quantum von Neumann entropy. Finally, the normal distribution saturates the inequality, and it is the only distribution with this property, because it is the maximum entropy probability distribution among those with fixed variance.

Entropic Uncertainty of the Normal Distribution

We demonstrate this method on the ground state of the QHO, which as discussed above saturates the usual uncertainty based on standard deviations. The length scale can be set to whatever is convenient, so we assign

$$x_0 = \sqrt{\frac{\hbar}{2m\omega}}$$

$$\begin{aligned}\psi(x) &= \left(\frac{m\omega}{\pi\hbar}\right)^{1/4} \exp\left(-\frac{m\omega x^2}{2\hbar}\right) \\ &= \left(\frac{1}{2\pi x_0^2}\right)^{1/4} \exp\left(-\frac{x^2}{4x_0^2}\right)\end{aligned}$$

The probability distribution is the normal distribution

$$|\psi(x)|^2 = \frac{1}{x_0\sqrt{2\pi}} \exp\left(-\frac{x^2}{2x_0^2}\right)$$

with Shannon entropy

$$\begin{aligned}H &= -\int |\psi(x)|^2 \ln(|\psi(x)|^2 \cdot x\ )dx \\ &= -\frac{1}{\sqrt{2\pi}} \int_{-\infty}^{\infty} \exp\left(-\frac{x^2}{2x_0^2}\right) \ln\left[\frac{1}{\sqrt{2\pi}} \exp\left(-\frac{x^2}{2x_0^2}\right)\right] dx \\ &= \frac{}{\sqrt{\ }} \int_{-\infty}^{\infty} \exp\left(-\frac{u^2}{2}\right) \left[\ln(\sqrt{\ }\pi) + \frac{u^2}{2}\right] du \\ &= \ln(\sqrt{\ }\pi) + \frac{}{}.\end{aligned}$$

A completely analogous calculation proceeds for the momentum distribution. Choosing a standard momentum of $p_0 = \hbar / x_0$:

$$\varphi(p) = \left(\frac{2x_0^2}{\pi\hbar^2}\right)^{1/4} \exp\left(-\frac{x_0^2 p^2}{\hbar^2}\right)$$

$$|\varphi(p)|^2 = \sqrt{\frac{2x_0^2}{\pi\hbar^2}} \exp\left(-\frac{2x_0^2 p^2}{\hbar^2}\right)$$

$$
\begin{aligned}
H_p &= -\int |\varphi(p)|^2 \ln(|\varphi(p)|^2 \cdot \hbar / x_0) dp \\
&= -\sqrt{\frac{2x_0^2}{\pi\hbar^2}} \int_{-\infty}^{\infty} \exp\left(-\frac{2x_0^2 p^2}{\hbar^2}\right) \ln\left[\sqrt{\frac{2}{\pi}} \exp\left(-\frac{2x_0^2 p^2}{\hbar^2}\right)\right] dp \\
&= \sqrt{\frac{2}{\pi}} \int_{-\infty}^{\infty} \exp(-2v^2) \left[\ln\left(\sqrt{\frac{\pi}{2}}\right) + 2v^2\right] dv \\
&= \ln\left(\sqrt{\frac{\pi}{2}}\right) + \frac{1}{2}.
\end{aligned}
$$

The entropic uncertainty is therefore the limiting value

$$
\begin{aligned}
H_x + H_p &= \ln(\sqrt{2\pi}) + \frac{1}{2} + \ln\left(\sqrt{\frac{\pi}{2}}\right) + \frac{1}{2} \\
&= 1 + \ln \pi = \ln(e\pi).
\end{aligned}
$$

A measurement apparatus will have a finite resolution set by the discretization of its possible outputs into bins, with the probability of lying within one of the bins given by the Born rule. We will consider the most common experimental situation, in which the bins are of uniform size. Let δx be a measure of the spatial resolution. We take the zeroth bin to be centered near the origin, with possibly some small constant offset c. The probability of lying within the jth interval of width δx is

$$
P[x_j] = \int_{(j-1/2)\delta x - c}^{(j+1/2)\delta x - c} |\psi(x)|^2 dx
$$

To account for this discretization, we can define the Shannon entropy of the wave function for a given measurement apparatus as

$$
H_x = -\sum_{j=-\infty}^{\infty} P[x_j] \ln P[x_j].
$$

Under the above definition, the entropic uncertainty relation is

$$
H_x + H_p > \ln\left(\frac{e}{2}\right) - \ln\left(\frac{\delta x \delta p}{h}\right).
$$

Here we note that $\delta x \, \delta p / h$ is a typical infinitesimal phase space volume used in the calculation of a partition function. The inequality is also strict and not saturated. Efforts to improve this bound are an active area of research.

Normal Distribution Example

We demonstrate this method first on the ground state of the QHO, which as discussed above saturates the usual uncertainty based on standard deviations.

$$
\psi(x) = \left(\frac{m\omega}{\pi\hbar}\right)^{1/4} \exp\left(-\frac{m\omega x^2}{2\hbar}\right)
$$

The probability of lying within one of these bins can be expressed in terms of the error function.

$$\mathrm{P}[x_j] = \sqrt{\frac{m\omega}{\pi\hbar}} \int_{(j-1/2)\delta x}^{(j+1/2)\delta x} \exp\left(-\frac{m\omega x^2}{\hbar}\right) dx$$

$$= \sqrt{\frac{1}{\pi}} \int_{(j-1/2)\delta x\sqrt{m\omega/\hbar}}^{(j+1/2)\delta x\sqrt{m\omega/\hbar}} e^{u^2}\, du$$

$$= \frac{1}{2}\left[\operatorname{erf}\left(\left(j+\frac{1}{2}\right)\delta x \cdot \sqrt{\frac{m\omega}{\hbar}}\right) - \operatorname{erf}\left(\left(j-\frac{1}{2}\right)\delta x \cdot \sqrt{\frac{m\omega}{\hbar}}\right)\right]$$

The momentum probabilities are completely analogous.

$$\mathrm{P}[p_j] = \frac{1}{2}\left[\operatorname{erf}\left(\left(j+\frac{1}{2}\right)\delta p \cdot \frac{1}{\sqrt{\hbar m\omega}}\right) - \operatorname{erf}\left(\left(j-\frac{1}{2}\right)\delta x \cdot \frac{1}{\sqrt{\hbar m\omega}}\right)\right]$$

For simplicity, we will set the resolutions to

$$\delta x = \sqrt{\frac{h}{m\omega}}$$

$$\delta p = \sqrt{hm\omega}$$

so that the probabilities reduce to

$$\mathrm{P}[x_j] = \mathrm{P}[p_j] = \frac{1}{2}\left[\operatorname{erf}\left(\left(j+\frac{1}{2}\right)\sqrt{2\pi}\right) - \operatorname{erf}\left(\left(j-\frac{1}{2}\right)\sqrt{2\pi}\right)\right]$$

The Shannon entropy can be evaluated numerically.

$$H_x = H_p = -\sum_{j=-\infty}^{\infty} \mathrm{P}[x_j] \ln \mathrm{P}[x_j]$$

$$= -\sum_{j=-\infty}^{\infty} \frac{1}{2}\left[\operatorname{erf}\left(\left(j+\frac{1}{2}\right)\sqrt{2\pi}\right) - \operatorname{erf}\left(\left(j-\frac{1}{2}\right)\sqrt{2\pi}\right)\right] \ln \frac{1}{2}\left[\operatorname{erf}\left(\left(j+\frac{1}{2}\right)\sqrt{2\pi}\right) - \operatorname{erf}\left(\left(j-\frac{1}{2}\right)\sqrt{2\pi}\right)\right] \approx 0.3226$$

The entropic uncertainty is indeed larger than the limiting value.

$$H_x + H_p \approx 0.3226 + 0.3226 = 0.6452 > \ln\left(\frac{e}{2}\right) - \ln 1 \approx 0.3069$$

Note that despite being in the optimal case, the inequality is not saturated.

Sinc Function Example

An example of a unimodal distribution with infinite variance is the sinc function. If the wave function is the correctly normalized uniform distribution,

$$\psi(x)=\begin{cases}\frac{1}{\sqrt{2a}} & \text{for } |x|\le a,\\ 0 & \text{for } |x|>a\end{cases}$$

then its Fourier transform is the sinc function,

$$\varphi(p)=\sqrt{\frac{a}{\pi\hbar}}\cdot\text{sinc}\left(\frac{ap}{\hbar}\right)$$

which yields infinite momentum variance despite having a centralized shape. The entropic uncertainty, on the other hand, is finite. Suppose for simplicity that the spatial resolution is just a two-bin measurement, $\delta x = a$, and that the momentum resolution is $\delta p = h/a$.

Partitioning the uniform spatial distribution into two equal bins is straightforward. We set the offset $c=1/2$ so that the two bins span the distribution.

$$\text{P}[x_0]=\int_{-a}^{0}\frac{1}{2a}dx=\frac{1}{2}$$

$$\text{P}[x_1]=\int_{0}^{a}\frac{1}{2a}dx=\frac{1}{2}$$

$$H_x=-\sum_{j=0}^{1}\text{P}[x_j]\ln\text{P}[x_j]=-\frac{1}{2}\ln\frac{1}{2}-\frac{1}{2}\ln\frac{1}{2}=\ln 2$$

The bins for momentum must cover the entire real line. As done with the spatial distribution, we could apply an offset. It turns out, however, that the Shannon entropy is minimized when the zeroth bin for momentum is centered at the origin. The probability of lying within an arbitrary momentum bin can be expressed in terms of the sine integral.

$$\begin{aligned}\text{P}[p_j]&=\frac{a}{\pi\hbar}\int_{(j-1/2)\delta p}^{(j+1/2)\delta p}\text{sinc}^2\left(\frac{ap}{\hbar}\right)dp\\ &=\frac{1}{\pi}\int_{2\pi(j-1/2)}^{2\pi(j+1/2)}\text{sinc}^2(u)du\\ &=\frac{1}{\pi}\left[\text{Si}((4j+2)\pi)-\text{Si}((4j-2)\pi)\right]\end{aligned}$$

The Shannon entropy can be evaluated numerically.

$$H_p=-\sum_{j=-\infty}^{\infty}\text{P}[p_j]\ln\text{P}[p_j]=-\text{P}[p_0]\ln\text{P}[p_0]-2\cdot\sum_{j=1}^{\infty}\text{P}[p_j]\ln\text{P}[p_j]\approx 0.53$$

The entropic uncertainty is indeed larger than the limiting value.

$$H_x+H_p\approx 0.69+0.53=1.22>\ln\left(\frac{e}{2}\right)-\ln 1\approx 0.31.$$

Harmonic Analysis

In the context of harmonic analysis, a branch of mathematics, the uncertainty principle implies that one cannot at the same time localize the value of a function and its Fourier transform. To wit, the following inequality holds,

$$\left(\int_{-\infty}^{\infty} x^2 |f(x)|^2 dx\right)\left(\int_{-\infty}^{\infty} \xi^2 |\hat{f}(\xi)|^2 d\xi\right) \geq \frac{\|f\|_2^4}{16\pi^2}.$$

Further mathematical uncertainty inequalities, including the above entropic uncertainty, hold between a function f and its Fourier transform $\hat{f}$:

$$H_x + H_\xi \geq \log(e/2)$$

Signal Processing

In the context of signal processing, and in particular time–frequency analysis, uncertainty principles are referred to as the Gabor limit, after Dennis Gabor, or sometimes the *Heisenberg–Gabor limit*. The basic result, which follows from "Benedicks's theorem", below, is that a function cannot be both time limited and band limited (a function and its Fourier transform cannot both have bounded domain).

$$\sigma_t \cdot \sigma_f \geq \frac{1}{4\pi} \approx 0.08 \text{ cycles}$$

where σ_t and σ_f are the standard deviations of the time and frequency estimates respectively.

Stated alternatively, "One cannot simultaneously sharply localize a signal (function f) in both the time domain and frequency domain ($\hat{f}$, its Fourier transform)".

When applied to filters, the result implies that one cannot achieve high temporal resolution and frequency resolution at the same time; a concrete example are the resolution issues of the short-time Fourier transform—if one uses a wide window, one achieves good frequency resolution at the cost of temporal resolution, while a narrow window has the opposite trade-off.

Alternate theorems give more precise quantitative results, and, in time–frequency analysis, rather than interpreting the (1-dimensional) time and frequency domains separately, one instead interprets the limit as a lower limit on the support of a function in the (2-dimensional) time–frequency plane. In practice, the Gabor limit limits the *simultaneous* time–frequency resolution one can achieve without interference; it is possible to achieve higher resolution, but at the cost of different components of the signal interfering with each other.

Benedicks'S Theorem

Amrein–Berthier and Benedicks's theorem intuitively says that the set of points where f is non-zero and the set of points where $\hat{f}$ is non-zero cannot both be small.

Specifically, it is impossible for a function f in $L^2(\mathbf{R})$ and its Fourier transform $\hat{f}$ to both be supported on sets of finite Lebesgue measure. A more quantitative version is

$$\| f \|_{L^2(\mathbf{R}^d)} \le Ce^{C|S||\Sigma|}\left(\| f \|_{L^2(S^c)} + \| \hat{f} \|_{L^2(\Sigma^c)}\right).$$

One expects that the factor $Ce^{C|S||\Sigma|}$ may be replaced by $Ce^{C(|S||\Sigma|)^{1/d}}$, which is only known if either S or Σ is convex.

Hardy's Uncertainty Principle

The mathematician G. H. Hardy formulated the following uncertainty principle: it is not possible for f and $\hat{f}$ to both be "very rapidly decreasing". Specifically, if f in $L^2(\mathrm{R})$ is such that

$$| f(x) | \le C(1+|x|)^N e^{-a\pi x^2}$$

and

$$| \hat{f}(\xi) | \le C(1+|\xi|)^N e^{-b\pi\xi^2} \quad (C > 0,\ N \text{ an integer}),$$

then, if $ab > 1$, $f = 0$, while if $ab = 1$, then there is a polynomial P of degree $\le N$ such that

$$f(x) = P(x)e^{-a\pi x^2}.$$

This was later improved as follows: if $f \in L^2(\mathrm{R}^d)$ is such that

$$\int_{\mathbf{R}^d}\int_{\mathbf{R}^d} |f(x)||\hat{f}(\xi)| \frac{e^{\pi i \langle x\ \xi \rangle}}{(1 \aleph |x|\ \ |\ \ |)^N} dx\, d\xi < +\infty,$$

then

$$f(x) = P(x)e^{-\pi\langle Ax, x\rangle},$$

where P is a polynomial of degree $(N-d)/2$ and A is a real $d \times d$ positive definite matrix.

This result was stated in Beurling's complete works without proof and proved in Hörmander (the case $d = 1$, $N = 0$) and Bonami, Demange, and Jamingfor the general case. Note that Hörmander–Beurling's version implies the case $ab > 1$ in Hardy's Theorem while the version by Bonami–Demange–Jaming covers the full strength of Hardy's Theorem.

Theorem: If a tempered distribution $f \in S'(\mathbb{R}^d)$ is such that

$$e^{\pi|x|^2} f \in S'(\mathbb{R}^d)$$

and

$$e^{\pi|\xi|^2} \hat{f} \in S'(\mathbb{R}^d),$$

then

$$f(x) = P(x)e^{-\pi\langle Ax, x\rangle},$$

for some convenient polynomial P and real positive definite matrix A of type $d \times d$.

Complementarity Principle

The principle of complementarity is related to Heisenberg's uncertainty principle, but they are not the same. Bohr's principle of complementarity lies at the central place of quantum mechanics.

Bohr had introduced complementarity in 1927: "The very nature of the quantum theory thus forces us to regard the space-time co-ordination and the claim of causality, the union of which characterizes the classical theories, as complementary but exclusive features of the description" (Bohr, p. 115). Thus, complementarity here means to keep distinct what has traditionally been merged. In contrast, the complementarity of the particle-like and the wave-like behavior of light brings together "contradictory" models that traditionally are regarded as excluding each other.

Complementarity refers to the possibility that the same entity/phenomenon manifests itself in distinct, categorically different ways. All the differing manifestations need to be described and explained, and be part of an overarching theory of the entity/phenomenon, but not all occur in the same spatial, temporal, or situational context, respectively.

Matter Wave

In quantum mechanics, a matter wave is when you think of matter as a wave. The concept of matter waves was first introduced by Louis de Broglie. Matter waves are hard to visualize, because we are used to thinking of matter as solid. De Broglie revolutionized quantum mechanics by producing the equation for matter waves.

Wavelength of De-Broglie Waves

Consider a photon whose energy is given by

$$E = h\upsilon = hc/\lambda$$

If a photon possesses mass (rest mass is zero), then according to the theory of relatively, its energy is given by

$$E = mc^2$$

From $E = h\upsilon = hc/\lambda$ and $E = mc^2$, we have

Mass of photon $m = h/c\lambda$

Therefore Momentum of photon

$$P = mc = hc/c\lambda = h/\lambda$$

$$\text{Or } \lambda = h/p$$

If instead of a photon, we consider a material particle of mass m moving with velocity v, then the momentum of the particle, $p = mv$. Therefore, the wavelength of the wave associated with this moving particle is given by:

$$h/mv$$

Or $\lambda = h/p$ (But here $p = mv$)

This wavelength is called De-Broglie wavelength.

Special Cases

- de-Broglie wavelength for material particle:

If E is the kinetic energy of the material particle of mass m moving with velocity v, then

$$E = 1/2\,mv^2 = 1/2\,m^2v^2 = p^2/2m$$

$$\text{Or } p = \sqrt{2mE}$$

Therefore the by putting above equation in equation $\lambda = h/p$, we get de-Broglie wavelength equation for material particle as:

$$\lambda = h/\sqrt{2mE}$$

- de-Broglie wavelength for particle in gaseous state:

According to kinetic theory of gases, the average kinetic energy of the material particle is given by

$$E = (3/2)kT$$

Where $k = 1.38 \times 10^{-23}\ J/K$ is the Boltzmann's constant and T is the absolute temperature of the particle.

$$\text{Also } E = p\ /2m$$

Comparing above two equations, we get:

$$p^2/2m = (3/2)kT$$

$$\text{or } p = /\sqrt{3mKT}$$

Therefore, equation $\lambda = h/p$ becomes

$$\lambda = h/\sqrt{3mKT}$$

This is the De-Broglie wavelength for particle in gaseous state.

- de-Broglie wavelength for an accelerated electron:

Suppose an electron accelerates through a potential difference of V volt. The work done by electric field on the electron appears as the gain in its kinetic energy

That is $E = eV$

Also $E = p^2 / 2m$

Where e is the charge on the electron, m is the mass of electron and v is the velocity of electron, then

Comparing above two equations, we get:

$$eV = p^2 / 2m$$

$$\text{or } p = \sqrt{2meV}$$

Thus by putting this equation in equation $\lambda = h/p$, we get the the de-Broglie wavelength of the electron as

$$\lambda = h/\sqrt{2meV}\ 6.63 \times 10^{-34} / \sqrt{2} \times 9.1 \times 10^{-31} \times 1.6 \times 10^{-19} V$$

$$\lambda = 12.27 / \sqrt{V}\ \AA$$

This is the de-Broglie wavelength for electron moving in a potential difference of V volt.

An Application of Matter Waves

Electrons behave like a waves, they can be used to "illuminate" objects in a manner similar to light. An electron microscope is an instrument that takes advantage of this situation. Electrons are given energy by accelerating them in a manner similar to the way a TV tube works. Then, using magnetic fields, they are directed at an object of interest. The electrons are focused to illuminate the object, and then to form the image of that object. A schematic diagram is shown in the diagram. This system can be used to look closely at very small objects.

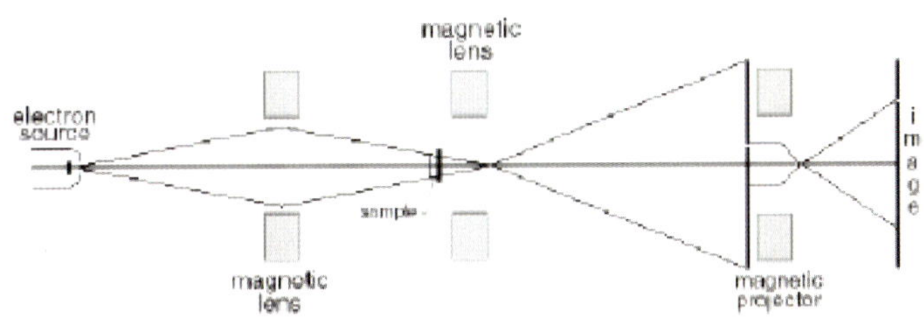

The wavelengths of the electrons are related to their kinetic energies. In electron microscopes, wavelengths as much as 100000 times smaller than those of visible light can be achieved. With such small wavelengths, electron microscopes can reveal features that are as small as 0.000000001 meters (1 nm).

Wave–particle Duality

Wave-particle duality is the term for the fact that fundamental objects in the universe such as photons or electrons appear to exhibit aspects of either waves or particles depending on the experiment.

Through the beginning of the twentieth century, light was widely accepted to be a wave while matter was understood to be comprised of atoms that themselves consisted of subatomic particles. However, with the development of quantum mechanics, experiments involving particles showed that they must possess some wave-like qualities, while experiments with light showed that it must possess some particle-like qualities. These apparent contradictions are explained within the theoretical framework of quantum mechanics.

The Particle Viewpoint

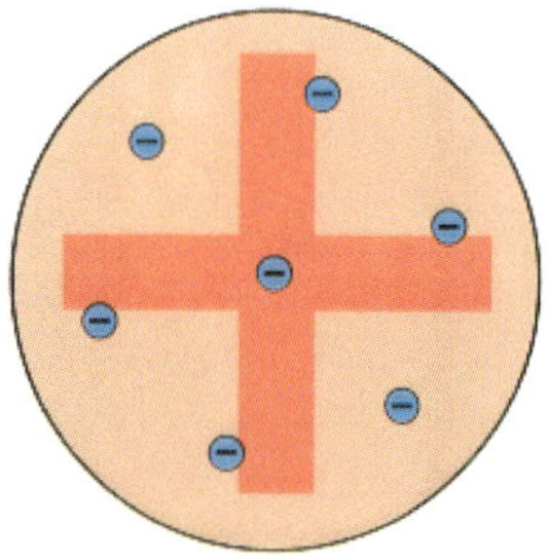

Plum pudding model of the atom where electrons were embedded in a solid positively charged sphere

The concept of matter composed of particles dates back to the Greeks, who were the first to conceive of atoms, though they did not know anything of their constituent parts. Not until the late 19th century did any understanding of individual particles come about, with J.J. Thomson's plum pudding model of the atom as a neutral composite consisting of individual negative particles or electrons embedded in a blob of positive charge.

This was motivated by Thomson's cathode ray tube experiments, where he showed that free electric charge could be carried across a vacuum. Thus, he proposed the existence of the electron as a particle carrying electron charge, and shortly after in 1904 devised the plum pudding model to account for chemical properties of the atom.

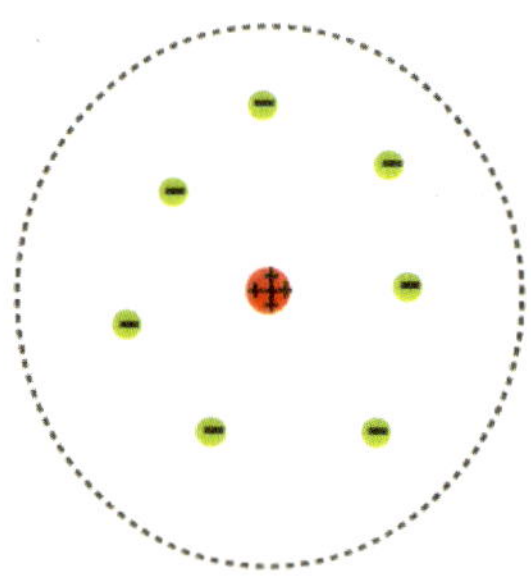

Rutherford model of the atom accounting for the small positively charged nucleus

Shortly after, in 1911, Rutherford overturned this model via his gold foil experiment which established the existence of the nucleus by bombarding a thin gold foil with alpha particles. Atoms were thus understood to consist of both positive and negative particles, but particles nonetheless.

With the discovery of quantum mechanics via the quantization of the emission lines of the hydrogen atom, Bohr updated Rutherford's model so that the electrons lay in orbits of integral angular momentum in units of $\hbar$. All of the elements in the periodic table that comprise all matter were thus ordered by the energies of their constituent electrons.

Throughout all of this scientific inquiry over several centuries, it was assumed and experimentally justified that essentially everything was composed of particles on a small enough level, with the exception of light (discussed below). Scientific investigation throughout the 19th century had put light on solid ground as a wave. In the early twentieth century, however, a number of ingenious experiments seemed unexplainable without light also taking on a particle nature:

The observation of light exhibiting particle qualities came as a result of the advent of quantum mechanics, with Planck's solution to the black-body radiation problem in 1900. When in thermal equilibrium, any physical object (a "black body") continuously emits electromagnetic radiation to its environment. If the spectrum of this radiation were continuous, the energy contained in this radiation would be infinite, a so called ultraviolet catastrophe.

The reason for this divergence was that classical statistical mechanics assumed every classical electromagnetic mode to have the same energy, dictated by the temperature. Planck's solution to this problem was to assume that radiation was quantized in discrete packets with energy $E = h\upsilon$ with υ the radiation frequency and h some constant. Einstein later interpreted this result as a demonstration of the particle nature of light, since the quantization of light energy into discrete packets suggests that each packet is a light particle or photon.

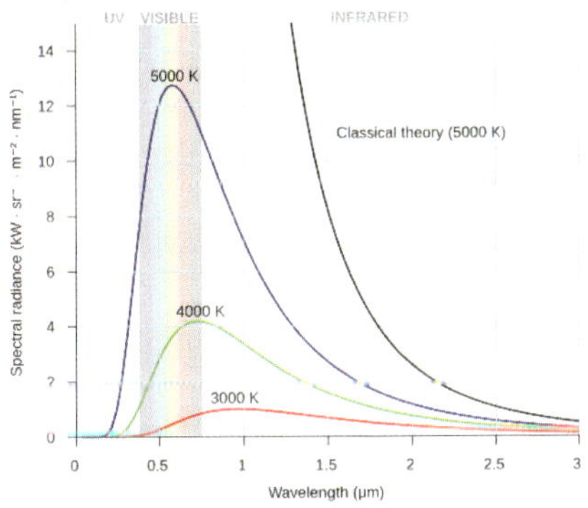

Power distribution of radiation from a black-body according to Planck's theory at different temperatures compared to the classical prediction.

A few years after Planck, Einstein's interpretation of a different experiment, the photoelectric effect, would further substantiate this claim. The photoelectric effect refers to the fact that high frequency light causes emission of electrons of metals regardless of how low the intensity of the light may be. Treating light as quantized packets or photons and using $E = h\upsilon$ explained this effect, because the intensity of light would only specify the number of photons, not their energy. Thus a small number of photons at high energy would still cause electronic emission.

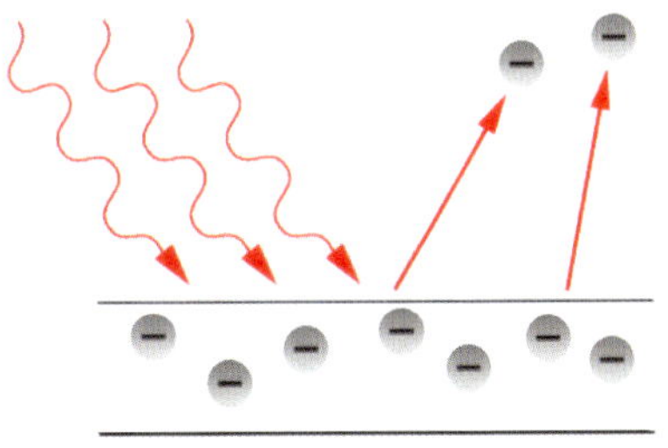

In the photoelectric effect, low frequency light does not eject electrons from a metal such as Potassium. The higher the frequency of the light, the more energetic the emitted electron

The Wave Viewpoint

The experiments shortly after the twentieth century demonstrating particle aspects of light were extremely shocking due to the nature of light being ingrained for centuries in the scientific community as a wave. The reasons light was universally accepted as a wave at the time began with the phenomenal success of optics centuries before. Describing light as waves whose rays bent in media like glass lenses was extremely useful for reproducing the correct experimental behavior.

At the turn of the twentieth century, further evidence for the wavelike nature of light came from Young's double-slit experiment.

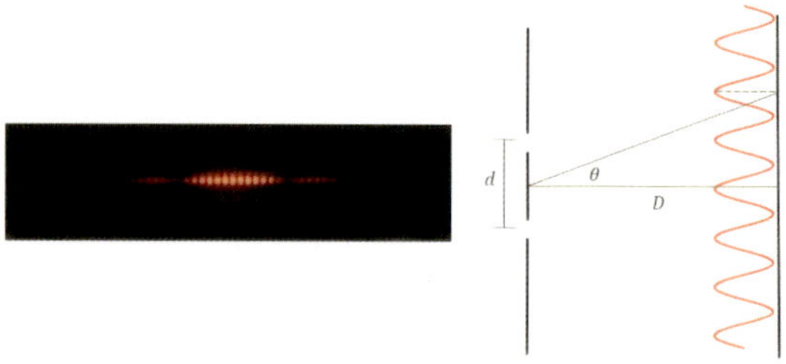

Left: actual experimental two-slit interference pattern of photons, exhibiting many small peaks and troughs. Right: schematic diagram of the experiment

In this experiment, light shining on a barrier with two narrow slits in it demonstrated an interference pattern on the screen behind the barrier as it diffracted through the slits. This was seemingly incontrovertible evidence that light was a wave, since such behavior seemed irreproducible from a particle viewpoint.

From a theoretical standpoint, Maxwell put the wavelike nature of light on a solid mathematical footing in the nineteenth century when he combined the equations governing electric and magnetic fields to yield the two equations:

$$\frac{1}{c^2}\frac{\partial^2 E}{\partial t^2} = \frac{\partial^2 E}{\partial x^2}, \qquad \frac{1}{c^2}\frac{\partial^2 B}{\partial t^2} = \frac{\partial^2 B}{\partial x^2},$$

i.e., the wave equation in each of the electric and magnetic fields. Maxwell thus established light as an electromagnetic wave, since the constant $\frac{1}{\sqrt{\mu_0 \varepsilon_0}}$ in the above equation was numerically exactly the speed of light in vacuum.

Modern Perspective: The Duality

As the nascent field of quantum mechanics developed in the early twentieth century, it would unify the wavelike treatment of light with the evidence that both matter and light were composed of particles.

As discussed above, the unification of particle and wave behavior for light was experimentally justified by the early twentieth century: centuries of optics and electromagnetism justified the wavelike nature of light, while the black-body problem and photoelectric effect supported the existence of the photon. The unification of particle and wave behavior for matter began in 1924, when de Broglie formulated his relation:

$$\lambda = \frac{h}{p}.$$

This relation says that matter particles with momentum p could be equally well described as waves of wavelength λ, with a proportionality constant equal to Planck's constant. The equation was motivated by the corresponding equation for light:

$$E = \frac{hc}{\lambda}$$

noting that $E = pc$ for light. Observing the wave-particle duality of light, de Broglie suggested that matter ought to obey the same relation. This hypothesis was experimentally justified in the following years with interference and diffraction experiments performed using electrons:

Even when fired one at a time, as in an experiment performed much later in the twentieth century, one found the same results for electrons in interference and diffraction experiments as predicted by wave mechanics. In recent decades the wave-particle duality of matter has been confirmed for even larger objects such as buckyballs, which are C_{60} allotropes.

Based on the de Broglie hypothesis, Heisenberg noticed that particles of large momenta ought to have small wavelengths and vice versa. He suggested that measuring the location of particles precisely would therefore require a high-energy photon to measure with, disturbing the momentum greatly upon measurement, and vice versa. In fact, this "measurement effect," while real, is not the full extent of the problem. In fact, particles themselves do not have definite positions or momenta, but rather obey statistical distributions given by their wave functions, which collapse to a single value upon measurement. These measurements have uncertainties which obey the Heisenberg uncertainty principle:

$$\sigma_x \sigma_p \geq \frac{\hbar}{2}.$$

That is, the momentum uncertainty is large if the position uncertainty is small, and vice versa.

This wave function formalism was promoted by Schrödinger, who devised a wave-like equation to govern its dynamics. In quantum mechanics, any particle (light or matter) can be described by a wave-like distribution, which gives the probability of measuring the particle to be in a particular location. This approach is able to capture both the particle and wave phenomena discussed above for both light and matter consistently, and also provides a formalism for computing many other important experimental results like the emission spectrum of hydrogen and the existence of quantum tunneling.

In quantum field theory, a modern and relativity-compatible version of quantum mechanics, wave-particle duality is elevated to a new level of mathematical abstraction in terms of group theory. In quantum field theory, all of light and matter is represented by fields that function as mathematical operators which create and destroy quanta of energy. These operators in turn obey wave-like equations that govern their dynamics. Particles in quantum field theory correspond to particular group representations under which these fields transform. The results of experiments as predicted by quantum field theory continue to provide support for the wave-particle duality of light and matter.

Wave Packet

A wave packet refers to the case where two (or more) waves exist simultaneously. A wave packet is often referred to as a wave group.

This situation is permitted by the principle of superposition. This principle states that if any two waves are a solution to the wave equation then the sum of the waves is also a solution. This principle holds only for linear systems.

Here we consider linear, one-dimensional sinusoidal disturbances. The first figure shows an example.

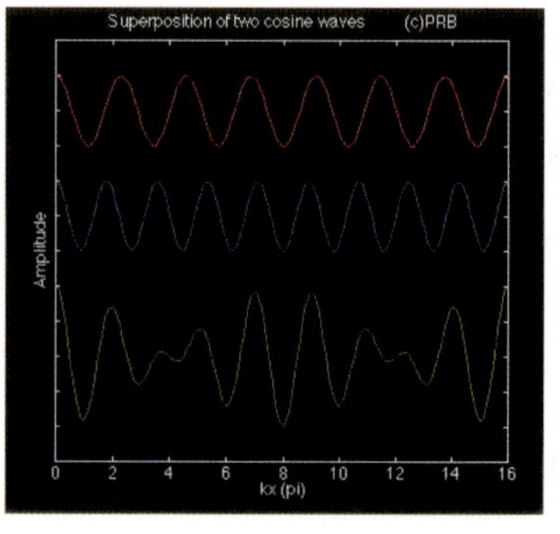

The red curve displays a simple wave form. The blue curve is another wave with a different wavelength. The wavelength is defined as the horizontal distance between two crests or between two troughs. (A crest corresponds to a wave maximum; a trough to a wave minimum). The wavelength of the red wave is longer than that of the blue wave.

The amplitude of the two waves is the same. Here the amplitude corresponds to the half the vertical distance between the crest and the trough. The amplitude of each wave equals the distance between the vertical tic marks in the diagram.

The green curve shows the sum of the two waves. It is the wave packet.

The amplitude of the packet is twice the amplitude of either the red or the blue wave. The crests and the troughs of either the red or blue wave have the same value anywhere along the wave. For the wave packet the value of the crest and the trough varies along the wave. This variation is a result of interference between the two waves. Recall that the packet is the sum of the two waves. At certain locations the two waves have a crest at the same point so the crest of the wave packet is twice the individual wave crest. This situation is called constructive interference since the two wave amplitudes add together. Constructive interference is also visible when the two wave troughs coincide. When a wave trough and a wave crest coincide, then the wave packet is zero. This situation is called destructive interference. A node or point of zero amplitude exists where there is complete destructive interference.

The wave packet can also be thought of as the product of two waves.

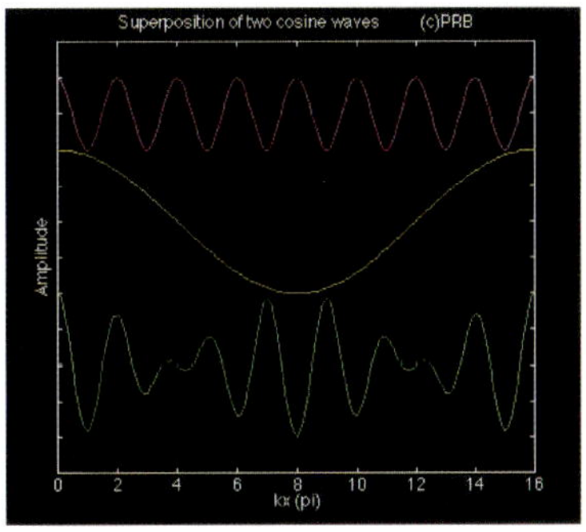

The top purple curve is called the average wave. It has the same amplitude as either the red or blue curve but its wavelength is the average of that of the red and blue wave. Count the number of crests in each of the waves and confirm this statement.

The average wave is also called the carrier wave.

The middle yellow curve is the wave envelope. It has an amplitude twice that of either the individual wave but its wavelength is half the difference of the red and the blue wave. The wavelength associated with the wave envelope is much larger than that of the carrier wave. The wave envelope is also called the wave modulation or wave amplitude.

The product of the carrier wave and the wave envelope appears as the bottom green curve. It is the same as the wave packet.

Basic Behaviors of Wave Packets

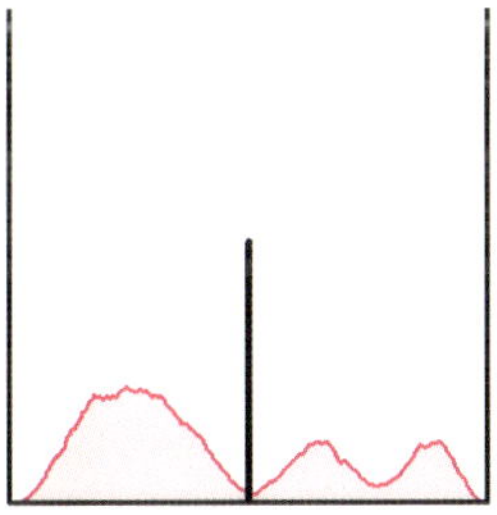

Position space probability density of an initially Gaussian state trapped in an infinite potential well experiencing periodic Quantum Tunneling in a centered potential wall.

Non-dispersive

As an example of propagation *without dispersion*, consider wave solutions to the following wave equation from classical physics

$$\frac{\partial^2 u}{\partial t^2} = c^2 \nabla^2 u,$$

where c is the speed of the wave's propagation in a given medium.

Using the physics time convention, $exp(-i\omega t)$, the wave equation has plane-wave solutions

$$u(\mathrm{x}, t) = e^{i(\mathrm{k}\cdot\mathrm{x}-\omega t)},$$

where

$$\omega^2 = |\mathrm{k}|^2 c^2, \text{ and } |\mathbf{k}|^2 = k_x^2 + k_y^2 + k_z^2.$$

This relation between ω and k should be valid so that the plane wave is a solution to the wave equation. It is called a dispersion relation.

To simplify, consider only waves propagating in one dimension (extension to three dimensions is straightforward). Then the general solution is

$$u(x, t) = Ae^{i(kx-\omega t)} + Be^{-i(kx+\omega t)},$$

in which we may take $\omega = kc$. The first term represents a wave propagating in the positive x-direction since it is a function of $x - ct$ only; the second term, being a function of $x + ct$, represents a wave propagating in the negative x-direction.

A wave packet is a localized disturbance that results from the sum of many different wave forms. If the packet is strongly localized, more frequencies are needed to allow the constructive superposition in the region of localization and destructive superposition outside the region. From the basic solutions in one dimension, a general form of a wave packet can be expressed as

$$u(x, t) = \frac{1}{\sqrt{2\pi}} \int_{-\infty}^{\infty} A(k) e^{i(kx-\omega(k)t)} dk.$$

As in the plane-wave case the wave packet travels to the right for $\omega(k) = kc$, since $u(x, t) = F(x - ct)$, and to the left for $\omega(k) = -kc$, since $u(x, t) = F(x + ct)$.

The factor $1/\sqrt{2}$ comes from Fourier transform conventions. The amplitude $A(k)$ contains the coefficients of the linear superposition of the plane-wave solutions. These coefficients can in turn be expressed as a function of $u(x, t)$ evaluated at $t = 0$ by inverting the Fourier transform relation above:

$$A(k) = \frac{1}{\sqrt{2\pi}} \int_{-\infty}^{\infty} u(x,0) e^{-ikx} dx.$$

For instance, choosing

$$u(x,0) = e^{-x^2 + ik_0 x},$$

we obtain

$$A(k) = \frac{1}{\sqrt{2}} e^{-\frac{(k-k_0)^2}{4}},$$

and finally

$$u(x,t) = e^{-(x-ct)^2 + ik_0(x-ct)}$$
$$= e^{-(x-ct)^2}\left[\cos\left(2\pi\frac{x-ct}{\lambda}\right) + i\sin\left(2\pi\frac{x-ct}{\lambda}\right)\right].$$

The imaginary part is a sine wave with perpendicular polarization to the cosine wave. The non-dispersive propagation of the real or imaginary part of this wave packet is presented in the above animation.

Dispersive

By contrast, as an example of propagation now *with dispersion*, consider instead solutions to the Schrödinger equation (with m and ħ set equal to one),

$$i\frac{\partial u}{\partial t} = -\frac{1}{2}\nabla^2 u,$$

yielding the dispersion relation

$$\omega = \frac{1}{2}|\mathrm{k}|^2 .$$

Once again, restricting attention to one dimension, the solution to the Schrödinger equation satisfying the initial condition $u(x,0) = exp(-x^2 + ik_0 x)$, representing a wave packet localized in space at the origin, is seen to be

$$u(x,t) = \frac{1}{\sqrt{1+2it}} e^{-\frac{1}{4}k_0^2} e^{-\frac{1}{1+2it}\left(x-\frac{ik_0}{2}\right)^2}$$
$$= \frac{1}{\sqrt{1+2it}} e^{-\frac{1}{1+4t^2}(x-k_0t)^2} e^{i\frac{1}{1+4t^2}\left((k_0+2tx)x-\frac{1}{2}tk_0^2\right)} .$$

An impression of the dispersive behavior of this wave packet is obtained by looking at the probability density,

$$|u(x,t)|^2 = \frac{1}{\sqrt{1+4t^2}} e^{-\frac{2(x-k_0t)^2}{1+4t^2}} .$$

It is evident that this dispersive wave packet, while moving with constant group velocity k_0, is delocalizing rapidly: it has a width increasing with time as $\sqrt{1+4t^2} \to 2t$, so eventually it diffuses to an unlimited region of space. The momentum profile $A(k)$ remains invariant.

Gaussian Wave Packets in Quantum Mechanics

The dispersive Gaussian wave packet, unnormalized and just centered at the origin, instead, at $t = 0$, can now be written in 3D, now in standard units:

$$\psi(\mathbf{r}, 0) = e^{-\mathbf{r}\cdot\mathbf{r}/2a},$$

where a is a positive real number, the square of the width of the wave packet,

$$a = 2\langle \mathbf{r}\cdot\mathbf{r}\rangle / 3\langle 1\rangle = 2(\Delta x)^2.$$

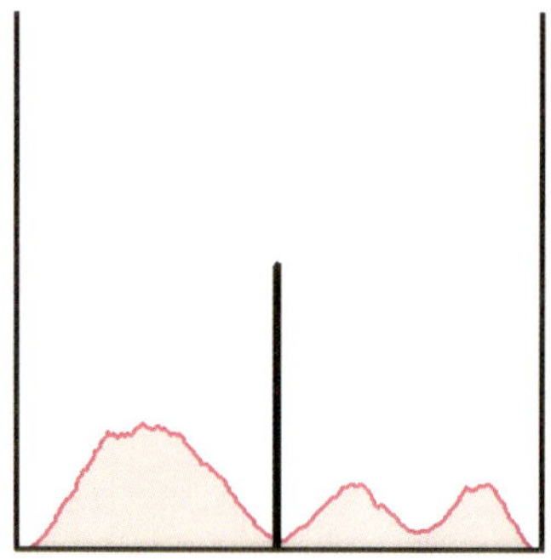

Position space probability density of an initially Gaussian state trapped in an infinite potential well experiencing periodic Quantum Tunneling in a centered potential wall.

The Fourier transform is also a Gaussian in terms of the wavenumber, $t = 0$, the $\mathbf{k}$-vector, (with inverse width,

$$1/a = 2\langle \mathbf{k}\cdot\mathbf{k}\rangle / 3\langle 1\rangle = 2(\Delta p_x/\hbar)^2,$$

so that

$$\Delta x \Delta p_x = \hbar/2,$$

i.e., it saturates the uncertainty relation),

$$\psi(\mathbf{k}, 0) = (2\pi a)^{3/2} e^{-a\mathbf{k}\cdot\mathbf{k}/2}.$$

Each separate wave only phase-rotates in time, so that the time dependent Fourier-transformed solution is

$$\begin{aligned}\Psi(\mathbf{k}, t) &= (2\pi a)^{3/2} e^{-a\mathbf{k}\cdot\mathbf{k}/2} e^{-iEt/\hbar} \\ &= (2\pi a)^{3/2} e^{-a\mathbf{k}\cdot\mathbf{k}/2 - i(\hbar^2\mathbf{k}\cdot\mathbf{k}/2m)t/\hbar} \\ &= (2\pi a)^{3/2} e^{-(a+i\hbar t/m)\mathbf{k}\cdot\mathbf{k}/2}.\end{aligned}$$

The inverse Fourier transform is still a Gaussian, but now the parameter a has become complex, and there is an overall normalization factor.

$$\Psi(\mathbf{r}, t) = \left(\frac{a}{a + i\hbar t/m}\right)^{3/2} e^{-\frac{\mathbf{r}\cdot\mathbf{r}}{2(a+i\hbar t/m)}}.$$

The integral of Ψ overall space is invariant, because it is the inner product of Ψ with the state of zero energy, which is a wave with infinite wavelength, a constant function of space. For any energy eigenstate $\eta(x)$, the inner product,

$$\langle\eta|\psi\rangle = \int \eta(\mathrm{r})\psi(\mathrm{r})d^3\mathrm{r},$$

only changes in time in a simple way: its phase rotates with a frequency determined by the energy of η. When η has zero energy, like the infinite wavelength wave, it doesn't change at all.

The integral $\int|\Psi|^2 d^3r$ is also invariant, which is a statement of the conservation of probability. Explicitly,

$$P(r) = |\Psi|^2 = \Psi^*\Psi = \left(\frac{a}{\sqrt{a^2+(\hbar t/m)^2}}\right)^3 e^{-\frac{a\mathbf{r}\cdot\mathbf{r}}{a^2+(\hbar t/m)^2}},$$

in which $\sqrt{a}$ is the width of $P(r)$ at $t = 0$; r is the distance from the origin; the speed of the particle is zero; and the time origin $t = 0$ can be chosen arbitrarily.

The width of the Gaussian is the interesting quantity which can be read off from the probability density, $|\Psi|^2$,

$$\sqrt{\frac{a^2+(\hbar t/m)^2}{a}}.$$

This width eventually grows linearly in time, as $\hbar t/(m\sqrt{a})$, indicating wave-packet spreading.

For example, if an electron wave packet is initially localized in a region of atomic dimensions (*i.e.*, $10^{-10}\,m$) then the width of the packet doubles in about 10^{-16} s. Clearly, particle wave packets spread out very rapidly indeed (in free space): For instance, after 1 ms, the width will have grown to about a kilometer.

This linear growth is a reflection of the (time-invariant) momentum uncertainty: the wave packet is confined to a narrow $\Delta x = \sqrt{a/2}$, and so has a momentum which is uncertain (according to the uncertainty principle) by the amount $\hbar/\sqrt{2a}$, a spread in velocity of $\hbar/m\sqrt{2a}$, and thus in the future position by $\hbar t/m\sqrt{2a}$. The uncertainty relation is then a strict inequality, very far from saturation, indeed. The initial uncertainty $\Delta x\Delta p = \hbar\sqrt{2}$ has now increased by a factor of $\hbar t/ma$ (for large t).

The Airy Wave Train

In contrast to the above Gaussian wave packet, it has been observed that a particular wave function based on Airy functions, propagates freely without envelope dispersion, maintaining its shape. It accelerates undistorted in the absence of a force field: $\psi = Ai(B(x - B^3t^2))\, exp(iB^3t(x - 2B^3t^2/3))$.

Nevertheless, there is no dissonance with Ehrenfest's theorem in this force-free situation, because the state is both non-normalizable and has an undefined (infinite) $\langle x\rangle$ for all times. (To the extent that it could be defined, $\langle p\rangle = 0$ for all times, despite the apparent acceleration of the front).

In phase space, this is evident in the pure state Wigner quasiprobability distribution of this wave train, whose shape in x and p is invariant as time progresses, but whose features accelerate to the right, in accelerating parabolas $B(x - B^3t^2) + (p/B - tB^2)^2 = 0$,

$$W(x, p; t) = W(x - B^3t^2, p - B^3t; 0)$$
$$= \frac{1}{2^{1/3}\pi B}\mathrm{Ai}\left(2^{2/3}\left(Bx + \frac{p^2}{B^2} - 2Bpt\right)\right).$$

Note the momentum distribution obtained by integrating over all x is constant. Since this is the probability density in momentum space, it is evident that the wave function itself is not normalizable.

Free Propagator

The narrow-width limit of the Gaussian wave packet solution discussed is the free *propagator kernel* K. For other differential equations, this is usually called the Green's function, but in quantum mechanics it is traditional to reserve the name Green's function for the time Fourier transform of K.

Returning to one dimension for simplicity, with m and $\hbar$ set equal to one, when a is the infinitesimal quantity ε, the Gaussian initial condition, rescaled so that its integral is one,

$$\psi_0(x) = \frac{1}{\sqrt{2\pi\epsilon}} e^{-\frac{x^2}{2\epsilon}}$$

becomes a delta function, $\delta(x)$,so that its time evolution,

$$K_t(x) = \frac{1}{\sqrt{2\pi(it+\epsilon)}} e^{\frac{-x^2}{2it+\epsilon}}$$

yields the propagator.

Note that a very narrow initial wave packet instantly becomes infinitely wide, but with a phase which is more rapidly oscillatory at large values of x. This might seem strange—the solution goes from being localized at one point to being "everywhere" at *all later times*, but it is a reflection of the enormous momentum uncertainty of a localized particle.

Further note that the norm of the wave function is infinite, which is also correct, since the square of a delta function is divergent in the same way.

The factor involving ε is an infinitesimal quantity which is there to make sure that integrals over K are well defined. In the limit that $\varepsilon \to 0$, K becomes purely oscillatory, and integrals of K are not absolutely convergent. It *will* be set to zero, but in order for all the integrations over intermediate states to be well defined, the limit $\varepsilon \to 0$ is to be only taken after the final state is calculated.

The propagator is the amplitude for reaching point x at time t, when starting at the origin, $x = 0$. By translation invariance, the amplitude for reaching a point x when starting at point y is the same function, only now translated,

$$K_t(x, y) = K_t(x-y) = \frac{1}{\sqrt{2\pi i t}} e^{\frac{i(x-y)^2}{2t}}.$$

In the limit when t is small, the propagator, of course, goes to a delta function,

$$\lim_{t \to 0} K_t(x-y) = \delta(x-y)$$

but only in the sense of distributions: The integral of this quantity multiplied by an arbitrary differentiable test function gives the value of the test function at zero.

To see this, note that the integral over all space of K equals 1 at all times,

$$\int K_t(x)dx = 1,$$

since this integral is the inner-product of K with the uniform wave function. But the phase factor in the exponent has a nonzero spatial derivative everywhere except at the origin, and so when the time is small there are fast phase cancellations at all but one point. This is rigorously true when the limit $\varepsilon \to 0$ is taken at the very end.

So the propagation kernel is the (future) time evolution of a delta function, and it is continuous, in a sense: it goes to the initial delta function at small times. If the initial wave function is an infinitely narrow spike at position y,

$$\psi_0(x) = \delta(x-y),$$

it becomes the oscillatory wave,

$$\psi_t(x) = \frac{1}{\sqrt{2\pi i t}} e^{i(x-y)^2/2t}.$$

Now, since every function can be written as a weighted sum of such narrow spikes,

$$\psi_0(x) = \int \psi_0(y)\delta(x-y)dy,$$

the time evolution of *every function* ψ_0 is determined by this propagation kernel K,

$$\psi_t(x) = \int \psi_0(y) \frac{1}{\sqrt{2\pi i t}} e^{i(x-y)^2/2t} dy.$$

Thus, this is a formal way to express the fundamental solution or ***general solution***. The interpretation of this expression is that the amplitude for a particle to be found at point x at time t is the amplitude that it started at y, times the amplitude that it went from y to x, *summed over all the possible starting points*. In other words, it is a convolution of the kernel K with the arbitrary initial condition ψ_0,

$$\psi_t = K * \psi_0.$$

Since the amplitude to travel from x to y after a time $t+t'$ can be considered in two steps, the propagator obeys the composition identity,

$$\int K(x-y;t)K(y-z;t')dy = K(x-z;t+t'),$$

which can be interpreted as follows: the amplitude to travel from x to z in time $t+t'$ is the sum of the amplitude to travel from x to y in time t, multiplied by the amplitude to travel from y to z in time t', summed over *all possible intermediate states* y. This is a property of an arbitrary quantum system, and by subdividing the time into many segments, it allows the time evolution to be expressed as a path integral.

Analytic Continuation to Diffusion

The spreading of wave packets in quantum mechanics is directly related to the spreading of probability densities in diffusion. For a particle which is randomly walking, the probability density function at any point satisfies the diffusion equation,

$$\frac{\partial}{\partial t}\rho = \frac{1}{2}\frac{\partial^2}{\partial x^2}\rho,$$

where the factor of 2, which can be removed by rescaling either time or space, is only for convenience.

A solution of this equation is the spreading Gaussian,

$$\rho_t(x) = \frac{1}{\sqrt{2\pi t}} e^{\frac{-x^2}{2t}},$$

and, since the integral of ρ_t *is constant while the width is becoming narrow at small times, this function approaches a delta function at* $t=0$,

$$\lim_{t\to 0}\rho_t(x) = \delta(x)$$

again only in the sense of distributions, so that

$$\lim_{t\to 0}\int_x f(x)\rho_t(x) = f(0)$$

for any smooth test function f.

The spreading Gaussian is the propagation kernel for the diffusion equation and it obeys the convolution identity,

$$K_{t+t'} = K_t * K_{t'},$$

which allows diffusion to be expressed as a path integral. The propagator is the exponential of an operator H,

$$K_t(x) = e^{-tH},$$

which is the infinitesimal diffusion operator,

$$H = -\frac{\nabla^2}{2}.$$

A matrix has two indices, which in continuous space makes it a function of x and x'. In this case, because of translation invariance, the matrix element K only depend on the difference of the position and a convenient abuse of notation is to refer to the operator, the matrix elements, and the function of the difference by the same name:

$$K_t(x, x') = K_t(x - x').$$

Translation invariance means that continuous matrix multiplication,

$$C(x, x'') = \int_{x'} A(x, x')B(x', x''),$$

is essentially convolution,

$$C(\Delta) = C(x - x'') = \int_{x'} A(x - x')B(x' - x'') = \int_y A(\Delta - y)B(y).$$

The exponential can be defined over a range of ts, which include complex values, so long as integrals over the propagation kernel stay convergent,

$$K_z(x) = e^{-zH}.$$

As long as the real part of z is positive, for large values of x, K is exponentially decreasing, and integrals over K are indeed absolutely convergent.

The limit of this expression for z approaching the pure imaginary axis is the above Schrödinger propagator encountered,

$$K_t^{\text{Schr}} = K_{it+\epsilon} = e^{-(it+\epsilon)H},$$

which illustrates the above time evolution of Gaussians.

From the fundamental identity of exponentiation, or path integration,

$$K_z * K_{z'} = K_{z+z'}$$

holds for all complex z values, where the integrals are absolutely convergent so that the operators are well defined.

Thus, quantum evolution of a Gaussian, which is the complex diffusion kernel K,

$$\psi_0(x) = K_a(x) = K_a * \delta(x)$$

amounts to the time-evolved state,

$$\psi_t = K_{it} * K_a = K_{a+it}.$$

This illustrates the above diffusive form of the complex Gaussian solutions,

$$\psi_t(x) = \frac{1}{\sqrt{2\pi(a+it)}} e^{-\frac{x^2}{2(a+it)}}.$$

Wave Function

Wave function in quantum mechanics is a variable quantity that mathematically describes the wave characteristics of a particle. The value of the wave function of a particle at a given point of space and time is related to the likelihood of the particle's being there at the time.

The wave function can have a positive or negative sign. The sign is important in calculations. It is also important when the wave functions of two or more atoms combine to form a molecule.

Wave functions with like signs (waves in phase) will interfere constructively, leading to the possibility of bonding. Wave functions with unalike signs (waves out of phase) will interfere destructively.

In 1926, Erwin Schrödinger deduced the wave function for the simplest of all atoms, hydrogen. Solving the Schrödinger equation enables scientists to determine wave functions for electrons in atoms and molecules.

The Schrödinger equation is an equation of quantum mechanics: calculated wave functions have discrete, allowed values for electrons bound in atoms and molecules; all other values are forbidden.

In addition to the importance of Ψ, its square Ψ^2 also has enormous significance in chemistry.

Ψ^2 is the probability density. It tells us where the electron is most likely to be found in the space around the nucleus.

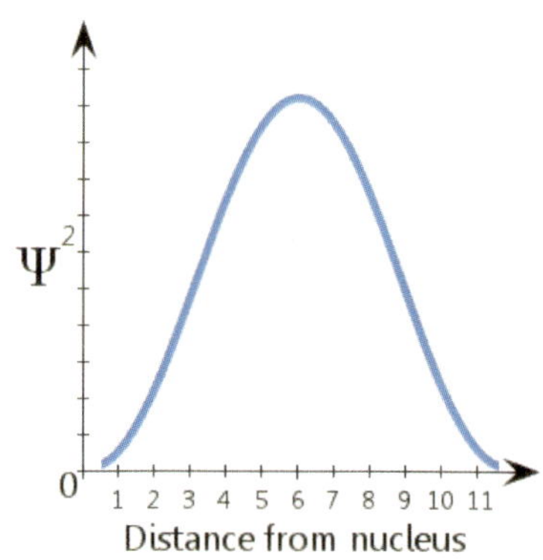

For example, in the (fictitious) schematic diagram plotted here, showing Ψ^2 against distance from nucleus, we see that the electron is most likely to be found between about 5-7 units from the nucleus. We also see that there is a vanishingly small likelihood that the electron will be at the nucleus or more than about 11½ units away from the nucleus.

There is a 100 percent probability that the electron is somewhere - in other words a probability of 1.

Ψ can be calculated exactly only for nuclei with one electron, i.e. H, He^{+}, Li^{2+}, Be^{3+}, B^{4+}, C^{5+}, etc. In mathematical language, we say that analytic solutions for Ψ are possible only for one-electron systems. One-electron systems are often described as hydrogenic - meaning "like hydrogen."

Now, a probability is a real number between 0 and 1. An outcome of a measurement which has a probability 0 is an impossible outcome, whereas an outcome which has a probability 1 is a certain outcome. The probability of a measurement of *x* yielding a result between $-\infty$ and $+\infty$ is

$$P_{x\in-\infty:\infty}(t)=\int_{-\infty}^{\infty}|\psi(x,t)|^2\,dx.$$

However, a measurement of *x must* yield a value between $-\infty$ and $+\infty$, since the particle has to be located somewhere. It follows that $P_{x\in-\infty:\infty}=1$, or

$$\int_{-\infty}^{\infty}|\psi(x,t)|^2\,dx=1,$$

which is generally known as the *normalization condition* for the wave function.

For example, suppose that we wish to normalize the wave function of a Gaussian wave packet, centered on $x=x_0$, and of characteristic width σ : *i.e.*,

$$\psi(x)=\psi_0\,e^{-(x-x_0)^2/(4\sigma^2)}.$$

In order to determine the normalization constant ψ_0, we simply substitute equation $\psi(x)=\psi_0\,e^{-(x-x_0)^2/(4\sigma^2)}$ into equation $\int_{-\infty}^{\infty}|\psi(x,t)|^2\,dx=1$, to obtain

$$|\psi_0|^2\int_{-\infty}^{\infty}e^{-(x-x_0)^2/(2\sigma^2)}\,dx=1.$$

Changing the variable of integration to $y=(x-x_0)(\sqrt{2}\sigma)$, we get

$$|\psi_0|^2\sqrt{2}\sigma\int_{-\infty}^{\infty}e^{-y^2}\,dy=1.$$

However,

$$\int_{-\infty}^{\infty}e^{-y^2}\,dy=\sqrt{\pi},$$

which implies that

$$|\psi_0|^2=\frac{1}{(2\pi\sigma^2)^{1/2}}$$

Hence, a general normalized Gaussian wave function takes the form

$$\psi(x)=\frac{e^{i\varphi}}{(2\pi\sigma^2)^{1/4}}e^{-(x-x_0)^2/(4\sigma^2)},$$

where φ is an arbitrary real phase-angle.

Now, it is important to demonstrate that if a wave function is initially normalized then it stays normalized as it evolves in time according to Schrödinger's equation. If this is not the case then the probability interpretation of the wave function is untenable, since it does not make sense for the probability that a measurement of x yields any possible outcome (which is, manifestly, unity) to change in time. Hence, we require that

$$\frac{d}{dt}\int_{-\infty}^{\infty}|\psi(x,t)|^2\,dx=0,$$

for wave functions satisfying Schrödinger's equation. The above equation gives

$$\frac{d}{dt}\int_{-\infty}^{\infty}\psi^*\psi dx=\int_{-\infty}^{\infty}\left(\frac{\partial\psi^*}{\partial t}\psi+\psi^*\frac{\partial\psi}{\partial t}\right)dx=0.$$

Now, multiplying Schrödinger's equation by $\psi^*/(i\hbar)$, we obtain

$$\psi^*\frac{\partial\psi}{\partial t}=\frac{i\hbar}{2m}\psi^*\frac{\partial^2\psi}{\partial x^2}-\frac{i}{\hbar}V|\psi|^2$$

The complex conjugate of this expression yields

$$\psi\frac{\partial\psi^*}{\partial t}=-\frac{i\hbar}{2m}\psi\frac{\partial^2\psi^*}{\partial x^2}+\frac{i}{\hbar}V|\psi|^2$$

[since $(AB)^*=A^*B^*$, $A^{**}=A$, and $i^*=-i$]. Summing the previous two equations, we get

$$\frac{\partial\psi^*}{\partial t}\psi+\psi^*\frac{\partial\psi}{\partial t}=\frac{i\hbar}{2m}\left(\psi^*\frac{\partial^2\psi}{\partial x^2}-\psi\frac{\partial^2\psi^*}{\partial x^2}\right)=\frac{i\hbar}{2m}\frac{\partial}{\partial x}\left(\psi^*\frac{\partial\psi}{\partial x}-\psi\frac{\partial\psi^*}{\partial x}\right).$$

Equations $\frac{d}{dt}\int_{-\infty}^{\infty}\psi^*\psi dx=\int_{-\infty}^{\infty}\left(\frac{\partial\psi^*}{\partial t}\psi+\psi^*\frac{\partial\psi}{\partial t}\right)dx=0.$ and $\frac{\partial\psi^*}{\partial t}\psi+\psi^*\frac{\partial\psi}{\partial t}=\frac{i\hbar}{2m}\left(\psi^*\frac{\partial^2\psi}{\partial x^2}-\psi\frac{\partial^2\psi^*}{\partial x^2}\right)$ $=\frac{i\hbar}{2m}\frac{\partial}{\partial x}\left(\psi^*\frac{\partial\psi}{\partial x}-\psi\frac{\partial\psi^*}{\partial x}\right)$ can be combined to produce

$$\frac{d}{dt}\int_{-\infty}^{\infty}|\psi|^2\,dx=\frac{i\hbar}{2m}\left[\psi^*\frac{\partial\psi}{\partial x}-\psi\frac{\partial\psi^*}{\partial x}\right]_{-\infty}^{\infty}=0.$$

The above equation is satisfied provided

$$|\psi|\to 0 \text{ as } |x|\to\infty.$$

However, this is a necessary condition for the integral on the left-hand side of eq. $\int_{-\infty}^{\infty}|\psi(x,t)|^2\,dx=1$ to converge. Hence, we conclude that all wave functions which are *square-integrable* [*i.e.*, are such that the integral in eq. $\int_{-\infty}^{\infty}|\psi(x,t)|^2\,dx=1$ converges] have the property that if the normalization condition $\int_{-\infty}^{\infty}|\psi(x,t)|^2\,dx=1$ is satisfied at one instant in time then it is satisfied at all subsequent times.

It is also possible to demonstrate, via very similar analysis to the above, that

$$\frac{dP_{x\in a:b}}{dt} + j(b,\, t) - j(a,t) = 0,$$

Where $P_{x\in a:b}$ is defined in equation $P_{x\in a:b}(t) = \int_a^b |\psi(x,t)|^2\, dx$ and

$$j(x,t) = \frac{i\hbar}{2m}\left(\psi\frac{\partial\psi^*}{\partial x} - \psi^*\frac{\partial\psi}{\partial x}\right)$$

is known as the *probability current*. Note that j is real. Equation $\frac{dP_{x\in a:b}}{dt} + j(b,\, t) - j(a,t) = 0$ is a probability conservation equation. According to this equation, the probability of a measurement of x lying in the interval a to b evolves in time due to the difference between the flux of probability into the interval [*i.e.*, $j(a,\, t)$] and that out of the interval [*i.e.*, $j(b,t)$]. Here, we are interpreting $j(x,\, t)$ as the *flux* of probability in the $+x$-direction at position x and time t.

Note, finally, that not all wave functions can be normalized according to the scheme set out in eq. $\int_{-\infty}^{\infty} |\psi\,(x,\, t)|^2\, dx = 1$. For instance, a plane wave wavefunction

$$\psi\,(x,\, t) = \psi_0\, e^{i(kx-wt)}$$

is not square-integrable and thus, cannot be normalized. For such wavefunctions, the best we can say is that

$$P_{x\in a:b}(t) \propto \int_a^b |\psi(x,t)|^2\, dx.$$

In the following, all wavefunctions are assumed to be square-integrable and normalized, unless otherwise stated.

Momentum-space Wave Functions

The particle also has a wave function in momentum space:

$$\Phi(p,t)$$

where p is the momentum in one dimension, which can be any value from $-\infty$ to $+\infty$, and t is time.

Analogous to the position case, the inner product of two wave functions $\Phi_1(p,t)$ and $\Phi_2(p,t)$ can be defined as:

$$(\Phi_1, \Phi_2) = \int_{-\infty}^{\infty} dp\Phi_1^*(p,t)\Phi_2(p,t).$$

One particular solution to the time-independent Schrödinger equation is

$$\Psi_p(x) = e^{ipx/\hbar},$$

a plane wave, which can be used in the description of a particle with momentum exactly p, since it is an eigenfunction of the momentum operator. These functions are not normalizable to unity (they aren't square-integrable), so they are not really elements of physical Hilbert space. The set

$$\{\Psi_p(x,t), -\infty \le p \le \infty\}$$

forms what is called the momentum basis. This "basis" is not a basis in the usual mathematical sense. For one thing, since the functions aren't normalizable, they are instead normalized to a delta function,

$$(\Psi_p, \Psi_{p'}) = \delta(p - p').$$

For another thing, though they are linearly independent, there are too many of them (they form an uncountable set) for a basis for physical Hilbert space. They can still be used to express all functions in it using Fourier transforms as described next.

Relations between Position and Momentum Representations

The x and p representations are

$$|\Psi\rangle = I|\Psi\rangle = \int |x\rangle\langle x|\Psi\rangle dx = \int \Psi(x)|x\rangle dx,$$
$$|\Psi\rangle = I|\Psi\rangle = \int |p\rangle\langle p|\Psi\rangle dp = \int \Phi(p)|p\rangle dp.$$

Now take the projection of the state Ψ onto eigenfunctions of momentum using the last expression in the two equations,

$$\int \Psi(x)\langle p|x\rangle\, dx = \int \Phi(p')\langle p|p'\rangle\, dp' = \int \Phi(p')\delta(p-p')dp' = \Phi(p).$$

Then utilizing the known expression for suitably normalized eigenstates of momentum in the position representation solutions of the free Schrödinger equation

$$\langle x|p\rangle = p(x) = \frac{1}{\sqrt{2\pi\hbar}} e^{\frac{i}{\hbar}px} \Rightarrow \langle p|x\rangle = \frac{1}{\sqrt{2\pi\hbar}} e^{-\frac{i}{\hbar}px},$$

one obtains

$$\Phi(p) = \frac{1}{\sqrt{2\pi\hbar}} \int \Psi(x) e^{-\frac{i}{\hbar}px} dx.$$

Likewise, using eigenfunctions of position,

$$\Psi(x) = \frac{1}{\sqrt{2\pi\hbar}} \int \Phi(p) e^{\frac{i}{\hbar}px} dp.$$

The position-space and momentum-space wave functions are thus found to be Fourier transforms of each other. The two wave functions contain the same information, and either one alone

is sufficient to calculate any property of the particle. As representatives of elements of abstract physical Hilbert space, whose elements are the possible states of the system under consideration, they represent the same state vector, hence *identical physical states*, but they are not generally equal when viewed as square-integrable functions.

In practice, the position-space wave function is used much more often than the momentum-space wave function. The potential entering the relevant equation (Schrödinger, Dirac, etc.) determines in which basis the description is easiest. For the harmonic oscillator, x and p enter symmetrically, so there it doesn't matter, which description one uses. The same equation (modulo constants) results. From this follows, with a little bit of afterthought, a factoid: The solutions to the wave equation of the harmonic oscillator are eigenfunctions of the Fourier transform in L^2.

Following are the general forms of the wave function for systems in higher dimensions and more particles, as well as including other degrees of freedom than position coordinates or momentum components.

One-particle States in 3D Position Space

The position-space wave function of a single particle without spin in three spatial dimensions is similar to the case of one spatial dimension above:

$$\Psi(\mathrm{r}, t)$$

where r is the position vector in three-dimensional space, and t is time. As always $\Psi(\mathrm{r}, t)$ is a complex-valued function of real variables. As a single vector in Dirac notation

$$|\Psi(t)\rangle = \int d^3\mathrm{r}\Psi(\mathrm{r}, t)|\mathrm{r}\rangle$$

For a particle with spin, ignoring the position degrees of freedom, the wave function is a function of spin only (time is a parameter);

$$(s\ , t)$$

where s_z is the spin projection quantum number along the z axis. (The z axis is an arbitrary choice; other axes can be used instead if the wave function is transformed appropriately.) The s_z parameter, unlike r and t, is a *discrete variable*. For example, for a spin-1/2 particle, s_z can only be +1/2 or −1/2, and not any other value. (In general, for spin s, s_z can be $s, s-1, \ldots, -s+1, -s$). Inserting each quantum number gives a complex valued function of space and time, there are $2s+1$ of them. These can be arranged into a column vector.

$$\xi \begin{bmatrix} \xi(s, t) \\ \xi(s-1, t) \\ \vdots \\ \xi(-(s-1), t) \\ \xi(-s, t) \end{bmatrix} = \xi(s, t)\begin{bmatrix} 1 \\ 0 \\ \vdots \\ 0 \\ 0 \end{bmatrix} + \xi(s-1, t)\begin{bmatrix} 0 \\ 1 \\ \vdots \\ 0 \\ 0 \end{bmatrix} + \ldots + \xi(-(s-1), t)\begin{bmatrix} 0 \\ 0 \\ \vdots \\ 1 \\ 0 \end{bmatrix} + \xi(-s, t)\begin{bmatrix} 0 \\ 0 \\ \vdots \\ 0 \\ 1 \end{bmatrix}$$

In bracket notation, these easily arrange into the components of a vector

$$|\xi(t)\rangle = \sum_{s_z=-s}^{s} \xi(s_z, t)|s_z\rangle$$

The entire vector ξ is a solution of the Schrödinger equation (with a suitable Hamiltonian), which unfolds to a coupled system of $2s+1$ ordinary differential equations with solutions $\xi(s, t), \xi(s-1, t), \ldots, \xi(-s, t)$. The term "spin function" instead of "wave function" is used by some authors. This contrasts the solutions to position space wave functions, the position coordinates being continuous degrees of freedom, because then the Schrödinger equation does take the form of a wave equation.

More generally, for a particle in 3d with any spin, the wave function can be written in "position–spin space" as:

$$\Psi(\mathrm{r}, s_z, t)$$

and these can also be arranged into a column vector

$$\Psi(\mathrm{r}, t) = \begin{bmatrix} \Psi(\mathrm{r}, s, t) \\ \Psi(\mathrm{r}, s-1, t) \\ \vdots \\ \Psi(\mathrm{r}, -(s-1), t) \\ \Psi(\mathrm{r}, -s, t) \end{bmatrix}$$

in which the spin dependence is placed in indexing the entries, and the wave function is a complex vector-valued function of space and time only.

All values of the wave function, not only for discrete but continuous variables also, collect into a single vector

$$|\Psi(t)\rangle = \sum_{s_z} \int d^3\mathrm{r}\,\Psi(\mathrm{r}, s_z, t)|\mathrm{r}, s_z\rangle$$

For a single particle, the tensor product $\otimes$ of its position state vector $|\psi\rangle$ and spin state vector $|\xi\rangle$ gives the composite position-spin state vector

$$|\psi(t)\rangle \otimes |\xi(t)\rangle = \sum_{s_z} \int d^3\mathrm{r}\ \psi(\mathrm{r}, t)\xi(s_z, t)|\mathrm{r}\rangle \otimes |s_z\rangle$$

with the identifications

$$|\Psi(t)\rangle = |\psi(t)\rangle \otimes |\xi(t)\rangle$$

$$\Psi(\mathrm{r}, s_z, t) = \psi(\mathrm{r}, t)\xi(s_z, t)$$

$$|\mathrm{r}, s_z\rangle = |\mathrm{r}\rangle \otimes |s_z\rangle$$

The tensor product factorization is only possible if the orbital and spin angular momenta of the particle are separable in the Hamiltonian operator underlying the system's dynamics (in other words, the Hamiltonian can be split into the sum of orbital and spin terms). The time dependence can be placed in either factor and time evolution of each can be studied separately. The factorization is not possible for those interactions where an external field or any space-dependent quantity couples to the spin; examples include a particle in a magnetic field, and spin-orbit coupling.

The preceding discussion is not limited to spin as a discrete variable, the total angular momentum J may also be used. Other discrete degrees of freedom, like isospin, can expressed similarly to the case of spin above.

Many Particle States in 3D Position Space

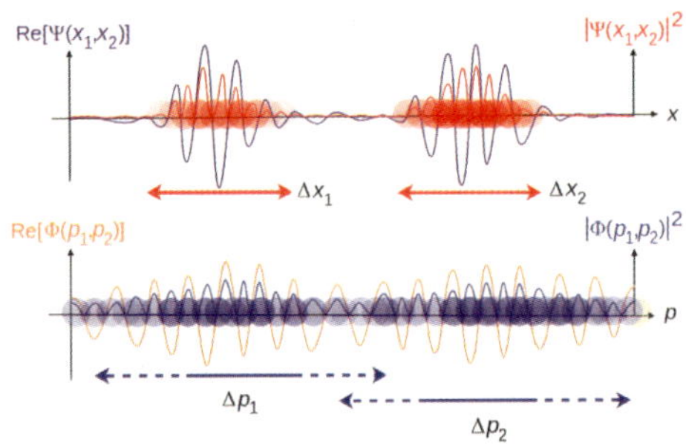

Traveling waves of two free particles, with two of three dimensions suppressed. Top is position space wave function, bottom is momentum space wave function, with corresponding probability densities.

If there are many particles, in general there is only one wave function, not a separate wave function for each particle. The fact that *one* wave function describes *many* particles is what makes quantum entanglement and the EPR paradox possible. The position-space wave function for N particles is written:

$$\Psi(\mathrm{r}_1, \mathrm{r}_2 \cdots \mathrm{r}_N, t)$$

where r_i is the position of the ith particle in three-dimensional space, and t is time. Altogether, this is a complex-valued function of $3N + 1$ real variables.

In quantum mechanics there is a fundamental distinction between identical particles and distinguishable particles. For example, any two electrons are identical and fundamentally indistinguishable from each other; the laws of physics make it impossible to "stamp an identification number" on a certain electron to keep track of it. This translates to a requirement on the wave function for a system of identical particles:

$$\Psi(\ldots \mathrm{r}_a, \ldots, \mathrm{r}_b, \ldots) = \pm\Psi(\ldots \mathrm{r}_b, \ldots, \mathrm{r}_a, \ldots)$$

where the + sign occurs if the particles are *all bosons* and – sign if they are *all fermions*. In other words, the wave function is either totally symmetric in the positions of bosons, or totally antisymmetric in the positions of fermions. The physical interchange of particles corresponds to mathematically switching arguments in the wave function. The antisymmetry feature of fermionic wave functions leads to the Pauli principle. Generally, bosonic and fermionic symmetry requirements are the manifestation of particle statistics and are present in other quantum state formalisms.

For N distinguishable particles (no two being identical, i.e. no two having the same set of quantum numbers), there is no requirement for the wave function to be either symmetric or antisymmetric.

For a collection of particles, some identical with coordinates $\mathbf{r}_1$, $\mathbf{r}_2$, ... and others distinguishable $\mathbf{x}_1$, $\mathbf{x}_2$, ... (not identical with each other and not identical to the aforementioned identical particles), the wave function is symmetric or antisymmetric in the identical particle coordinates $\mathbf{r}_i$ only:

$$\Psi(\ldots \mathbf{r}_a, \ldots, \mathbf{r}_b, \ldots, \mathbf{x}_1, \mathbf{x}_2, \ldots) - \pm\Psi(\ldots \mathbf{r}_b, \ldots, \mathbf{r}_a, \ldots, \mathbf{x}_1, \mathbf{x}_2, \ldots)$$

Again, there is no symmetry requirement for the distinguishable particle coordinates $\mathbf{x}_i$.

The wave function for N particles each with spin is the complex-valued function

$$\Psi(\mathbf{r}_1, \mathbf{r}_2 \cdots \mathbf{r}_N, s_{z1}, s_{z2} \cdots s_{zN}, t)$$

Accumulating all these components into a single vector,

$$\underbrace{|\Psi\rangle}_{\text{state vector (ket)}} = \underbrace{\overbrace{\sum_{s_{z1},\ldots,s_{zN}}}^{\text{discrete lables}} \overbrace{\int_{R_N} d^3r_N \cdots \int_{R_1} d^3 r_1}^{\text{continuous labels}}}_{\text{adding up}} \underbrace{\Psi(r_1, \ldots, r_N, s_{z1}, \ldots, s_{zN})}_{\text{wavefunctiom (component of state vector along basis state)}} \underbrace{|r_1, \ldots, r_N, s_{z1}, \ldots, s_{zN}\rangle}_{\text{basis state (basis ket)}}$$

For identical particles, symmetry requirements apply to both position and spin arguments of the wave function so it has the overall correct symmetry.

The formulae for the inner products are integrals over all coordinates or momenta and sums over all spin quantum numbers. For the general case of N particles with spin in 3d,

$$(\Psi_1, \Psi_2) = \sum_{s_{zN}} \cdots \sum_{s_{z2}} \sum_{s_{z1}} \int_{\text{all space}} d^3\mathbf{r}_1 \int_{\text{all space}} d^3\mathbf{r}_2 \cdots \int_{\text{all space}} d^3\mathbf{r}_N \Psi_1^*(\mathbf{r}_1 \cdots \mathbf{r}_N, s_{z1} \cdots s_{zN}, t)\Psi_2(\mathbf{r}_1 \cdots \mathbf{r}_N, s_{z1} \cdots s_{zN}, t)$$

this is altogether N three-dimensional volume integrals and N sums over the spins. The differential volume elements $d^3\mathbf{r}_i$ are also written "dV_i" or "$dx_i\, dy_i\, dz_i$".

The multidimensional Fourier transforms of the position or position–spin space wave functions yields momentum or momentum–spin space wave functions.

Probability Interpretation

For the general case of N particles with spin in 3d, if Ψ is interpreted as a probability amplitude, the probability density is

$$\rho(\mathbf{r}_1 \cdots \mathbf{r}_N, s_{z1} \cdots s_{zN}, t) = |\Psi(\mathbf{r}_1 \cdots \mathbf{r}_N, s_{z1} \cdots s_{zN}, t)|^2$$

and the probability that particle 1 is in region R_1 with spin $s_{z1} = m_1$ *and* particle 2 is in region R_2 with spin $s_{z2} = m_2$ etc. at time t is the integral of the probability density over these regions and evaluated at these spin numbers:

$$P_{\mathbf{r}_1 \in R_1, s_{z1} = m_1, \ldots, \mathbf{r}_N \in R_N, s_{zN} = m_N}(t) = \int_{R_1} d^3\mathbf{r}_1 \int_{R_2} d^3\mathbf{r}_2 \cdots \int_{R_N} d^3\mathbf{r}_N |\Psi(\mathbf{r}_1 \cdots \mathbf{r}_N, m_1 \cdots m_N, t)|^2$$

Time Dependence

For systems in time-independent potentials, the wave function can always be written as a function of the degrees of freedom multiplied by a time-dependent phase factor, the form of which is given by the Schrödinger equation. For N particles, considering their positions only and suppressing other degrees of freedom,

$$\Psi(\mathrm{r}_1, \mathrm{r}_2, \ldots, \mathrm{r}_N, t) = e^{-iEt/\hbar}\psi(\mathrm{r}_1, \mathrm{r}_2, \ldots, \mathrm{r}_N),$$

where E is the energy eigenvalue of the system corresponding to the eigenstate Ψ. Wave functions of this form are called stationary states.

The time dependence of the quantum state and the operators can be placed according to unitary transformations on the operators and states. For any quantum state $|\Psi\rangle$ and operator O, in the Schrödinger picture $|\Psi(t)\rangle$ changes with time according to the Schrödinger equation while O is constant. In the Heisenberg picture it is the other way round, $|\Psi\rangle$ is constant while $O(t)$ evolves with time according to the Heisenberg equation of motion. The Dirac (or interaction) picture is intermediate, time dependence is places in both operators and states which evolve according to equations of motion. It is useful primarily in computing S-matrix elements.

Non-relativistic Examples

The following are solutions to the Schrödinger equation for one non-relativistic spinless particle.

Finite Potential Barrier

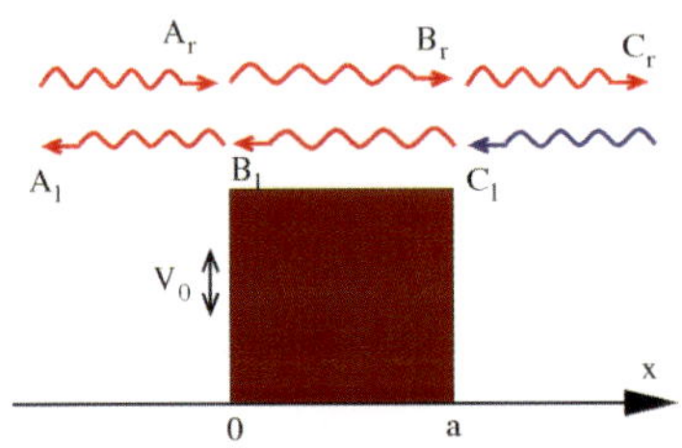

Scattering at a finite potential barrier of height V_0. The amplitudes and direction of left and right moving waves are indicated. In red, those waves used for the derivation of the reflection and transmission amplitude. $E > V_0$ for this illustration.

One of most prominent features of the wave mechanics is a possibility for a particle to reach a location with a prohibitive (in classical mechanics) force potential. A common model is the "potential barrier", the one-dimensional case has the potential

$$V(x) = \begin{cases} V_0 & |x| < a \\ 0 & |x| \geq L \end{cases}$$

and the steady-state solutions to the wave equation have the form (for some constants k, κ)

$$\Psi(x) = \begin{cases} A_r e^{ikx} + A_l e^{-ikx} & x < -a, \\ B_r e^{\kappa x} + B_l e^{-\kappa x} & |x| \leq a, \\ C_r e^{ikx} + C_l e^{-ikx} & x > a. \end{cases}$$

Note that these wave functions are not normalized.

The standard interpretation of this is as a stream of particles being fired at the step from the left (the direction of negative x): setting $A_r = 1$ corresponds to firing particles singly; the terms containing A_r and C_r signify motion to the right, while A_l and – to the left. Under this beam interpretation, put $C_l = 0$ since no particles are coming from the right. By applying the continuity of wave functions and their derivatives at the boundaries, it is hence possible to determine the constants above.

In a semiconductor crystallite whose radius is smaller than the size of its exciton Bohr radius, the excitons are squeezed, leading to quantum confinement. The energy levels can then be modeled using the particle in a box model in which the energy of different states is dependent on the length of the box.

Quantum harmonic oscillator

The wave functions for the quantum harmonic oscillator can be expressed in terms of Hermite polynomials H_n, they are

$$\Psi_n(x) = \sqrt{\frac{1}{2^n n!}} \cdot \left(\frac{m\omega}{\pi\hbar}\right)^{1/4} \cdot e^{-\frac{m\omega x^2}{2\hbar}} \cdot H_n\left(\sqrt{\frac{m\omega}{\hbar}}x\right)$$

where $n = 0, 1, 2, \ldots$.

Hydrogen Atom

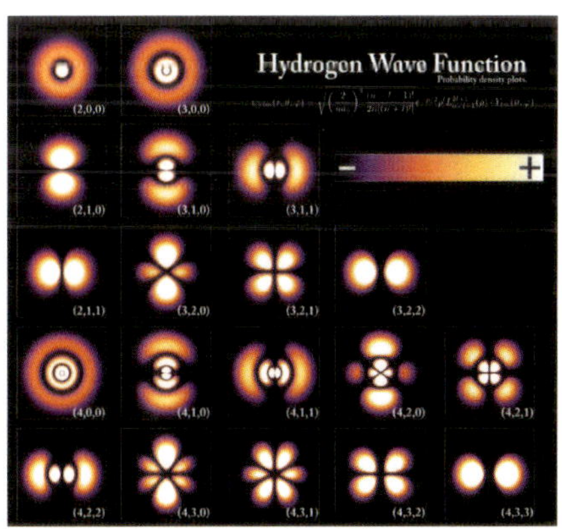

The electron probability density for the first few hydrogen atomelectron orbitals shown as cross-sections. These orbitals form an orthonormal basis for the wave function of the electron. Different orbitals are depicted with different scale.

The wave functions of an electron in a Hydrogen atom are expressed in terms of spherical harmonics and generalized Laguerre polynomials.

It is convenient to use spherical coordinates, and the wave function can be separated into functions of each coordinate,

$$\Psi_{nlm}(r, \theta, \phi) = R(r) Y_\ell^m(\theta, \phi)$$

where R are radial functions and $Y_\ell^m(\theta, \varphi)$ are spherical harmonics of degree ℓ and order m. This is the only atom for which the Schrödinger equation has been solved for exactly. Multi-electron atoms require approximative methods. The family of solutions are:

$$\Psi_{n\ell m}(r, \theta, \phi) = \sqrt{\left(\frac{2}{na_0}\right)^3 \frac{(n-\ell-1)!}{2n[(n+\ell)!]}} e^{-r/na_0} \left(\frac{2r}{na_0}\right)^\ell L_{n-\ell-1}^{2\ell+1}\left(\frac{2r}{na_0}\right) \cdot Y_\ell^m(\theta, \phi)$$

where $a_0 = 4\pi\varepsilon_0\hbar^2 / m_e e^2$ is the Bohr radius, $L_{n-\ell-1}^{2\ell+1}$ are the generalized Laguerre polynomials of degree $n-\ell-1$, $n = 1, 2, \ldots$ is the principal quantum number, $\ell = 0, 1, \ldots n-1$ the azimuthal quantum number, $m = -\ell, -\ell+1, \ldots, \ell-1, \ell$ the magnetic quantum number. Hydrogen-like atoms have very similar solutions.

This solution does not take into account the spin of the electron.

In the figure of the hydrogen orbitals, the 19 sub-images are images of wave functions in position space (their norm squared). The wave functions each represent the abstract state characterized by the triple of quantum numbers (n, l, m), in the lower right of each image. These are the principal quantum number, the orbital angular momentum quantum number, and the magnetic quantum number. Together with one spin-projection quantum number of the electron, this is a complete set of observables.

The figure can serve to illustrate some further properties of the function spaces of wave functions.

- In this case, the wave functions are square integrable. One can initially take the function space as the space of square integrable functions, usually denoted L^2.
- The displayed functions are solutions to the Schrödinger equation. Obviously, not every function in L^2 satisfies the Schrödinger equation for the hydrogen atom. The function space is thus a subspace of L^2.
- The displayed functions form part of a basis for the function space. To each triple (n, l, m), there corresponds a basis wave function. If spin is taken into account, there are two basis functions for each triple. The function space thus has a countable basis.
- The basis functions are mutually orthonormal.

Wave Functions and Function Spaces

The concept of function spaces enters naturally in the discussion about wave functions. A function space is a set of functions, usually with some defining requirements on the functions (in the present case that they are square integrable), sometimes with an algebraic structure on the set (in the present case a vector space structure with an inner product), together with a topology on the set. The latter will sparsely be used here, it is only needed to obtain a precise definition of what it means for a subset of a function space to be closed. It will be concluded below that the function space of wave functions is a Hilbert space. This observation is the foundation of the predominant mathematical formulation of quantum mechanics.

Vector Space Structure

A wave function is an element of a function space partly characterized by the following concrete and abstract descriptions.

- The Schrödinger equation is linear. This means that the solutions to it, wave functions, can be added and multiplied by scalars to form a new solution. The set of solutions to the Schrödinger equation is a vector space.
- The superposition principle of quantum mechanics. If Ψ and Φ are two states in the abstract space of states of a quantum mechanical system, and a and b are any two complex numbers, then $a\Psi + b\Phi$ is a valid state as well. (Whether the null vector counts as a valid state ("no system present") is a matter of definition. The null vector does *not* at any rate describe the vacuum state in quantum field theory.) The set of allowable states is a vector space.

This similarity is of course not accidental. There are also a distinctions between the spaces to keep in mind.

Representations

Basic states are characterized by a set of quantum numbers. This is a set of eigenvalues of a maximal set of commuting observables. Physical observables are represented by linear operators, also called observables, on the vectors space. Maximality means that there can be added to the set no further algebraically independent observables that commute with the ones already present. A choice of such a set may be called a choice of representation.

- It is a postulate of quantum mechanics that a physically observable quantity of a system, such as position, momentum, or spin, is represented by a linear Hermitian operator on the state space. The possible outcomes of measurement of the quantity are the eigenvalues of the operator. At a deeper level, most observables, perhaps all, arise as generators of symmetries.
- The physical interpretation is that such a set represents what can – in theory – be simultaneously be measured with arbitrary precision. The Heisenberg uncertainty relation prohibits simultaneous exact measurements of two non-commuting observables.
- The set is non-unique. It may for a one-particle system, for example, be position and spin z-projection, (x, S_z) or it may be momentum and spin y-projection (p, S_y). In this case, the operator corresponding to position (a multiplication operator in the position representation) and the operator corresponding to momentum (a differential operator in the position representation) do not commute.
- Once a representation is chosen, there is still arbitrariness. It remains to choose a coordinate system. This may, for example, correspond to a choice of x, y- and z-axis, or a choice of curvilinear coordinates as exemplified by the spherical coordinates used for the Hydrogen atomic wave functions. This final choice also fixes a basis in abstract Hilbert space. The basic states are labeled by the quantum numbers corresponding to the maximal set of commuting observables and an appropriate coordinate system.

The abstract states are "abstract" only in that an arbitrary choice necessary for a particular *explicit* description of it is not given. This is the same as saying that no choice of maximal set of commuting observables has been given. This is analogous to a vector space without a specified basis. Wave functions corresponding to a state are accordingly not unique. This non-uniqueness reflects the non-uniqueness in the choice of a maximal set of commuting observables. For one spin particle in one dimension, to a particular state there corresponds two wave functions, $\Psi(x, S_z)$ and $\Psi(p, S_y)$, both describing the *same* state.

- For each choice of maximal commuting sets of observables for the abstract state space, there is a corresponding representation that is associated to a function space of wave functions.
- Between all these different function spaces and the abstract state space, there are one-to-one correspondences (here disregarding normalization and unobservable phase factors), the common denominator here being a particular abstract state. The relationship between the momentum and position space wave functions, for instance, describing the same state is the Fourier transform.

Each choice of representation should be thought of as specifying a unique function space in which wave functions corresponding to that choice of representation lives. This distinction is best kept even if one could argue that two such function spaces are mathematically equal, e.g. being the set of square integrable functions. One can then think of the function spaces as two distinct copies of that set.

Inner Product

There is an additional algebraic structure on the vector spaces of wave functions and the abstract state space.

- Physically, different wave functions are interpreted to overlap to some degree. A system in a state Ψ that does *not* overlap with a state Φ cannot be found to be in the state Φ upon measurement. But if $\Phi_1, \Phi_2,...$ overlap Ψ to *some* degree, there is a chance that measurement of a system described by Ψ will be found in states $\Phi_1, \Phi_2,....$ Also selection rules are observed apply. These are usually formulated in the preservation of some quantum numbers. This means that certain processes allowable from some perspectives (e.g. energy and momentum conservation) do not occur because the initial and final *total* wave functions don't overlap.
- Mathematically, it turns out that solutions to the Schrödinger equation for particular potentials are orthogonal in some manner, this is usually described by an integral

$$\int \Psi_m^* \Psi_n w dV = \delta_{nm},$$

where m, n are (sets of) indices (quantum numbers) labeling different solutions, the strictly positive function w is called a weight function, and δ_{mn} is the Kronecker delta. The integration is taken over all of the relevant space.

This motivates the introduction of an inner product on the vector space of abstract quantum states, compatible with the mathematical observations above when passing to a representation. It is denoted (Ψ, Φ), or in the Bra–ket notation $\langle\Psi|\Phi\rangle$. It yields a complex number. With the inner product, the function space is an inner product space. The explicit appearance of the inner product (usually an integral or a sum of integrals) depends on the choice of representation, but the complex number (Ψ, Φ) does not. Much of the physical interpretation of quantum mechanics stems from the Born rule. It states that the probability p of finding upon measurement the state Φ given the system is in the state Ψ is

$$p = |(\Phi, \Psi)|^2,$$

where Φ and Ψ are assumed normalized. Consider a scattering experiment. In quantum field theory, if Φ_{out} describes a state in the "distant future" (an "out state") after interactions between scattering particles have ceased, and Ψ_{in} an "in state" in the "distant past", then the quantities (Φ_{out}, Ψ_{in}), with Φ_{out} and Ψ_{in} varying over a complete set of in states and out states respectively, is called the S-matrix or scattering matrix. Knowledge of it is, effectively, having *solved* the theory at hand, at least as far as predictions go. Measurable quantities such as decay rates and scattering cross sections are calculable from the S-matrix.

Hilbert Space

The above observations encapsulate the essence of the function spaces of which wave functions are elements. However, the description is not yet complete. There is a further technical requirement on the function space, that of completeness, that allows one to take limits of sequences in the function space, and be ensured that, if the limit exists, it is an element of the function space. A complete inner product space is called a Hilbert space. The property of completeness is crucial in advanced treatments and applications of quantum mechanics. For instance, the existence of projection operators or orthogonal projections relies on the completeness of the space. These projection operators, in turn, are essential for the statement and proof of many useful theorems, e.g. the spectral theorem. It is not very important in introductory quantum mechanics, and technical details and links may be found in footnotes like the one that follows. The space L^2 is a Hilbert space, with inner product presented later. The function space of the example of the figure is a subspace of L^2. A subspace of a Hilbert space is a Hilbert space if it is closed.

In summary, the set of all possible normalizable wave functions for a system with a particular choice of basis, together with the null vector, constitute a Hilbert space.

Not all functions of interest are elements of some Hilbert space, say L^2. The most glaring example is the set of functions $e^{2\pi i p \cdot x/h}$. These are plane wave solutions of the Schrödinger equation for a free particle, but are not normalizable, hence not in L^2. But they are nonetheless fundamental for the description. One can, using them, express functions that *are* normalizable using wave packets. They are, in a sense, a basis (but not a Hilbert space basis, nor a Hamel basis) in which wave functions of interest can be expressed. There is also the artifact "normalization to a delta function" that is frequently employed for notational convenience, see further down. The delta functions themselves aren't square integrable either.

The above description of the function space containing the wave functions is mostly mathematically motivated. The function spaces are, due to completeness, very *large* in a certain sense. Not all functions are realistic descriptions of any physical system. For instance, in the function space L^2

one can find the function that takes on the value 0 for all rational numbers and $-i$ for the irrationals in the interval [0, 1]. This *is* square integrable but can hardly represent a physical state.

Common Hilbert Spaces

While the space of solutions as a whole is a Hilbert space there are many other Hilbert spaces that commonly occur as ingredients.

- Square integrable complex valued functions on the interval $[0, 2\pi]$. The set $\{e^{int} / 2\pi, n \in \mathbb{Z}\}$ is a Hilbert space basis, i.e. a maximal orthonormal set.
- The Fourier transform takes functions in the above space to elements of $l^2(\mathbb{Z})$, the space of *square summable* functions $\mathbb{Z} \to \mathbb{C}$. The latter space is a Hilbert space and the Fourier transform is an isomorphism of Hilbert spaces. Its basis is $\{e_i, i \in \mathbb{Z}\}$ with $e_i(j) = \delta_{ij}$, i, $j \in \mathbb{Z}$.
- The most basic example of spanning polynomials is in the space of square integrable functions on the interval [−1, 1] for which the Legendre polynomials is a Hilbert space basis (complete orthonormal set).
- The square integrable functions on the unit sphere S^2 is a Hilbert space. The basis functions in this case are the spherical harmonics. The Legendre polynomials are ingredients in the spherical harmonics. Most problems with rotational symmetry will have "the same" (known) solution with respect to that symmetry, so the original problem is reduced to a problem of lower dimensionality.
- The associated Laguerre polynomials appear in the hydrogenic wave function problem after factoring out the spherical harmonics. These span the Hilbert space of square integrable functions on the semi-infinite interval $[0, \infty)$.

More generally, one may consider a unified treatment of all second order polynomial solutions to the Sturm–Liouville equations in the setting of Hilbert space. These include the Legendre and Laguerre polynomials as well as Chebyshev polynomials, Jacobi polynomials and Hermite polynomials. All of these actually appear in physical problems, the latter ones in the harmonic oscillator, and what is otherwise a bewildering maze of properties of special functions becomes an organized body of facts.

There occurs also finite-dimensional Hilbert spaces. The space $\mathbb{C}^n$ is a Hilbert space of dimension n. The inner product is the standard inner product on these spaces. In it, the "spin part" of a single particle wave function resides.

- In the non-relativistic description of an electron one has $n = 2$ and the total wave function is a solution of the Pauli equation.
- In the corresponding relativistic treatment, $n = 4$ and the wave function solves the Dirac equation.

With more particles, the situations is more complicated. One has to employ tensor products and use representation theory of the symmetry groups involved (the rotation group and the Lorentz group respectively) to extract from the tensor product the spaces in which the (total) spin wave functions reside. Corresponding remarks apply to the concept of isospin, for which the symmetry group is SU(2). The models of the nuclear forces of the sixties used the symmetry group SU(3). In this case, as well, the part of the wave functions corresponding to the inner symmetries reside in some $\mathbb{C}^n$ or subspaces of tensor products of such spaces.

- In quantum field theory the underlying Hilbert space is Fock space. It is built from free single-particle states, i.e. wave functions when a representation is chosen, and can accommodate any finite, not necessarily constant in time, number of particles. The interesting (or rather the *tractable*) dynamics lies not in the wave functions but in the field operators that are operators acting on Fock space. Thus the Heisenberg picture is the most common choice (constant states, time varying operators).

Due to the infinite-dimensional nature of the system, the appropriate mathematical tools are objects of study in functional analysis.

Simplified Description

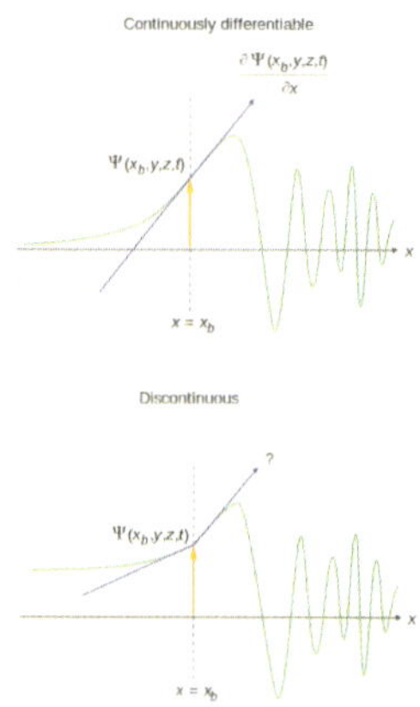

Continuity of the wave function and its first spatial derivative (in the x direction, y and z coordinates not shown), at some time t.

Not all introductory textbooks take the long route and introduce the full Hilbert space machinery, but the focus is on the non-relativistic Schrödinger equation in position representation for certain standard potentials. The following constraints on the wave function are sometimes explicitly formulated for the calculations and physical interpretation to make sense:

- The wave function must be square integrable. This is motivated by the Copenhagen interpretation of the wave function as a probability amplitude.
- It must be everywhere continuous and everywhere continuously differentiable. This is motivated by the appearance of the Schrödinger equation for most physically reasonable potentials.

It is possible to relax these conditions somewhat for special purposes. If these requirements are not met, it is not possible to interpret the wave function as a probability amplitude.

This does not alter the structure of the Hilbert space that these particular wave functions inhabit, but it should be pointed out that the subspace of the square-integrable functions L^2, which is a Hilbert space, satisfying the second requirement *is not closed* in L^2, hence not a Hilbert space in itself. The functions that does not meet the requirements are still needed for both technical and practical reasons.

More on Wave Functions and Abstract State Space

As has been demonstrated, the set of all possible wave functions in some representation for a system constitute an in general infinite-dimensional Hilbert space. Due to the multiple possible choices of representation basis, these Hilbert spaces are not unique. One therefore talks about an abstract Hilbert space, state space, where the choice of representation and basis is left undetermined. Specifically, each state is represented as an abstract vector in state space. A quantum state $|\Psi\rangle$ in any representation is generally expressed as a vector

$$|\Psi\rangle = \sum_{\alpha} \int d^m \omega \psi(\alpha, \omega, t) |\alpha, \omega\rangle$$

where $\alpha = (\alpha_1, \alpha_2, ..., \alpha_n)$ are (dimensionless) discrete quantum numbers, and $\grave{u} = (\omega_1, \omega_2, ..., \omega_m)$ are continuous variables (not necessarily dimensionless). All of them index the components of the vector, and $|\alpha, \omega\rangle$ are the basis vectors in this representation. All α are in an n-dimensional set $A = A_1 \times A_2 \times ... A_n$ where each A_i is the set of allowed values for α_i, likewise all ω are in an m-dimensional "volume" $\Omega \subseteq \mathbb{R}^m$ where $\Omega = \Omega_1 \times \Omega_2 \times ... \Omega_m$ and each $\Omega_i \subseteq \mathbb{R}$ is the set of allowed values for ω_i, a subset of the real numbers $\mathbb{R}$. For generality n and m are not necessarily equal.

For example, for a single particle in 3d with spin s, neglecting other degrees of freedom, using Cartesian coordinates, we could take $\alpha = (s_z)$ for the spin quantum number of the particle along the z direction, and $\omega = (x, y, z)$ for the particle's position coordinates. Here $A = \{-s, -s+1, ..., s-1, s\}$ is the set of allowed spin quantum numbers and $\Omega = R^3$ is the set of all possible particle positions throughout 3d position space. An alternative choice is $\alpha = (s_y)$ for the spin quantum number along the y direction and $\omega = (p_x, p_y, p_z)$ for the particle's momentum components. In this case A and Ω are the same as before.

Then, a component $\psi(\alpha, \omega, t)$ of the vector $|\Psi\rangle$ is referred to as the "wave function" of the system.

When interpreted as a probability amplitude (non-relativistic systems with constant number of particles), the probability density of finding the system at α, ω is

$$\rho_{\alpha,\omega} t = |\psi(\alpha, \omega, t)|^2$$

The probability of finding system with α in some or all possible discrete-variable configurations, $D \subseteq A$, and ω in some or all possible continuous-variable configurations, $C \subseteq \Omega$, is the sum and integral over the density,

$$P(t) = \sum_{\alpha \in D} \int_C \rho_{\alpha,\omega}(t)\, d^m \omega$$

where $d^m\omega = d\omega_1 d\omega_2 ... d\omega_m$ is a "differential volume element" in the continuous degrees of freedom. Since the sum of all probabilities must be 1, the normalization condition

$$1 = \sum_{\alpha \in A} \int_{\Omega} \rho_{\alpha,\omega}(t)\, d^m \omega$$

must hold at all times during the evolution of the system.

The normalization condition requires $\rho\, d^m\omega$ to be dimensionless, by dimensional analysis Ψ must have the same units as $(\omega_1\omega_2...\omega_m)^{-1/2}$.

Ontology

Whether the wave function really exists, and what it represents, are major questions in the interpretation of quantum mechanics. Many famous physicists of a previous generation puzzled over this problem, such as Schrödinger, Einstein and Bohr. Some advocate formulations or variants of the Copenhagen interpretation (e.g. Bohr, Wigner and von Neumann) while others, such as Wheeler or Jaynes, take the more classical approach and regard the wave function as representing information in the mind of the observer, i.e. a measure of our knowledge of reality. Some, including Schrödinger, Bohm and Everett and others, argued that the wave function must have an objective, physical existence. Einstein thought that a complete description of physical reality should refer directly to physical space and time, as distinct from the wave function, which refers to an abstract mathematical space.

Wave Function Collapse

Consider an extended wave function $\psi(x, t)$. According to our usual interpretation, $|\psi(x, t)|^2$ is proportional to the probability density of a measurement of the particle's position yielding the value x at time t. If the wave function is extended then there is a wide range of likely values that this measurement could give. Suppose that we make such a measurement, and obtain the value x_0. We now know that the particle is located at $x = x_0$. If we make another measurement immediately after the first one then what value do we expect to obtain? Well, common sense tells us that we must obtain the same value x_0, since the particle cannot have shifted position appreciably in an infinitesimal time interval. Thus, immediately after the first measurement, a measurement of the particle's position is certain to give the value x_0, and has no chance of giving any other value. This implies that the wave function must have *collapsed* to some sort of "spike" function located at $x = x_0$. This is illustrated in figure. Of course, as soon as the wave function has collapsed, it starts to expand again. Thus, the second measurement must be made reasonably quickly after the first, in order to guarantee that the same result will be obtained.

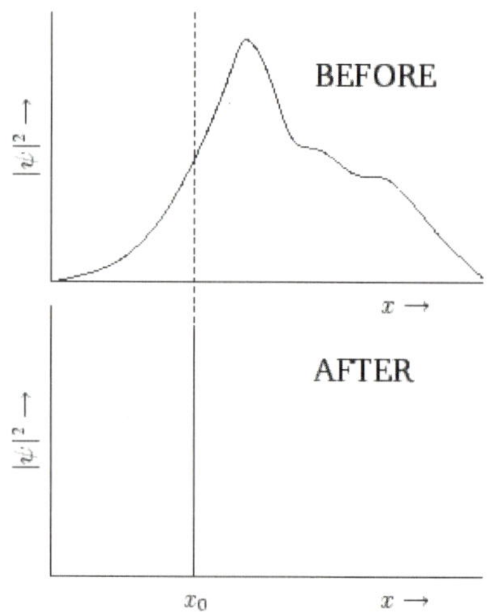

Figure: Collapse of the wave function upon measurement of x.

The above discussion illustrates an important point in quantum mechanics. Namely, that the wave function of a particle changes *discontinuously* (in time) whenever a measurement is made. We conclude that there are two types of time evolution of the wave function in quantum mechanics. First, there is a smooth evolution which is governed by Schrödinger's equation. This evolution takes place between measurements. Second, there is a discontinuous evolution which takes place each time a measurement is made.

Decoherence

Wave function collapse is the mechanism in which a system, by interacting with its environment (which includes a measuring apparatus), is transformed from a superposition of states into a definite (classical) state with a well-defined value of a given measurable quantity. This mechanism is not understood, although significant progress has been made in recent years. It is the property of a wave that leads to interference which gives rise to the strange phenomenon that permits a quantum system to simultaneously possess several (often an infinite number) of values for a particular measurable property. This property is called coherence and is related to the fact that the phases of two waves that interfere have to be "locked" so that there is a constant phase difference between them which generates this interference. The process of unlocking this phase so that the phase difference of the two waves is randomized is known as "decoherence" and this is now believed to be the mechanism that leads to wave function collapse. When decoherence occurs, the quantum property of superposition of states, with many different possible values for a given quantity, is lost - the wave function has collapsed into the wave function for a state in which that quantity is indeed well-defined and has a unique value.

Monochromatic (single wavelength) light, or other monochromatic electromagnetic radiation, or particles with the same momentum, used in the Young's double slit experiment, can interfere (giving rise to interference patterns) because the incident wave is coherent, that is to say the wave front at the double slits has peaks and troughs at the same time (or at least with a fixed time interval between them) - they are phase locked. If something happens to randomize the phases of the wave incident on the two slits, the interference is lost (decoherence has occurred). The attempt to determine through which slit a photon or electron has passed is an example of a disturbance of the system which leads to decoherence and consequently loss of the interference pattern.

Decoherence was introduced in Quantum Physics by Dieter Zeh and developed in the 1980s. The idea is that when a sub-microscopic system which is in a superposition of states interacts with a macroscopic environment, the evolution in time, determined by Quantum Mechanics, will change the entire system (the original sub-microscopic system plus the environment) into a different superposition of states of the entire system, i.e., the sub-microscopic system plus the environment. Subsequently a measurement which selects a particular state of the environment automatically selects a particular state of the sub microscopic system. As an example, suppose the sub-microscopic system is the spin of an electron which is in a state which is a superposition of spin-up and spin-down,

$$a\Psi_{\uparrow} + b\Psi_{\downarrow}$$

and the environment is a detector which has two states which we call "red" and "green". If the environment is in state "red" then a light shines red and if it is the state "green" the light shines green. Now suppose that the sub-microscopic system (the electron spin) interacts with the detector in such a way that if the electron has spin-up the detector is transformed into the state "red" and if the electron is spin-down the detector is transformed into the state "green". After this interaction the combined wave function for the electron-spin and the detector is

$$a\Psi_{\uparrow}\Psi_{red} + b\Psi_{\downarrow}\Psi_{green}.$$

This is still a superposition of states, but it is a superposition of states of a system consisting of the electron-spin plus the detector.

Observation of the detector (observing whether the light is red or green) then determines if the electron is spin-up or spin-down. The probability that the result is found to be spin-up, i.e., the probability that the light is red is a^2 - and similarly the probability that the result is found to be spin-down, i.e., the probability that the light is green is b^2.

The rate at which this decoherence takes place depends on a number of properties of the environment - such as its temperature - but estimates of the decoherence time have suggested that for macroscopic systems, the time-scale for such a decoherence process is too short for any interference to occur (two waves can only interfere if they remain coherent for a period which is much longer than the period of the waves) which is why we do not observe quantum behaviour in macroscopic objects.

Decoherence does not lead directly to wave function collapse. It does not directly explain why a measurement returns a particular value of a property of a quantum system, despite the fact that the system was initially in a state for which that value was not defined. It does, however, provide us with a mechanism which could explain how wave function collapse is observed.

Wave Equation

The wave equation is a linear second-order partial differential equation which describes the propagation of oscillations at a fixed speed in some quantity y:

$$\frac{1}{v^2}\frac{\partial^2 y}{\partial t^2} = \frac{\partial^2 y}{\partial x^2},$$

where v is the velocity of the wave.

The equation is a good description for a wide range of phenomena because it is typically used to model small oscillations about an equilibrium, for which systems can often be well approximated by Hooke's law. Solutions to the wave equation are of course important in fluid dynamics, but also play an important role in electromagnetism, optics, gravitational physics, and heat transfer. Especially important are the solutions to the Fourier transform of the wave equation, which define Fourier series, spherical harmonics, and their generalizations.

Derivation of the Wave Equation

The derivation of the wave equation varies depending on context. A particularly simple physical setting for the derivation is that of small oscillations on a piece of string obeying Hooke's law.

Consider the below diagram showing a piece of string displaced by a small amount from equilibrium:

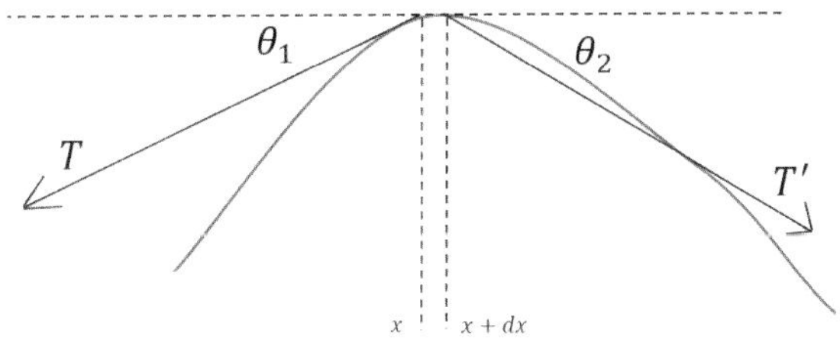

Small oscillations of a string (blue). On a small element of mass contained in a small interval dx, tensions T and T' pull the element downwards.

Consider the forces acting on a small element of mass dm contained in a small interval dx. If the displacement is small, the horizontal force is approximately zero. The vertical force is

$$\sum F_y = -T'\sin\theta_2 - T\sin\theta_1 = (dm)\alpha = \mu dx\frac{\partial^2 y}{\partial t^2},$$

Where μ is the mass density $\mu = \frac{\partial m}{\partial x}$ of the string.

On the other hand, since the horizontal force is approximately zero for small displacements, $T\cos\theta_1 \approx T'\cos\theta_2 \approx T$. Therefore,

$$-\frac{\mu dx\frac{\partial^2 y}{\partial t^2}}{T} \approx \frac{T'\sin\theta_2 + T\sin\theta_1}{T} = \frac{T'\sin\theta_2}{T} + \frac{T\sin\theta_1}{T} \approx \frac{T'\sin\theta_2}{T'\cos\theta_2} + \frac{T\sin\theta_1}{T\cos\theta_1} = \tan\theta_1 + \tan\theta_2$$

However, $\tan\theta_1 + \tan\theta_2 = -\Delta\frac{\partial y}{\partial x}$, where the difference is between x and $x + dx$. This is because the tangent is equal to thc slopc gcometrically. Dividing over dx, one finds

$$-\frac{\mu\frac{\partial^2 y}{\partial t^2}}{T} = \frac{\tan\theta_1 + \tan\theta_2}{dx} = \frac{\Delta\frac{\partial y}{\partial x}}{dx}.$$

The rightmost term above is the definition of the derivative with respect to x since the difference is over an interval dx, and therefore one has

$$\frac{\mu}{T}\frac{\partial^2 y}{\partial t^2} = \frac{\partial^2 y}{\partial x^2},$$

which is exactly the wave equation in one dimension for velocity $v = \sqrt{\frac{T}{\mu}}$.

Many derivations for physical oscillations are similar. Below, a derivation is given for the wave equation for light which takes an entirely different approach.

Another derivation can be performed providing the assumption that the definition of an entity is the same as the description of an entity.

So, a wave is a squiggly thing, with a speed, and when it moves it does not change shape:

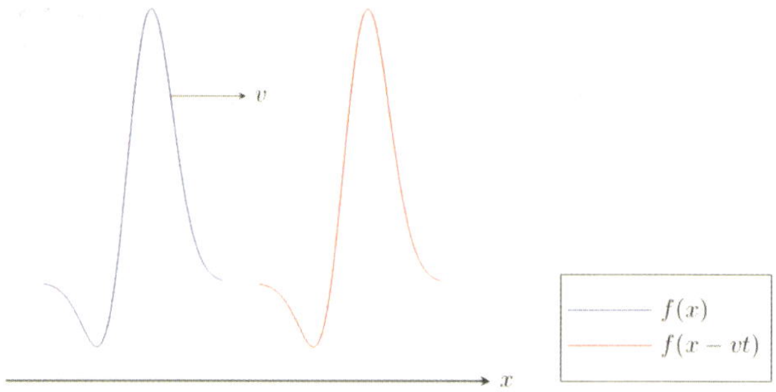

The squiggly thing is $f(x)$, the speed is v, and the red graph is the wave after time t given by a graph transformation of a translation in the -axis in the positive direction by the distance vt (the distance travelled by the wave travelling at constant speed v over time t): $f(x-vt)$.

Now, since the wave can be translated in either the positive or the negative x direction, change $f(x-vt)$ to $f(x \pm vt)$. Now, we are going to let $u = x \pm vt$, so differentiating with respect to x, keeping t constant,

$$\partial u = \partial x,$$

and differentiating with respect to t, keeping x constant,

$$\partial u = \pm\, v \partial t.$$

Now, we have a $\pm$ sign. Since the negative sign is not feasible, the second derivative of f with respect to u will be taken which will introduce a square value of v^2.

So, substituting the various ∂u:

$$\frac{\partial}{\partial u}\left(\frac{\partial f}{\partial u}\right) = \frac{\partial}{\partial x}\left(\frac{\partial f}{\partial x}\right) = \pm\frac{1}{v}\frac{\partial}{\partial t}\left(\pm\frac{1}{v}\frac{\partial f}{\partial t}\right) \Rightarrow \frac{\partial^2 f}{\partial u^2} = \frac{\partial^2 f}{\partial x^2} = \frac{1}{v^2}\frac{\partial^2 f}{\partial t^2}.$$

So we finally have the wave equation:

$$\frac{\partial^2 f}{\partial x^2} = \frac{1}{v^2}\frac{\partial^2 f}{\partial t^2}.$$

Solution of the Wave Equation

All solutions to the wave equation are super positions of "left-traveling" and "right-traveling" waves, $f(x+vt)$ and $g(x-vt)$. These are called left-traveling and right-traveling because while the overall shape of the wave remains constant, the wave translates to the left or right in time. Furthermore, any super positions of solutions to the wave equation are also solutions, because the equation is linear.

The fact that solutions to the wave equation are super positions of "left-traveling" and "right-traveling" waves.

Example:

Prove that all solutions to the wave equation are super positions of "left-traveling" and "right-traveling" waves.

Define new coordinates $a = x - vt$ and $b = x + vt$ representing right and left propagation of waves, respectively. Then the partial derivatives can be rewritten as

$$\frac{\partial}{\partial x} = \frac{1}{2}(\frac{\partial}{\partial a} + \frac{\partial}{\partial b}) \Rightarrow \frac{\partial^2}{\partial x^2} = \frac{1}{4}\left(\frac{\partial^2}{\partial a^2} + 2\frac{\partial^2}{\partial a \partial b} + \frac{\partial^2}{\partial b^2}\right)$$

$$\frac{\partial}{\partial t} = \frac{v}{2}(\frac{\partial}{\partial b} - \frac{\partial}{\partial a}) \Rightarrow \frac{\partial^2}{\partial t^2} = \frac{v^2}{4}\left(\frac{\partial^2}{\partial a^2} - 2\frac{\partial^2}{\partial a \partial b} + \frac{\partial^2}{\partial b^2}\right).$$

Since the wave equation is

$$\frac{\partial^2 y}{\partial x^2} - \frac{1}{v^2}\frac{\partial^2 y}{\partial t^2} = 0,$$

substituting in for the partial derivatives yields the equation in the coordinates a and b:

$$\frac{\partial^2 y}{\partial a \partial b} = 0.$$

This is solved in general by $y = f(a) + g(b) = f(x - vt) + g(x + vt)$ as claimed.

The most commonly used examples of solutions are harmonic waves:

$$y(x, t) = A \sin(x - vt) + B \sin(x + vt),$$

Where y_0 is the amplitude of the wave and A and B are some constants depending on initial conditions.

If the boundary conditions are such that the solutions take the same value at both endpoints, the solutions can lead to standing waves as seen above. These take the functional form

$$y(x, t) = y_0 \sin(x - vt) + y_0 \sin(x + vt) = 2y_0 \sin(x) \cos(vt),$$

where y_0 is the amplitude of the wave. The solution has constant amplitude and the spatial part $\sin(x)$ has no time dependence.

Formally, there are two major types of boundary conditions for the wave equation:

Dirichlet boundary conditions: The amplitude is fixed at the boundary point x^*. This can be written as

$$y(x^*) = y^*,$$

- where y^* is some fixed constant. A more useful way of writing this condition is

$$\left.\frac{dy}{dt}\right|_{x=x^*} = 0,$$

which says that the amplitude at the endpoint does not change over time.

- Neumann boundary conditions: The derivative of the amplitude is specified at the endpoints. Often, this derivative is taken to be zero. For an approximately massless (very light) string, this means the endpoints are free rather than fixed; for a pipe with pressure waves inside of it, this means that the endpoints are open to the atmosphere:

$$\left.\frac{dy}{dx}\right|_{x=x^*} = 0.$$

A string with Dirichlet boundary conditions at the left end, where the string is fixed to a wall, and Neumann boundary conditions at the right end, where the string is attached to a freely sliding ring

Example:

A string of tension T and mass density per unit length μ is attached to a small massless ring which slides on a slippery rod. The rod exerts a damping force $F_d = -b\frac{\partial y}{\partial t}$ on the ring. Find the boundary condition on the oscillations of the string at the end attached to the ring.

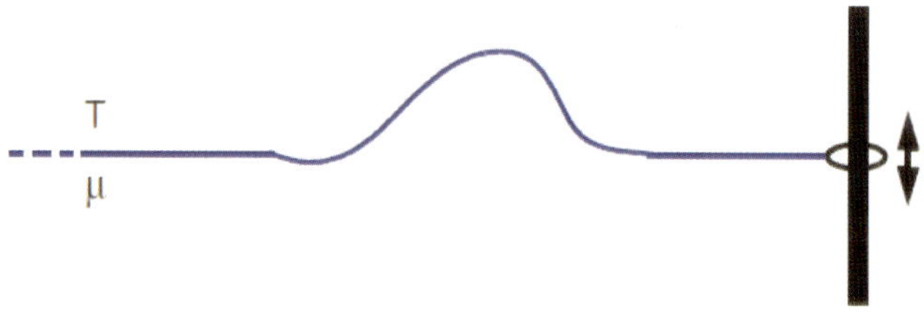

A string attached to a ring sliding on a slippery rod.

The ring is free to slide, so the boundary conditions are Neumann and since the ring is massless the total force on the ring must be zero. Consider the following free body diagram:

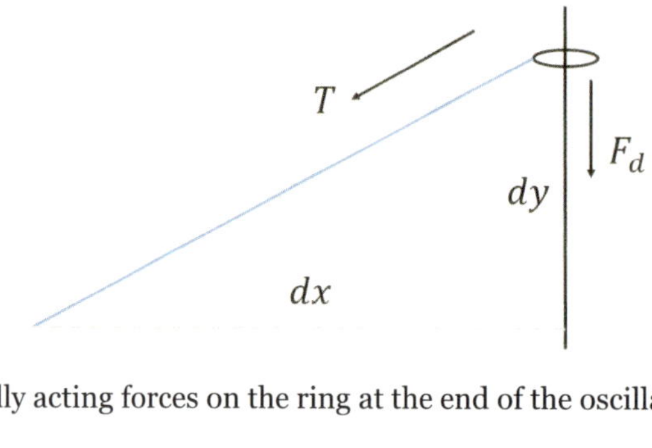

All vertically acting forces on the ring at the end of the oscillating string.

Using the fact that the wave equation holds for *small* oscillations only, $dx \gg dy$. Balancing the forces in the vertical direction thus yields

$$-T\frac{\partial y}{\partial x}-b\frac{\partial y}{\partial t}=0 \Rightarrow \frac{\partial y}{\partial x}=-\frac{b}{T}\frac{\partial y}{\partial t}.$$

This slope condition is the Neumann boundary condition on the oscillations of the string at the end attached to the ring.

One way of writing down solutions to the wave equation generates Fourier series which may be used to represent a function as a sum of sinusoidals. This method uses the fact that the complex exponentials e^{-iwt} are eigenfunctions of the operator $\frac{\partial^2}{\partial t^2}$. Using this fact, ansatz a solution for a particular w:

$$y(x,t)=e^{-iwt}f(x),$$

where the exponential has essentially factored out the time dependence. Plugging into the wave equation, one finds

$$\frac{\partial^2 y}{\partial t^2}=-w^2 y(x,t)=v^2\frac{\partial^2 y}{\partial x^2}=v^2 e^{-iwt}\frac{\partial^2 f}{\partial x^2}.$$

The function f therefore satisfies the equation

$$\frac{\partial^2 f}{\partial x^2}=-\frac{w^2}{v^2}f.$$

This is solved by the plane waves

$$f(x)=f_0 e^{\perp iwx/v}.$$

Therefore, the general solution for a particular w can be written as

$$y(x,t)=f_0 e^{i\frac{w}{v}(x\pm vt)}.$$

This is consistent with the assertion above that solutions are written as super positions of $f(x-vt)$ and $g(x+vt)$ for some functions f and g. By the linearity of the wave equation, an arbitrary solution can be built up in terms of super positions of the above solutions that have w fixed. This is exactly the statement of existence of the Fourier series.

Scalar Wave Equation in Three Space Dimensions

A solution of the initial-value problem for the wave equation in three space dimensions can be obtained from the corresponding solution for a spherical wave. The result can then be also used to obtain the same solution in two space dimensions.

Swiss mathematician and physicist Leonhard Euler discovered the wave equation in three space dimensions.

Spherical Waves

The wave equation can be solved using the technique of separation of variables. To obtain a solution with constant frequencies, let us first Fourier transform the wave equation in time as

$$\Psi(\vec{r},t)=\int_{-\infty}^{\infty}\Psi(\vec{r},\omega)e^{-i\omega t}d\omega.$$

So we get,

$$\left(\nabla^2+\frac{\omega^2}{c^2}\right)\Psi(\vec{r},\omega)=0$$

This is the Helmholtz equation and can be solved using separation of variables. If spherical coordinates are used to describe a problem, then the solution to the angular part of the Helmholtz equation is given by spherical harmonics and the radial equation now becomes

$$\left[\frac{d^2}{dr^2}+\frac{2}{r}\frac{d}{dr}+k^2-\frac{l(l+1)}{r^2}\right]f_{lm}(r)=0$$

Here $k\equiv\frac{\omega}{c}$ and the complete solution is now given by

$$\Psi(\vec{r},\omega)=\sum_{lm}\left[A_{lm}^{(1)}h_l^{(1)}(kr)+A_{lm}^{(2)}h_l^{(2)}(kr)\right]Y_{lm}(\theta,\phi),$$

where $h_l^{(1)}(kr)$ and $h_l^{(2)}(kr)$ are the spherical Hankel functions. To gain a better understanding of the nature of these spherical waves, let us go back and look at the case when $l=0$. In this case, there is no angular dependence and the amplitude depends only on the radial distance i.e. $\Psi(\vec{r},t)\rightarrow u(r,t)$. In this case, the wave equation reduces to

$$\left(\nabla^2-\frac{1}{c^2}\frac{\partial^2}{\partial t^2}\right)\Psi(\vec{r},t)=0\rightarrow\left(\frac{\partial^2}{\partial r^2}+\frac{2}{r}\frac{\partial}{\partial r}-\frac{1}{c^2}\frac{\partial^2}{\partial t^2}\right)u(r,t)=0$$

This equation can be rewritten as

$$\frac{\partial^2(ru)}{\partial t^2} - c^2 \frac{\partial^2(ru)}{\partial r^2} = 0;$$

where the quantity ru satisfies the one-dimensional wave equation. Therefore, there are solutions in the form

$$u(r, t) = \frac{1}{r} F(r - ct) + \frac{1}{r} G(r + ct),$$

where F and G are general solutions to the one-dimensional wave equation, and can be interpreted as respectively an outgoing or incoming spherical wave. Such waves are generated by a point source, and they make possible sharp signals whose form is altered only by a decrease in amplitude as r increases. Such waves exist only in cases of space with odd dimensions.

For physical examples of non-spherical wave solutions to the 3D wave equation that do possess angular dependence.

Monochromatic Spherical Wave

Although the word "monochromatic" is not exactly accurate since it refers to light or electromagnetic radiation with well-defined frequency, the spirit is to discover the eigenmode of the wave equation in three-dimensions. Following the derivation on Plane wave eigenmodes, if we again restrict our solutions to spherical waves that oscillate in time with well-defined *constant* angular frequency ω, then the transformed function $ru(r, t)$ has simply plane wave solutions,

$$ru(r, t) \quad Ae^{i(\omega t \pm kr)},$$

or

$$u(r, t) = \frac{A}{r} e^{i(\omega t \pm kr)}.$$

From this we can observe that the peak intensity of the spherical wave oscillation, characterized as the squared wave amplitude

$$I = |u(r, t)|^2 = \frac{|A|^2}{r^2}.$$

drops at the rate proportional to $1/r^2$, an example of the inverse-square law.

Solution of a General Initial-value Problem

The wave equation is linear in u and it is left unaltered by translations in space and time. Therefore, we can generate a great variety of solutions by translating and summing spherical waves. Let $\varphi(\xi, \eta, \zeta)$ be an arbitrary function of three independent variables, and let the spherical wave form F be a delta-function: that is, let F be a weak limit of continuous functions whose integral is unity, but whose support (the region where the function is non-zero) shrinks to the origin. Let a family of spherical waves have center at (ξ, η, ζ), and let r be the radial distance from that point. Thus

$$r^2 = (x-\xi)^2 + (y-\eta)^2 + (z-\zeta)^2.$$

If u is a superposition of such waves with weighting function φ, then

$$u(t, x, y, z) = \frac{1}{4\pi c} \iiint \varphi(\xi, \eta, \zeta) \frac{\delta(r-ct)}{r} d\xi d\eta d\zeta;$$

the denominator $4\pi c$ is a convenience.

From the definition of the delta-function, u may also be written as

$$u(t, x, y, z) = \frac{t}{4\pi} \iint_S \varphi(x + ct\alpha, y + ct\beta, z + ct\gamma) d\omega,$$

where α, β, and γ are coordinates on the unit sphere S, and ω is the area element on S. This result has the interpretation that $u(t, x)$ is t times the mean value of ϕ on a sphere of radius ct centered at x:

$$u(t, x, y, z) = tM_{ct}[\phi].$$

It follows that

$$u(0, x, y, z) = 0, \quad u_t(0, x, y, z) = \phi(x, y, z).$$

The mean value is an even function of t, and hence if

$$v(t, x, y, z) = \frac{\partial}{\partial t}\left(tM_{ct}[\psi]\right),$$

then

$$v(0, x, y, z) = \psi(x, y, z), \quad v_t(0, x, y, z) = 0.$$

These formulas provide the solution for the initial-value problem for the wave equation. They show that the solution at a given point P, given (t, x, y, z) depends only on the data on the sphere of radius ct that is intersected by the light cone drawn backwards from P. It does *not* depend upon data on the interior of this sphere. Thus the interior of the sphere is a lacuna for the solution. This phenomenon is called Huygens' principle. It is true for odd numbers of space dimension, where for one dimension the integration is performed over the boundary of an interval with respect to the Dirac measure. It is not satisfied in even space dimensions.

Scalar Wave Equation in Two Space Dimensions

In two space dimensions, the wave equation is

$$u_{tt} = c^2\left(u_{xx} + u_{yy}\right).$$

We can use the three-dimensional theory to solve this problem if we regard u as a function in three dimensions that is independent of the third dimension. If

$$u(0, x, y) = 0, \quad u_t(0, x, y) = \phi(x, y),$$

then the three-dimensional solution formula becomes

$$u(t, x, y) = tM_{ct}[\phi] = \frac{t}{4\pi}\iint_S \phi(x + ct\alpha,\ y + ct\beta)d\omega,$$

where α and β are the first two coordinates on the unit sphere, and $d\omega$ is the area element on the sphere. This integral may be rewritten as a double integral over the disc D with center (x, y) and radius ct:

$$u(t, x, y) = \frac{1}{2\pi c}\iint_D \frac{\phi(x+\xi, y+\eta)}{\sqrt{(ct)^2 - \xi^2 - \eta^2}} d\xi d\eta.$$

It is apparent that the solution at (t, x, y) depends not only on the data on the light cone where

$$(x-\xi)^2 + (y-\eta)^2 = c^2t^2,$$

but also on data that are interior to that cone.

Scalar Wave Equation in General Dimension and Kirchhoff's Formulae

We want to find solutions to $u_{tt} - \Delta u = 0$ for $u : \mathbb{R}^n \times (0, \infty) \to \mathbb{R}$ with $u(x, 0) = g(x)$ and $u_t(x, 0) = h(x)$.

Odd Dimensions

Assume $n \geq 3$ is an odd integer and $g \in C^{m+1}(\mathbb{R}^n), h \in C^m(\mathbb{R}^n)$ *for* $m = (n+1)/2$. Let $\gamma_n = 1\cdot 3\cdot 5\cdot\ldots\cdot(n-3)$ and let

$$u(x, t) = \frac{1}{\gamma_n}\left[\partial_t\left(\frac{1}{t}\partial_t\right)^{\frac{n-3}{2}}\left(t^{n-2}\frac{1}{|\partial B_t(x)|}\int_{\partial B_t(x)} gdS\right) + \left(\frac{1}{t}\partial_t\right)^{\frac{n-3}{2}}\left(t^{n-2}\frac{1}{|\partial B_t(x)|}\int_{\partial B_t(x)} hdS\right)\right]$$

then

$$u \in C^2\left(\mathbb{R}^n \times [0, \infty)\right)$$
$$u_{tt} - \Delta u = 0 \text{ in } \mathbb{R}^n \times (0, \infty)$$
$$\lim_{(x,t)\to(x^0,0)} u(x, t) = g(x^0)$$
$$\lim_{(x,t)\to(x^0,0)} u_t(x, t) = h(x^0)$$

Even Dimensions

Assume $n \geq 2$ is an even integer and $g \in C^{m+1}(\mathbb{R}^n)$, $h \in C^m(\mathbb{R}^n)$, for $m = (n+2)/2$. Let $\gamma_n = 2\cdot 4\cdot\ldots\cdot n$ and let

$$u(x,t)=\frac{1}{\gamma_n}\left[\partial_t\left(\frac{1}{t}\partial_t\right)^{\frac{n-2}{2}}\left(t^n\frac{1}{|B_t(x)|}\int_{B_t(x)}\frac{g}{(t^2-|y-x|^2)^{\frac{1}{2}}}dy\right)+\left(\frac{1}{t}\partial_t\right)^{\frac{n-2}{2}}\left(t^n\frac{1}{|B_t(x)|}\int_{B_t(x)}\frac{h}{(t^2-|y-x|^2)^{\frac{1}{2}}}dy\right)\right]$$

then

$$\begin{aligned}
&u\in C^2\left(\mathrm{R}^n\times[0,\infty)\right)\\
&u_{tt}-\Delta u=0 \text{ in } \mathrm{R}^n\times(0,\infty)\\
&\lim_{(x,t)\to(x^0,0)} u(x,t)=g(x^0)\\
&\lim_{(x,t)\to(x^0,0)} u_t(x,t)=h(x^0)
\end{aligned}$$

One Space Dimension

The Sturm-Liouville Formulation

A flexible string that is stretched between two points $x=0$ and $x=L$ satisfies the wave equation for $t>0$ and $0<x<L$. On the boundary points, u may satisfy a variety of boundary conditions. A general form that is appropriate for applications is

$$-u_x(t,0)+au(t,0)=0,$$

$$u_x(t,L)+bu(t,L)=0,$$

where a and b are non-negative. The case where u is required to vanish at an endpoint is the limit of this condition when the respective a or b approaches infinity. The method of separation of variables consists in looking for solutions of this problem in the special form

$$u(t,x)=T(t)v(x).$$

A consequence is that

$$\frac{T''}{c^2T}=\frac{v''}{v}=-\lambda.$$

The eigenvalue λ must be determined so that there is a non-trivial solution of the boundary-value problem

$$\begin{aligned}
&v''+\lambda v=0,\\
&-v'(0)+av(0)=0,\quad v'(L)+bv(L)=0.
\end{aligned}$$

This is a special case of the general problem of Sturm–Liouville theory. If a and b are positive, the eigenvalues are all positive, and the solutions are trigonometric functions. A solution that satisfies square-integrable initial conditions for u and u_t can be obtained from expansion of these functions in the appropriate trigonometric series.

Investigation by Numerical Methods

Approximating the continuous string with a finite number of equidistant mass points one gets the following physical model:

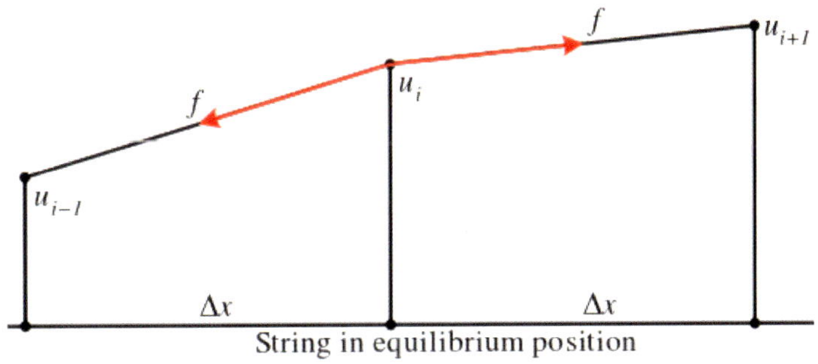

Figure: Three consecutive mass points of the discrete model for a string

If each mass point has the mass m, the tension of the string is f, the separation between the mass points is Δx and u_i, $i = 1, \ldots, n$ are the offset of these n points from their equilibrium points (i.e. their position on a straight line between the two attachment points of the string) the vertical component of the force towards point $i+1$ is

$$\frac{u_{i+1} - u_i}{\Delta x} f$$

and the vertical component of the force towards point $i-1$ is

$$\frac{u_{i-1} - u_i}{\Delta x} f$$

Taking the sum of these two forces and dividing with the mass m one gets for the vertical motion:

$$\ddot{u}_i = \left(\frac{f}{m\Delta x}\right)(u_{i+1} + u_{i-1} - 2u_i)$$

As the mass density is

$$\rho = \frac{m}{\Delta x}$$

this can be written

$$\ddot{u}_i = \left(\frac{f}{\rho\Delta x^2}\right)(u_{i+1} + u_{i-1} - 2u_i)$$

The wave equation is obtained by letting $\Delta x \to 0$ in which case $u_i(t)$ takes the form $u(x, t)$ where $u(x, t)$ is continuous function of two variables, $\ddot{u}_i$ takes the form $\frac{\partial^2 u}{\partial t^2}$ and

$$\frac{u_{i+1} + u_{i-1} - 2u_i}{\Delta x^2} \to \frac{\partial^2 u}{\partial x^2}$$

But the discrete formulation $\ddot{u}_i = \left(\frac{f}{m\Delta x}\right)(u_{i+1} + u_{i-1} - 2u_i)$ of the equation of state with a finite number of mass point is just the suitable one for a numerical propagation of the string motion. The boundary condition

$$u(0,t) = u(L,t) = 0$$

where L is the length of the string takes in the discrete formulation the form that for the outermost points u_1 and u_n the equation of motion are

$$\ddot{u}_1 = \left(\frac{c}{\Delta x}\right)^2 (u_2 - 2u_1)$$

and

$$\ddot{u}_n = \left(\frac{c}{\Delta x}\right)^2 (u_{n-1} - 2u_n)$$

while for $1 < i < n$

$$\ddot{u}_i = \left(\frac{c}{\Delta x}\right)^2 (u_{i+1} + u_{i-1} - 2u_i)$$

where $c = \sqrt{\frac{f}{\rho}}$.

If the string is approximated with 100 discrete mass points one gets the 100 coupled second order differential equations $\ddot{u}_1 = \left(\frac{c}{\Delta x}\right)^2 (u_2 - 2u_1)$, $\ddot{u}_n = \left(\frac{c}{\Delta x}\right)^2 (u_{n-1} - 2u_n)$ and $\ddot{u}_i = \left(\frac{c}{\Delta x}\right)^2 (u_{i+1} + u_{i-1} - 2u_i)$ or equivalently 200 coupled first order differential equations.

Propagating these up to the times

$$\frac{L}{c} k 0.05 k = 0, \cdots, 5$$

using an 8th order multistep method the 6 states displayed in figure is found:

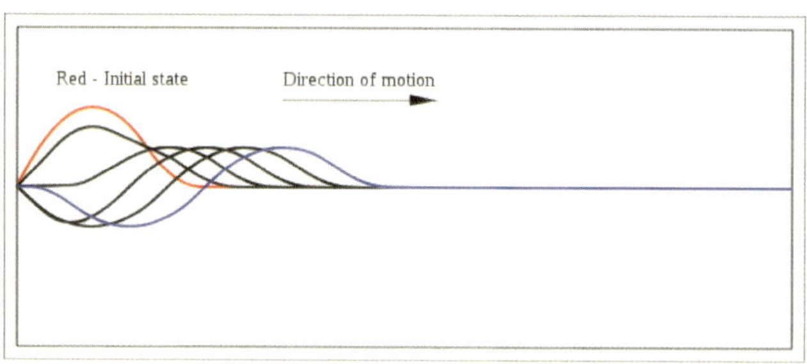

Figure: The string at 6 consecutive epochs, the first (red) corresponding to the initial time with the string in rest

The red curve is the initial state at time zero at which the string is "let free" in a predefined shape with all $\dot{u}_i = 0$. The blue curve is the state at time $\frac{L}{c}(0.25)$, i.e. after a time that corresponds to the time a wave that is moving with the nominal wave velocity $c=\sqrt{\frac{f}{\rho}}$ would need for one fourth of the length of the string.

Figure displays the shape of the string at the times $\frac{L}{c}k(0.05),\ k=6,\cdots,11$. The wave travels in direction right with the speed $c=\sqrt{\frac{f}{\rho}}$ without being actively constraint by the boundary conditions at the two extremes of the string. The shape of the wave is constant, i.e. the curve is indeed of the form $f(x-ct)$.

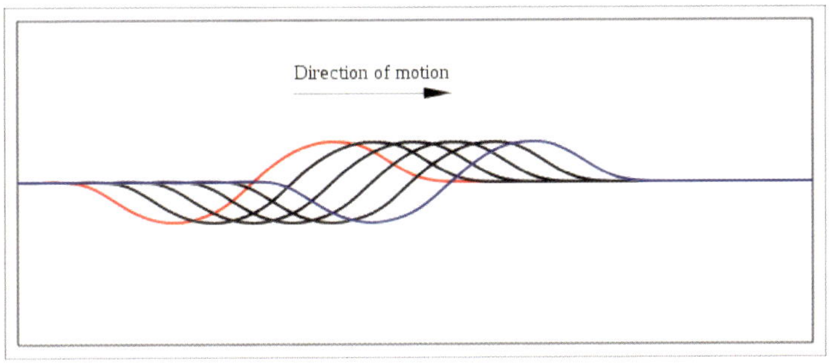

Figure displays the shape of the string at the times $\frac{L}{c}k(0.05),\ k=12,\cdots,17$. The constraint on the right extreme starts to interfere with the motion preventing the wave to raise the end of the string.

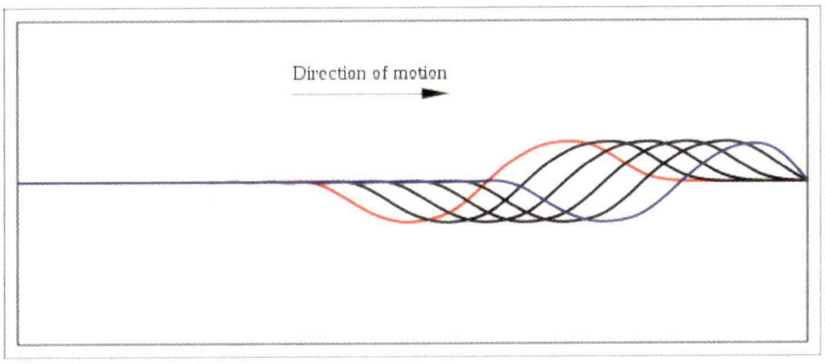

Figure displays the shape of the string at the times $\frac{L}{c}k(0.05),\ k=18,\cdots,23$ when the direction of motion is reversed. The red, green and blue curves are the states at the times $\frac{L}{c}k(0.05),\ k=18,\cdots,20$ while the 3 black curves correspond to the states at times $\frac{L}{c}k(0.05),\ k=21,\cdots,23$ with the wave starting to move back towards left.

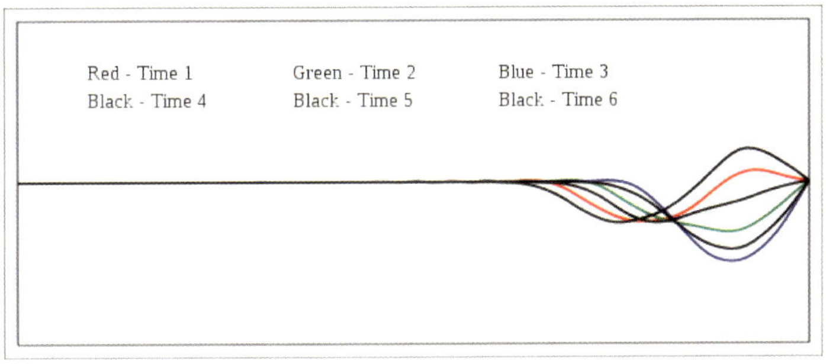

Figure (a) and figure (b) finally display the shape of the string at the times $\frac{L}{c}k(0.05),\ k=24,\cdots,29$ and $\frac{L}{c}k(0.05),\ k=30,\cdots,35$. The wave now travels towards left and the constraints at the end points are not active any more. When finally the other extreme of the string the direction will again be reversed in a way similar to what is displayed in figure (a).

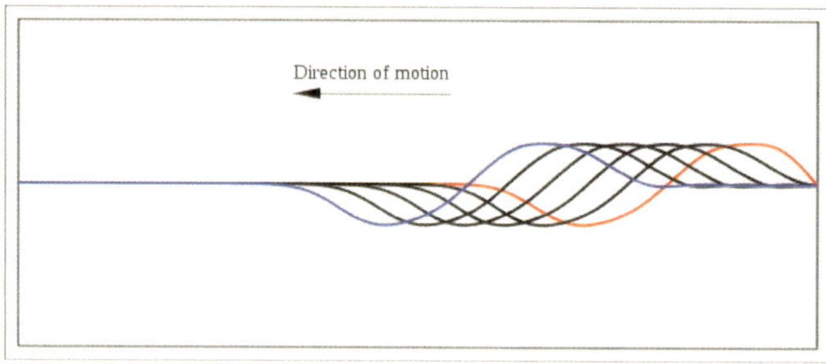

Figure (a)

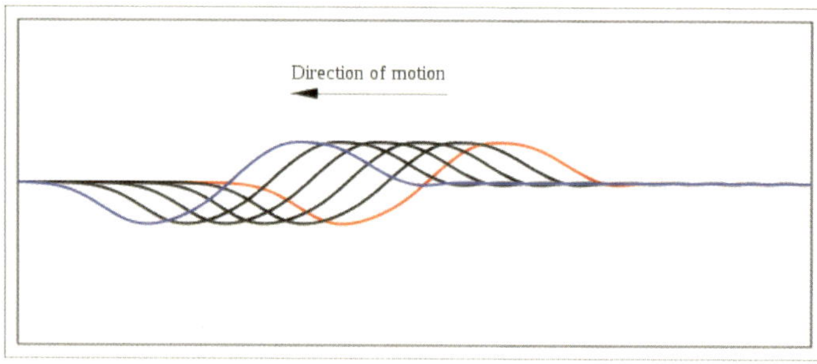

Figure(b)

Several Space Dimensions

The one-dimensional initial-boundary value theory may be extended to an arbitrary number of space dimensions. Consider a domain D in m-dimensional x space, with boundary B. Then the wave equation is to be satisfied if x is in D and $t > 0$. On the boundary of D, the solution u shall satisfy

$$\frac{\partial u}{\partial n} + au = 0,$$

where n is the unit outward normal to B, and a is a non-negative function defined on B. The case where u vanishes on B is a limiting case for a approaching infinity. The initial conditions are

$$u(0, x) = f(x), \quad u_t(0, x) = g(x),$$

where f and g are defined in D. This problem may be solved by expanding f and g in the eigenfunctions of the Laplacian in D, which satisfy the boundary conditions. Thus the eigenfunction v satisfies

$$\nabla \cdot \nabla v + \lambda v = 0,$$

in D, and

$$\frac{\partial v}{\partial n} + av = 0,$$

on B.

In the case of two space dimensions, the eigenfunctions may be interpreted as the modes of vibration of a drumhead stretched over the boundary B. If B is a circle, then these eigenfunctions have an angular component that is a trigonometric function of the polar angle θ, multiplied by a Bessel function (of integer order) of the radial component.

If the boundary is a sphere in three space dimensions, the angular components of the eigenfunctions are spherical harmonics, and the radial components are Bessel functionsof half-integer order.

Inhomogeneous Wave Equation in One Dimension

The inhomogeneous wave equation in one dimension is the following:

$$c^2u_{xx}(x,t)-u_{tt}(x,t)=s(x,t)$$

with initial conditions given by

$$u(x,0)=f(x)$$
$$u_t(x,0)=g(x)$$

The function $s(\mathrm{x},t)$ is often called the source function because in practice it describes carrying them. Physical examples of source functions include the force driving a wave on a string, or the charge or current density in the Lorenz gauge of electromagnetism.

One method to solve the initial value problem (with the initial values as posed above) is to take advantage of a special property of the wave equation in an odd number of space dimensions, namely that its solutions respect causality. That is, for any point (x_i, t_i), the value of $u(x_i, t_i)$ depends only on the values of $f(x_i+ct_i)$ and $f(x_i-ct_i)$ and the values of the function $g(x)$ between (x_i-ct_i) and (x_i+ct_i). This can be seen in d'Alembert's formula, stated above, where these quantities are the only ones that show up in it. Physically, if the maximum propagation speed is c, then no part of the wave that can't propagate to a given point by a given time can affect the amplitude at the same point and time.

In terms of finding a solution, this causality property means that for any given point on the line being considered, the only area that needs to be considered is the area encompassing all the points that could causally affect the point being considered. Denote the area that casually affects point (x_i, t_i) *as* R_C. Suppose we integrate the inhomogeneous wave equation over this region.

$$\iint_{R_C}\left(c^2u_{xx}(x,t)-u_{tt}(x,t)\right)dxdt=\iint_{R_C}s(x,t)dxdt.$$

To simplify this greatly, we can use Green's theorem to simplify the left side to get the following:

$$\int_{L_0+L_1+L_2}\left(-c^2u_x(x,t)dt-u_t(x,t)dx\right)=\iint_{R_C}s(x,t)dxdt.$$

The left side is now the sum of three line integrals along the bounds of the causality region. These turn out to be fairly easy to compute

$$\int_{x_i-ct_i}^{x_i+ct_i}-u_t(x,0)dx=-\int_{x_i-ct_i}^{x_i+ct_i}g(x)dx.$$

In the above, the term to be integrated with respect to time disappears because the time interval involved is zero, thus $dt = 0$.

For the other two sides of the region, it is worth noting that $x \pm ct$ is a constant, namingly $x_i \pm ct_i$, where the sign is chosen appropriately. Using this, we can get the relation $dx \pm cdt = 0$, again choosing the right sign:

$$\begin{aligned}\int_{L_1}\left(-c^2u_x(x,t)dt - u_t(x,t)dx\right) &= \int_{L_1}(cu_x(x,t)dx + cu_t(x,t)dt)\\ &= c\int_{L_1} du(x,t)\\ &= cu(x_i,t_i) - cf(x_i + ct_i).\end{aligned}$$

And similarly for the final boundary segment:

$$\begin{aligned}\int_{L_2}\left(-c^2u_x(x,t)dt - u_t(x,t)dx\right) &= -\int_{L_2}(cu_x(x,t)dx + cu_t(x,t)dt)\\ &= -c\int_{L_2} du(x,t)\\ &= cu(x_i,t_i) - cf(x_i - ct_i).\end{aligned}$$

Adding the three results together and putting them back in the original integral:

$$\begin{aligned}\iint_{RC} s(x,t)dxdt &= -\int_{x_i - ct_i}^{x_i + ct_i} g(x)dx + cu(x_i,t_i) - cf(x_i + ct_i) + cu(x_i,t_i) - cf(x_i - ct_i)\\ &= 2cu(x_i,t_i) - cf(x_i + ct_i) - cf(x_i - ct_i) - \int_{x_i - ct_i}^{x_i + ct_i} g(x)dx\end{aligned}$$

Solving for $u(x_i, t_i)$ we arrive at

$$u(x_i,t_i) = \frac{f(x_i + ct_i) + f(x_i - ct_i)}{2} + \frac{1}{2c}\int_{x_i - ct_i}^{x_i + ct_i} g(x)dx + \frac{1}{2c}\int_0^{t_i}\int_{x_i - c(t_i - t)}^{x_i + c(t_i - t)} s(x,t)dxdt.$$

In the last equation of the sequence, the bounds of the integral over the source function have been made explicit. Looking at this solution, which is valid for all choices (x_i, t_i) compatible with the wave equation, it is clear that the first two terms are simply d'Alembert's formula, as stated above as the solution of the homogeneous wave equation in one dimension. The difference is in the third term, the integral over the source.

Other Coordinate Systems

In three dimensions, the wave equation, when written in elliptic cylindrical coordinates, may be solved by separation of variables, leading to the Mathieu differential equation.

Elastic Waves

The elastic wave equation in three dimensions describes the propagation of waves in an isotropic homogeneous elastic medium. Most solid materials are elastic, so this equation describes such phenomena as seismic waves in the Earth and ultrasonic waves used to detect flaws in materials. While linear, this equation has a more complex form than the equations given above, as it must account for both longitudinal and transverse motion:

$$\rho \ddot{u} = f + (\lambda + 2\mu)\nabla(\nabla \cdot u) - \mu\nabla \times (\nabla \times u)$$

where:

- λ and μ are the so-called Lamé parameters describing the elastic properties of the medium,
- ρ is the density,
- f is the source function (driving force),
- and u is the displacement vector.

Note that in this equation, both force and displacement are vector quantities. Thus, this equation is sometimes known as the vector wave equation. As an aid to understanding, the reader will observe that if f and $\nabla \cdot u$ are set to zero, this becomes (effectively) Maxwell's equation for the propagation of the electric field E, which has only transverse waves.

Dispersion Relation

In dispersive wave phenomena, the speed of wave propagation varies with the wavelength of the wave, which is reflected by a dispersion relation

$$\omega = \omega(k),$$

where ω is the angular frequency and k is the wavevector describing plane wave solutions. For light waves, the dispersion relation is $\omega = \pm c|\mathbf{k}|$, but in general, the constant speed *c* gets replaced by a variable phase velocity:

$$v_p = \frac{\omega(k)}{k}.$$

Double-Slit Experiment

The double-slit experiment is a nineteenth-century investigation into the properties of lightthat has since been found to demonstrate both the duality of photons and the concepts of superposition and quantum interference.

Debate over whether light is made up of particles or waves dates back over three hundred years. In the seventeenth century, Isaac Newton proclaimed that light consisted of a stream of particles. In the early nineteenth century, Thomas Young devised the double-slit experiment to prove that it consisted of waves. While the implications of Young's experiment are difficult to accept, they are illustrative of quantum theory. The noted physicist Richard Feynman claimed that the essentials of quantum mechanics could be grasped from an exploration of the results of the experiment.

In the double-slit experiment, a beam of light is aimed at a barrier with two vertical slits. After the light passes through the slits, the resulting pattern is recorded on a photographic plate. When one slit is covered, a single line of light is displayed, aligned with whichever slit is open.

Intuitively, one might hypothesize that if both slits were open, the resulting pattern would display as two lines of light, aligned with the slits. What occurs in practice, however, is that light passing through the slits and displayed on the photographic plate is entirely separated into multiple lines of lightness and darkness in varying degrees. The result illustrates that interference is taking place between the waves and particles going through the slits in what a layman might expect to be two non-crossing trajectories.

If the beam of photons is slowed enough to ensure that individual photons are hitting the plate, one might expect there to be no interference and a pattern of light would be two lines of light, aligned with the slits. The results of the experiment, however, indicate the presence of interference. Somehow, the single particles are interfering with themselves. On the face of things, this might seem impossible: we expect that a single photon will go through one slit or the other and will end up in one of two possible light line areas.

That expectation, however, is invalidated by the results of the double-slit experiment. What actually occurs is that each photon not only goes through both slits, but also simultaneously traverses every possible trajectory en route to the target. Research into this phenomenon has demonstrated that other elementary particles, such as electrons, exhibit the same behavior.

References

- Chafaï, D. (2003), Gaussian maximum of entropy and reversed log-Sobolev inequality, pp. 194–200, arXiv:math/0102227 , doi:10.1007/978-3-540-36107-7_5, ISBN 978-3-540-00072-3
- Uncertainty-principle, science: britannica.com, Retrieved 12 May 2018
- Curtright, T.; Zachos, C. (2001). "Negative Probability and Uncertainty Relations". Modern Physics Letters A. 16 (37): 2381–2385. arXiv:hep-th/0105226 . Bibcode:2001MPLA...16.2381C. doi:10.1142/S021773230100576X.
- De-broglie-concept-of-matter-waves-dual-nature-of-matter: winnerscience.com, Retrieved 19 June 2018
- Steiger, Nathan. "Quantum Uncertainty and Conservation Law Restrictions on Gate Fidelity". Brigham Young University. Retrieved 19 June 2011
- Wave-function: chemicool.com, Retrieved 11 July 2018
- Folland, Gerald; Sitaram, Alladi (May 1997), "The Uncertainty Principle: A Mathematical Survey", Journal of Fourier Analysis and Applications, 3 (3): 207–238, doi:10.1007/BF02649110, MR 1448337
- Wave-equation: brilliant.org, Retrieved 31 March 2018
- Sitaram, A (2001) [1994], "Uncertainty principle, mathematical", in Hazewinkel, Michiel, Encyclopedia of Mathematics, Springer Science+Business Media B.V. / Kluwer Academic Publishers, ISBN 978-1-55608-010-4
- Double-slit-experiment: techtarget.com, Retrieved 11 April 2018
- Busch, P.; Lahti, P.; Werner, R. F. (2013). "Proof of Heisenberg's Error-Disturbance Relation". Physical Review Letters. 111 (16): 160405. arXiv:1306.1565 . Bibcode:2013PhRvL.111p0405B. doi:10.1103/PhysRevLett.111.160405. PMID 24182239

6

Theories in Quantum Mechanics

This chapter closely examines the key theories in quantum mechanics such as quantum field theory, perturbation theory and scattering theory for an extensive understanding of the subject. All the principles and results of these theories have been carefully analyzed in this chapter.

Quantum Field Theory

Quantum Field Theory (QFT) is the mathematical and conceptual framework for contemporary elementary particle physics. In a rather informal sense QFT is the extension of quantum mechanics (QM), dealing with particles, over to fields, i.e. systems with an infinite number of degrees of freedom.

Two examples of modern quantum field theories are quantum electrodynamics, describing the interaction of electrically charged particles and the electromagnetic force, and quantum chromodynamics, representing the interactions of quarks and the strong force. Designed to account for particle-physics phenomena such as high-energy collisions in which subatomic particles may be created or destroyed, quantum field theories have also found applications in other branches of physics.

The prototype of quantum field theories is quantum electrodynamics (QED), which provides a comprehensive mathematical framework for predicting and understanding the effects of electromagnetism on electrically charged matter at all energy levels. Electric and magnetic forces are regarded as arising from the emission and absorption of exchange particles called photons. These can be represented as disturbances of electromagnetic fields, much as ripples on a lake are disturbances of the water. Under suitable conditions, photons may become entirely free of charged particles; they are then detectable as light and as other forms of electromagnetic radiation. Similarly, particles such as electrons are themselves regarded as disturbances of their own quantized fields. Numerical predictions based on QED agree with experimental data to within one part in 10 million in some cases.

There is a widespread conviction among physicists that other forces in nature—the weak force responsible for radioactive beta decay; the strong force, which binds together the constituents of atomic nuclei; and perhaps also the gravitational force—can be described by theories similar to QED. These theories are known collectively as gauge theories. Each of the forces is mediated by its own set of exchange particles, and differences between the forces are reflected in the properties of these particles. For example, electromagnetic and gravitational forces operate over long distances, and their exchange particles—the well-studied photon and the as-yet-undetected graviton, respectively—have no mass.

In contrast, the strong and weak forces operate only over distances shorter than the size of an atomic nucleus. Quantum chromodynamics (QCD), the modern quantum field theory describing the effects of the strong force among quarks, predicts the existence of exchange particles called gluons, which are also massless as with QED but whose interactions occur in a way that essentially confines quarks to bound particles such as the proton and the neutron. The weak force is carried by massive exchange particles—the W and Z particles—and is thus limited to an extremely short range, approximately 1 percent of the diameter of a typical atomic nucleus.

The current theoretical understanding of the fundamental interactions of matter is based on quantum field theories of these forces. Research continues, however, to develop a single unified field theory that encompasses all the forces. In such a unified theory, all the forces would have a common origin and would be related by mathematical symmetries. The simplest result would be that all the forces would have identical properties and that a mechanism called spontaneous symmetry breaking would account for the observed differences. A unified theory of electromagnetic and weak forces, the electroweak theory, has been developed and has received considerable experimental support. It is likely that this theory can be extended to include the strong force. There also exist theories that include the gravitational force, but these are more speculative.

The Basic Structure of the Conventional Formulation

The Lagrangian Formulation of QFT

The crucial step towards *quantum* field theory is in some respects analogous to the corresponding quantization in quantum mechanics, namely by imposing commutation relations, which leads to operator valued quantum fields. The starting point is the classical Lagrangian formulation of mechanics, which is a so-called analytical formulation as opposed to the standard version of Newtonian mechanics. A generalized notion of momentum (the conjugate or canonical momentum) is defined by setting $p = \partial L / \partial \dot{q}$, where L is the Lagrange function $L = T - V$ (T is the kinetic energy and V the potential) and $\dot{q} \equiv dq/dt$. This definition can be motivated by looking at the special case of a Lagrange function with a potential V which depends only on the position so that (using Cartesian coordinates) $\partial\dot{x} = (\partial/\partial\dot{x})(m\dot{x}^2/2) = m\dot{x} = p_x$. Under these conditions the generalized momentum coincides with the usual mechanical momentum. In classical Lagrangian *field* theory one associates with the given field φ a second field, namely the conjugate field

$$\pi = \partial\mathcal{L}/\partial\dot{\phi}$$

where $\mathcal{L}$ is a Lagrangian density. The field ϕ and its conjugate field π are the direct analogues of the canonical coordinate q and the generalized (canonical or conjugate) momentum p in classical mechanics of point particles.

In both cases, QM and QFT, requiring that the canonical variables satisfy certain commutation relations implies that the basic quantities become operator valued. From a physical point of view this shift implies a restriction of possible measurement values for physical quantities some (but not all) of which can have their values only in discrete steps now. In QFT the canonical commutation relations for a field ϕ and the corresponding conjugate field π are:

$$[\phi(\mathrm{x},t),\ \pi(\mathrm{y},t)] = i\delta^3(\mathrm{x}-\mathrm{y})$$
$$[\phi(\mathrm{x},t),\ \phi(\mathrm{y},t)] = [\pi(\mathrm{x},t),\ \pi(\mathrm{y},t)] = 0$$

which are equal-time commutation relations, i.e., the commutators always refer to fields at the same time. It is not obvious that the equal-time commutation relations are Lorentz invariant but one can formulate a manifestly covariant form of the canonical commutation relations. If the field to be quantized is not a bosonic field, like the Klein-Gordon field or the electromagnetic field, but a fermionic field, like the Dirac field for electrons one has to use anticommutation relations.

While there are close analogies between quantization in QM and in QFT there are also important differences. Whereas the commutation relations in QM refer to a quantum object with three degrees of freedom, so that one has a set of 15 equations, the commutation relations in QFT do in fact comprise an infinite number of equations, namely for each of the infinitely many space-time 4-tuples (x, t) there is a new set of commutation relations. This infinite number of degrees of freedom embodies the field character of QFT.

It is important to realize that the operator valued field $\varphi(x, t)$ in QFT is *not* analogous to the wave function $\psi(x, t)$ in QM, i.e., the quantum mechanical state in its position representation. While the wave function in QM is acted upon by observables/operators, in QFT it is the (operator valued) field itself which acts on the space of states. In a certain sense the single particle wave functions have been transformed, via their reinterpretation as operator valued quantum fields, into observables. This step is sometimes called 'second quantization' because the single particle wave equations in relativistic QM already came about by a quantization procedure, e.g., in the case of the Klein-Gordon equation by replacing position and momentum by the corresponding quantum mechanical operators. Afterwards the solutions to these single particle wave equations, which are states in relativistic QM, are considered as classical fields, which can be subjected to the canonical quantization procedure of QFT. The term 'second quantization' has often been criticized partly because it blurs the important fact that the single particle wave function φ in relativistic QM and the operator valued quantum field φ are fundamentally different kinds of entities despite their connection in the context of discovery.

In conclusion, it must be emphasized that both in QM and QFT states *and* observables are equally important. However, to some extent their roles are switched. While states in QM can have a concrete spatio-temporal meaning in terms of probabilities for position measurements, in QFT states are abstract entities and it is the quantum field operators that seem to allow for a spatio-temporal interpretation.

Interaction

Up to this point, the aim was to develop a free field theory. Doing so does not only neglect interaction with other particles (fields), it is even unrealistic for one free particle because it interacts with the field that it generates itself. For the description of interactions—such as scattering in particle colliders—we need certain extensions and modifications of the formalism. The immediate contact between scattering experiments and QFT is given by the scattering or S-matrix which contains all the relevant predictive information about, e.g., scattering cross sections. In order to calculate the S-matrix the interaction Hamiltonian is needed. The Hamiltonian can in turn be derived from the Lagrangian density by means of a Legendre transformation.

In order to discuss interactions one introduces a new representation, the interaction picture, which is an alternative to the Schrödinger and the Heisenberg picture. For the interaction picture one splits up the Hamiltonian, which is the generator of time-translations, into two parts $H = H_0 + H_{int}$, where H_0 describes the free system, i.e., without interaction, and gets absorbed in the definition of the fields and H_{int} is the interaction part of the Hamiltonian, or short the 'interaction Hamiltonian'. Using the interaction picture is advantageous because the equations of motion as well as, under certain conditions, the commutation relations are the same for interacting fields as for free fields. Therefore, various results that were established for free fields can still be used in the case of interacting fields. The central instrument for the description of interaction is again the S-matrix, which expresses the connection between in and out states by specifying the transition amplitudes. In QED, for instance, a state $|in\rangle$ describes one particular configuration of electrons, positrons and photons, i.e., it describes how many of these particles there are and which momenta, spins and polarizations they have before the interaction. The S-matrix supplies the probability that this state goes over to a particular $|out\rangle$ state, e.g., that a particular counter responds after the interaction. Such probabilities can be checked in experiments.

The canonical formalism of QFT is only applicable in the case of free fields since the inclusion of interaction leads to infinities. For this reason perturbation theory makes up a large part of most publications on QFT. The importance of perturbative methods is understandable realizing that they establish the immediate contact between theory and experiment. Although the techniques of perturbation theory have become ever more sophisticated it is somewhat disturbing that perturbative methods could not be avoided even in principle. One reason for this unease is that perturbation theory is felt to be rather a matter of (highly sophisticated) craftsmanship than of understanding nature.

Gauge Invariance

Some theories are distinguished by being gauge invariant, which means that gauge transformations of certain terms do not change any observable quantities. Requiring gauge invariance provides an elegant and systematic way of introducing terms for interacting fields. Moreover, gauge invariance plays an important role in selecting theories. The prime example of an intrinsically gauge invariant theory is electrodynamics. In the potential formulation of Maxwell's equations one introduces the vector potential A and the scalar potential ϕ, which are linked to the magnetic field $\mathrm{B}(\mathrm{x}, t)$ and the electric field $\mathrm{E}(\mathrm{x}, t)$ by

$$\mathrm{B} = \nabla \times \mathrm{A}$$
$$E = -(\partial \mathrm{A} / \partial t) - \nabla\phi$$

or covariantly

$$F^{\mu\nu} = \partial^{\mu} A^{\nu} - \partial^{\nu} A^{\mu} t$$

where $F^{\mu\nu}$ is the electromagnetic field tensor and $A^{\mu} = (\phi, \mathrm{A})$ the 4-vector potential. The important point in the present context is that given the identification $\mathrm{B} = \nabla \times \mathrm{A}$ or $F^{\mu\nu} = \partial^{\mu} A^{\nu} - \partial^{\nu} A^{\mu}$, there remains a certain flexibility or freedom in the choice of A and ϕ, or A^{μ}. In order to see that, consider the so-called gauge transformations

$$A \rightarrow \quad \mathrm{A} - \nabla\psi$$
$$\varphi \rightarrow \quad \varphi + \partial\chi / \partial$$

or covariantly

$$A^{\mu} \to A^{\mu} + \partial^{\mu}\chi$$

where χ is a scalar function (of space and time or of space-time) which can be chosen arbitrarily. Inserting the transformed potential(s) into equation(s) $B = \nabla \times A$ or $F^{\mu\nu} = \partial^{\mu} A^{\nu} - \partial^{\nu} A^{\mu}$, one can see that the electric field E and the magnetic field B, or covariantly the electromagnetic field tensor $F^{\mu\nu}$ are not affected by a gauge transformation of the potential(s). Since only the electric field E and the magnetic field B, and quantities constructed from them, are observable, whereas the vector potential itself is not, nothing physical seems to be changed by a gauge transformation because it leaves E and B unaltered. Note that gauge invariance is a kind of symmetry that does not come about by space-time transformations.

In order to link the notion of gauge invariance to the Lagrangian formulation of QFT one needs a more general form of gauge transformations which applies to the field operator φ and which is supplied by

$$\varphi \to e^{-i\Lambda}\varphi$$

$$\varphi^{*} \to e^{i\Lambda}\varphi^{*}$$

where Λ is an arbitrary real constant. Equations $\varphi \to e^{-i\Lambda}\varphi$ describe a *global gauge* transformation where as a local gauge transformation

$$\varphi(x) \to e^{-i\alpha(x)}\varphi(x)$$

varies with x.

It turned out that requiring invariance under local gauge transformations supplies a systematic way for finding the equations describing fundamental interactions. For instance, starting with the Lagrangian for a free electron, the requirement of local gauge invariance can only be fulfilled by introducing additional terms, namely those for the electromagnetic field. Gauge invariance can be captured by certain symmetry groups: $U(1)$ for electromagnetic, $SU(2)\otimes U(1)$ for electroweak and $SU(3)$ for strong interaction. This is an important basis for unification programs, as is the analogy to general relativity where a local gauge symmetry is associated with the gravitational field. Moreover, it turned out that only gauge invariant quantum field theories are renormalizable. All this can be taken to show that a mathematically rich theory, with surplus structures, can be very valuable in the construction of theories.

Quantization (Physics)

In physics, quantization is a procedure for constructing a quantum field theory starting from a classical field theory. This is a generalization of the procedure for building quantum mechanics from classical mechanics.

Field quantization can also be considered, as in the quantization of the electromagnetic field, where one refers to photons as field quanta (for instance as light quanta).

This procedure is basic to theories of particle physics, nuclear physics, condensed matter physics, and quantum optics.

Quantization converts classical fields into operators acting on quantum states of the field theory. The lowest energy state is called the vacuum state and may be very complicated.

The reason for quantizing a theory is to deduce properties of materials, objects or particles through the computation of quantum amplitudes. Such computations have to deal with certain subtleties called renormalization, which, if neglected, can often lead to nonsense results, such as the appearance of infinities in various amplitudes.

The first method to be developed for quantization of field theories was canonical quantization.

Canonical quantization of a field theory is analogous to the construction of quantum mechanics from classical mechanics.

Second Quantization

Second Quantization is a body of physical analysis in quantum field theory that focuses on the occupation numbers of states rather than the physical states of particular particles. It talks of operations which create and which annihilate a particle in a field. When this is applied to the quantum analysis of a harmonic oscillator it is seen that what the term particle refers to second quantization is entirely different from what particle refers to in other contexts. Here is what the probability density function looks like for a particular level of energy.

The number of peaks of the probability density function is 2n+1, where n is the principal quantum number. What an increase in n adds to the solution is roughly depicted below.

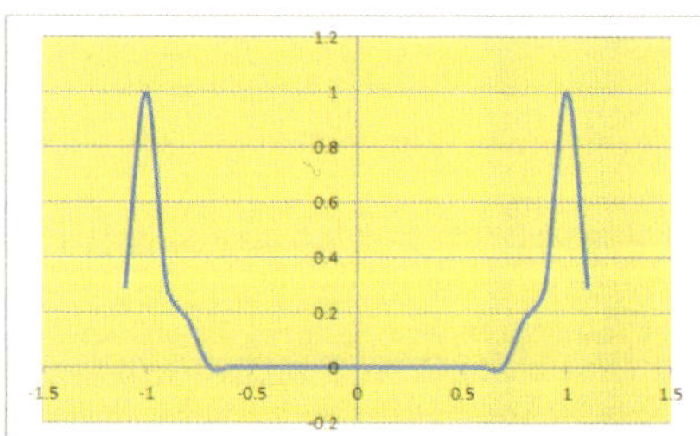

It is n which the creation and annihilation operators change.

Mathematical Background

Let V be a vector space of complex elements. A function that maps V onto V is called a transformation. A linear transformation is called an *operator*. Operators can be added and substracted. The concatenation of two operators is defined and will be referred to as the multiplication of operators. This multiplication of operators is *associative* $(AB)C = A(BC)$ but not communicative $AB \neq BA$.

The complex conjugate of an element u of V is denoted as u*. If an operator A maps v to u then the operator that maps v to u* is denoted as A^* and is called the adjoint of A.

If V is finite dimensional then an operator is just a matrix. When V consists of functions it is infinite dimensional.

Let A and B be operators on V. The commutator [A, B] is defined as

$$[A, B] = AB - BA$$

This leads to

$$AB = BA + [A, B]$$
$$BA = AB + [B, A] - AB - [A, B]$$

Note that the commutator is anti-symmetric, $AB = -BA$.

Some General Relationships for Operators on a Function Space

Let A, B and C be operators. Then for the commutator $[AB, C]$

$$[AB, C] = A[B, C] - [A, C]B$$

Proof:

$$[AB, C] = (AB)C - C(AB)$$

and from the associativity
of operator applications

$$[AB, C] = A(BC) - (CA)B$$

and from subtracting and
adding A(CB) on the right

$$[AB, C] = A(BC) - A(CB) + A(CB) - (CA)B$$

and hence

$$[AB, C] = A[B, C] + (AC)B - (CA)B$$

and finally

$$[AB, C] = A[B, C] + [A, C]B$$

Now let A be any operator and A^* its adjoint (conjugate operation). Applying the above indentity to $[A^* A, A]$ gives

$$[A^* A, A] = A^*[A, A] + [A^*, A]A$$

but $[A, A] = 0$, so

$$[A^* A, A] = [A^*, A]A$$

and

since commutation
is antisymmetric

$$[A^* A, A] = -[A, A^*]A$$

Likewise

$$[A^* A, A^*] = A^*[A, A^*] + [A^*, A^*]A = A^*[A, A^*]$$

Canonical Quantization

Canonical quantization is defined as

$$[A, A^*] = I$$

where I is the identity operator. If A satisfies the canonical quantization condition then the above two relationships reduce to

$$[A^* A, A] = -A$$

and

$$[A^* A, A^*] = A^*$$

These can also be expressed as

$$(A^* A)A = A(A^* A) - A = A(A^* A - I)$$

and

$$(A^* A)A^* = A^*(A^* A) + A^* = A^*(A^* A + I)$$

An Eigenvector for an Operator and its Eigenvalue

For and operator A, any element v of V such that

$$Av = \alpha v$$

is called an eigenvector of A and α is its eigenvalue.

Quantum Field Theory in Curved Spacetime

Quantum Field Theory (QFT) in curved spacetime is a hybrid approximate theory in which quantum matter fields are assumed to propagate in a fixed classical background gravitational field. Its basic physical prediction is that strong gravitational fields can polarize the vacuum and, when time dependent, lead to pair creation just as a strong and/or time-dependent electromagnetic field can polarize the vacuum and/or give rise to pair-creation of charged particles. One expects it to be a good approximation to full quantum gravity provided the typical frequencies of the gravitational background are very much less than the Planck frequency $(c^5/G\hbar)^{1/2} \sim 10^{43} s^{-1}$) and provided, with a suitable measure for energy, the energy of created particles is very much less than the energy of the background gravitational field or of its matter sources. Undoubtedly the most important prediction of the theory is the Hawking effect, according to which a, say spherically symmetric, classical black hole of mass M will emit thermal radiation at the Hawking temperature $T = (8\pi M)^{-1}$ (here and from now on, we use Planck units where G, c, $\hbar$ *and* k [Boltzmann's constant] are all taken to be 1.)

On the mathematical side, the need to formulate the laws and derive the general properties of QFT on non-flat space times forces one to state and prove results in local terms and, as a byproduct, thereby leads to an improved perspective on flat-space-time QFT too. It's also interesting to formulate QFT on idealized space times with particular global geometrical features. Thus, QFT on space times with bifurcate Killing horizons is intimately related to the Hawking effect; QFT on space times with closed time like curves is intimately related to the question whether the laws of physics permit the manufacture of a time-machine.

As is standard in General Relativity, a curved spacetime is modelled mathematically as a manifold $\mathcal{M}$ equipped with a pseudo-Riemannian metric g of signature $(-, + + +)$. We shall also assume, except where otherwise stated, our spacetime to be globally hyperbolic, i.e. that M admits a global time coordinate, by which we mean a global coordinate t such that each constant-t surface is a smooth Cauchy surface i.e. a smooth spacelike 3-surface cut exactly once by each inextendible causal curve. In view of this definition, globally hyperbolic spacetimes are clearly time orientable and we shall assume a choice of time-orientation has been made so we can talk about the "future" and "past" directions. Modern formulations of the subject take, as the fundamental mathematical structure modelling the quantum field, a $*$-algebra $\mathcal{A}$ (with identity I) together with a family of local sub $*$-algebras $\mathcal{A}(O)$ labelled by bounded open regions O of the spacetime $(\mathcal{M}, g)$ and satisfying the isotony or net condition that $O_1 \subset O_2$ implies $\mathcal{A}(O_1)$ is a subalgebra of $\mathcal{A}(O_2)$, as well as the condition that whenever O_1 and O_2 are spacelike separated, then $\mathcal{A}(O_1)$ and $\mathcal{A}(O_2)$ commute.

Standard concepts and techniques from algebraic quantum theory are then applicable: In particular, states are defined to be positive (this means $\omega(A^*A) \geq 0 \ \forall A \in A$) normalized (this means $\omega(I) = 1$) linear functionals on A. One distinguishes between pure states and mixed states, only the latter being writeable as non-trivial convex combinations of other states. To each state, ω, the GNS-construction associates a representation, $\rho\omega$, of A on a Hilbert space $\mathcal{H}_\omega$ together with a cyclic vector $\Omega \in H_\omega$ such that

$$w(A) = \langle \Omega | \rho_w (A) \Omega \rangle$$

(and the GNS triple $(\rho_\omega, \mathcal{H}, \Omega)$ is unique up to equivalence). There are often technical advantages in formulating things so that the $*$-algebra is a C^*-algebra. Then the GNS representation is as everywhere-defined bounded operators and is irreducible if and only if the state is pure. A useful concept, due to Haag, is the folium of a given state w which may be defined to be the set of all states w_σ which arise in the form $Tr(\sigma\rho_\omega(\cdot))$ where σ ranges over the density operators (trace-class operators with unit trace) on $\mathcal{H}w$.

Given a state, w, and an automorphism, α, which preserves the state $(i.e.\ \omega \circ \alpha = \omega)$ then there will be a unitary operator, U, on which implements α in the sense that $\rho_\omega(\alpha(A)) = U^{-1}\rho_\omega(A)U$ and U is chosen uniquely by the condition $U\,\Omega = \Omega$.

On a stationary spacetime, i.e. one which admits a one-parameter group of isometries whose integral curves are everywhere time like, the algebra will inherit a one-parameter group (i.e. satisfying $\alpha(t_1) \circ \alpha(t_2) = \alpha(t_1 + t_2)$) of time-translation automorphisms, $\alpha(t)$, and, given any stationary state (i.e. one which satisfies $\omega \circ \alpha(t) = \omega \forall t \in R$) these will be implemented by a one-parameter group of unitaries, $U(t)$, on its GNS Hilbert space satsifying $U(t)\Omega = \Omega$. If

$U(t)$ is strongly continuous so that it takes the form e^{-iHt} and if the Hamiltonian, H, is positive, then ω is said to be a ground state. Typically one expects ground states to exist and often be unique.

Another important class of stationary states for the algebra of a stationary spacetime is the class of KMS states, ω^{β}, at inverse temperature β; these have the physical interpretation of thermal equilibrium states. In the GNS representation of one of these, the automorphisms are also implemented by a strongly-continuous unitary group, e^{-iHt}, which preserves Ω but (in place of H positive) there is a complex conjugation, J, on $H\omega$ such that for all $A \in A$.

$$e^{-\beta H/2}\ \rho_{\omega}(A)\Omega = J\rho_{\omega}(A^{*})\Omega$$

An attractive feature of the subject is that its main qualitative features are already present for linear field theories and, unusually in comparison with other questions in QFT, these are susceptible of a straightforward explicit and rigorous mathematical formulation. In fact, as our principal example, a construction for the field algebra for the quantized real linear Klein Gordon equation

$$(\Box_g - m^2 - V)\phi = 0$$

of mass m on a globally hyperbolic spacetime $(\mathcal{M}, g)$. Here, $\Box_g$ denotes the Laplace Beltrami operator g $g^{ab}\nabla_a\partial_b\left((|\det(g)|)^{-1/2}\partial_a\left((|\det(g)|^{1/2} g^{ab}\partial_b)\right)\right)$. We include a scalar external background classical field, V in addition to the external gravitational field represented by g. In case m is zero, taking V to equal $R/6$ where R denotes the Riemann scalar, makes the equation conformally invariant.

The main new feature of quantum field theory in curved spacetime (present already for linear field theories) is that, in a general (neither flat, nor stationary) spacetime there will not be any single preferred state but rather a family of preferred states, members of which are best regarded as on an equal footing with one-another. It is this feature which makes the above algebraic framework particularly suitable, indeed essential, to a clear formulation of the subject. Conceptually it is this feature which takes the most getting used to. In particular, one must realize that, the interpretation of a state as having a particular "particle-content" is in general problematic because it can only be relative to a particular choice of "vacuum" state and, depending on the spacetime of interest, there may be one state or several states or, frequently, no states at all which deserve the name "vacuum" and even when there are states which deserve this name, they will often only be defined in some approximate or asymptotic or transient sense or only on some subregion of the spacetime.

Concomitantly, one does not expect global observables such as the "particle number" or the quantum Hamiltonian of flat-spacetime free field theory to generalize to a curved spacetime context, and for this reason local observables play a central role in the theory. The quantized stress-energy tensor is a particularly natural and important such local observable and the theory of this is central to the whole subject.

Perturbation Theory

Perturbation Theory is an extremely important method of seeing how a Quantum System will be affected by a small change in the potential and as such the Hamiltonian.

The basic idea of perturbation theory is very simple: we split the Hamiltonian into a piece we know how to solve (the "reference" or "unperturbed" Hamiltonian) and a piece we don't know how to solve (the "perturbation"). As long as the perturbation is small compared to the unperturbed Hamiltonian, perturbation theory tells us how to correct the solutions to the unperturbed problem to approximately account for the influence of the perturbation. For example, perturbation theory can be used to approximately solve an anharmonic oscillator problem with the Hamiltonian

$$\hat{H} = -\frac{\hbar^2}{2\mu}\frac{d^2}{dx^2} + \frac{1}{2}kx^2 + \frac{1}{6}\gamma x^3.$$

Here, since we know how to solve the harmonic oscillator problem, we make that part the unperturbed Hamiltonian (denoted $\hat{H}^{(0)}$), and the new, anharmonic term is the perturbation (denoted $\hat{H}^{(1)}$):

$$\hat{H}^{(0)} = -\frac{\hbar^2}{2\mu}\frac{d^2}{dx^2} + \frac{1}{2}kx^2,$$

$$\hat{H}^{(1)} = +\frac{1}{6}\gamma x^3.$$

Perturbation theory solves such a problem in two steps. First, obtain the eigenfunctions and eigenvalues of the unperturbed Hamiltonian, $\hat{H}^{(0)}$:

$$\hat{H}^{(0)}\Psi_n^{(0)} = E_n^{(0)}\Psi_n^{(0)}.$$

Second, correct these eigenvalues and/or eigenfunctions to account for the perturbation's influence. Perturbation theory gives these corrections as an infinite series of terms, which become smaller and smaller for well-behaved systems:

$$E_n = E_n^{(0)} + E_n^{(1)} + E_n^{(2)} + \cdots$$

$$\Psi_n = \Psi_n^{(0)}\Psi_n^{(1)} + \Psi_n^{(2)} + \cdots$$

Quite frequently, the corrections are only taken through first or second order (i.e., superscripts (1) or (2)). According to perturbation theory, the first-order correction to the energy is

$$E_n^{(1)} = \int \Psi_n^{(0)*}\hat{H}^{(1)}\Psi_n^{(0)},$$

and the second-order correction is

$$E_n^{(2)} = \int \Psi_n^{(0)*} \hat{H}^{(1)} \Psi_n^{(1)}.$$

One can see that the first-order correction to the wave function, $\Psi_n^{(1)}$, , seems to be needed to compute the second-order energy correction. However, it turns out that the correction $\Psi_n^{(1)}$ can be written in terms of the zeroth-order wave function as

$$\Psi_n^{(1)} = \sum_{i \neq n} \Psi_i^{(0)} \frac{\int \Psi_i^{(0)*} \hat{H}^{(1)} \Psi_n^{(0)}}{E_n^{(0)} - E_i^{(0)}}.$$

Substituting this in the expression for $E_n^{(2)}$, we obtain

$$E_n^{(2)} = \sum_{i \neq n} \frac{|\int \Psi_n^{(0)*} \hat{H}^{(1)} \Psi_i^{(0)}|^2}{E_n^{(0)} - E_i^{(0)}}.$$

Going back to the anharmonic oscillator example, the ground state wave function for the unperturbed problem is just

$$E_0^{(0)} = \frac{1}{2}\hbar w,$$

$$\Psi_0^{(0)}(x) = N_0 H_0(\alpha^{1/2} x) e^{-\alpha x^2/2}$$

$$= \left(\frac{\alpha}{\pi}\right)^{1/4} e^{-\alpha x^2/2}.$$

The first-order correction to the ground state energy would be

$$E_0^{(1)} = \left(\frac{\alpha}{\pi}\right)^{1/2} \int_{-\infty}^{\infty} \frac{1}{6} \gamma x^3 e^{-\alpha x^2} dx.$$

It turns out in this case that $E_0^{(1)} = 0$, since the integrand is odd. Does this mean that the anharmonic energy levels are the same as for the harmonic oscillator? No, because there are higher-order corrections such as $E_0^{(2)}$, which is not necessarily zero.

Time-independent Perturbation Theory

Time-independent perturbation theory is one of two categories of perturbation theory, the other being time-dependent perturbation. In time-independent perturbation theory the perturbation Hamiltonian is static (i.e., possesses no time dependence). Time-independent perturbation theory was presented by Erwin Schrödinger in a 1926 paper, shortly after he produced his theories in wave mechanics. In this paper Schrödinger referred to earlier work of Lord Rayleigh, who investigated harmonic vibrations of a string perturbed by small inhomogeneities. This is why this perturbation theory is often referred to as Rayleigh–Schrödinger perturbation theory.

First Order Corrections

We begin with an unperturbed Hamiltonian H_0, which is also assumed to have no time dependence. It has known energy levels and eigenstates, arising from the time-independent Schrödinger equation:

$$H_0 \left| n^{(0)} \right\rangle = E_n^{(0)} \left| n^{(0)} \right\rangle, \qquad n = 1, 2, 3, \cdots$$

For simplicity, we have assumed that the energies are discrete. The (0) superscripts denote that these quantities are associated with the unperturbed system. Note the use of bra–ket notation.

We now introduce a perturbation to the Hamiltonian. Let V be a Hamiltonian representing a weak physical disturbance, such as a potential energy produced by an external field. (Thus, V is formally a Hermitian operator.) Let λ be a dimensionless parameter that can take on values ranging continuously from 0 (no perturbation) to 1 (the full perturbation). The perturbed Hamiltonian is

$$H = H_0 + \lambda V$$

The energy levels and eigenstates of the perturbed Hamiltonian are again given by the Schrödinger equation:

$$(H_0 + \lambda V) | n \rangle = E_n | n \rangle.$$

Our goal is to express E_n and $|n\rangle$ in terms of the energy levels and eigenstates of the old Hamiltonian. If the perturbation is sufficiently weak, we can write them as a (Maclaurin) power series in λ:

$$E_n = E_n^{(0)} + \lambda E_n^{(1)} + \lambda^2 E_n^{(2)} + \cdots$$
$$| n \rangle = \left| n^{(0)} \right\rangle + \lambda \left| n^{(1)} \right\rangle + \lambda^2 \left| n^{(2)} \right\rangle + \cdots$$

where

$$E_n^{(k)} = \frac{1}{k!} \frac{d^k E_n}{d\lambda^k} \Big|_{\lambda=0}$$
$$\left| n^{(k)} \right\rangle = \frac{1}{k!} \frac{d^k |n\rangle}{d\lambda^k} \Big|_{\lambda=0}$$

When $k = 0$, these reduce to the unperturbed values, which are the first term in each series. Since the perturbation is weak, the energy levels and eigenstates should not deviate too much from their unperturbed values, and the terms should rapidly become smaller as we go to higher order.

Substituting the power series expansion into the Schrödinger equation, we obtain

$$(H_0 + \lambda V)\left(\left| n^{(0)} \right\rangle + \lambda \left| n^{(1)} \right\rangle + \cdots \right) = \left(E_n^{(0)} + \lambda E_n^{(1)} + \cdots \right)\left(\left| n^{(0)} \right\rangle + \lambda \left| n^{(1)} \right\rangle + \cdots \right)$$

Expanding this equation and comparing coefficients of each power of λ results in an infinite series of simultaneous equations. The zeroth-order equation is simply the Schrödinger equation for the unperturbed system. The first-order equation is

$$H_0|n^{(1)}\rangle + V|n^{(0)}\rangle = E_n^{(0)}|n^{(1)}\rangle + E_n^{(1)}|n^{(0)}\rangle$$

Operating through by $\langle n^{(0)}|$, the first term on the left-hand side cancels the first term on the right-hand side. Hermitian. This leads to the first-order energy shift:

$$E_n^{(1)} = \langle n^{(0)}|V|n^{(0)}\rangle$$

This is simply the expectation value of the perturbation Hamiltonian while the system is in the unperturbed state. This result can be interpreted in the following way: suppose the perturbation is applied, but we keep the system in the quantum state $|n^{(0)}\rangle$, which is a valid quantum state though no longer an energy eigenstate. The perturbation causes the average energy of this state to increase by $\langle n^{(0)}|V|n^{(0)}\rangle$. However, the true energy shift is slightly different, because the perturbed eigenstate is not exactly the same as $|n^{(0)}\rangle$. These further shifts are given by the second and higher order corrections to the energy.

Before we compute the corrections to the energy eigenstate, we need to address the issue of normalization. We may suppose

$$\langle n^{(0)}|n^{(0)}\rangle = 1,$$

but perturbation theory assumes we also have $\langle n|n\rangle = 1$. It follows that at first order in λ, we must have

$$\left(\langle n^{(0)}| + \lambda\langle n^{(1)}|\right)\left(|n^{(0)}\rangle + \lambda|n^{(1)}\rangle\right) = 1$$

$$\langle n^{(0)}|n^{(0)}\rangle + \lambda\langle n^{(0)}|n^{(1)}\rangle + \lambda\langle n^{(1)}|n^{(0)}\rangle + \lambda^2\cancel{\langle n^{(1)}|n^{(1)}\rangle} = 1$$

$$\langle n^{(0)}|n^{(1)}\rangle + \langle n^{(1)}|n^{(0)}\rangle = 0.$$

Since the overall phase is not determined in quantum mechanics, without loss of generality, we may assume $\langle n^{(0)}|n^{(1)}\rangle$ is purely real. Therefore,

$$\langle n^{(0)}|n^{(1)}\rangle = -\langle n^{(1)}|n^{(0)}\rangle,$$

and we deduce

$$\langle n^{(0)}|n^{(1)}\rangle = 0.$$

To obtain the first-order correction to the energy eigenstate, we insert our expression for the first-order energy correction back into the result shown above of equating the first-order coefficients of λ. We then make use of the resolution of the identity,

$$\begin{aligned} V|n^{(0)}\rangle &= \left(\sum_{k\neq n}|k^{(0)}\rangle\langle k^{(0)}|\right)V|n^{(0)}\rangle + \left(|n^{(0)}\rangle\langle n^{(0)}|\right)V|n^{(0)}\rangle \\ &= \sum_{k\neq n}|k^{(0)}\rangle\langle k^{(0)}|V|n^{(0)}\rangle + E_n^{(1)}|n^{(0)}\rangle, \end{aligned}$$

where the $|k^{(0)}\rangle$ are in the orthogonal complement of $|n^{(0)}\rangle$. The first-order equation may thus be expressed as

$$\left(E_n^{(0)}-H_0\right)|n^{(1)}\rangle=\sum_{k\neq n}|k^{(0)}\rangle\langle k^{(0)}|V|n^{(0)}\rangle$$

For the moment, suppose that the zeroth-order energy level is not degenerate, i.e. there is no eigenstate of H_0 in the orthogonal complement of $|n^{(0)}\rangle$ with the energy $E_n^{(0)}$. After renaming the summation dummy index above as k', we can pick any $k\neq n$,, and multiply through by $\langle k^{(0)}|$ giving

$$\left(E_n^{(0)}-E_k^{(0)}\right)\langle k^{(0)}|n^{(1)}\rangle=\langle k^{(0)}|V|n^{(0)}\rangle$$

We see that the above $\langle k^{(0)}|n^{(1)}\rangle$ also gives us the component of the first-order correction along $|k^{(0)}\rangle$.

Thus in total we get,

$$|n^{(1)}\rangle=\sum_{k\neq n}\frac{\langle k^{(0)}|V|n^{(0)}\rangle}{E_n^{(0)}-E_k^{(0)}}|k^{(0)}\rangle$$

The first-order change in the n-th energy eigenket has a contribution from each of the energy eigenstates $k\neq n$. Each term is proportional to the matrix element $\langle k^{(0)}|V|n^{(0)}\rangle$, which is a measure of how much the perturbation mixes eigenstate n with eigenstate k; it is also inversely proportional to the energy difference between eigenstates k and n, which means that the perturbation deforms the eigenstate to a greater extent if there are more eigenstates at nearby energies. The expression is singular if any of these states have the same energy as state n, which is why we assumed that there is no degeneracy.

Second-order and Higher Corrections

We can find the higher-order deviations by a similar procedure, though the calculations become quite tedious with our current formulation. Our normalization prescription gives that

$$2\langle n^{(0)}|n^{(1)}\rangle+\langle n^{(2)}|n^{(1)}\rangle=0.$$

Up to second order, the expressions for the energies and (normalized) eigenstates are:

$$E_n(\lambda)=E_n^{(0)}+\lambda\langle n^{(0)}|V|n^{(0)}\rangle+\lambda^2\sum_{k\neq n}\frac{\left|\langle k^{(0)}|V|n^{(0)}\rangle\right|^2}{E_n^{(0)}-E_k^{(0)}}+O(\lambda^3)$$

$$|n(\lambda)\rangle=|n^{(0)}\rangle+\lambda\sum_{k\neq n}|k^{(0)}\rangle\frac{\langle k^{(0)}|V|n^{(0)}\rangle}{E_n^{(0)}-E_k^{(0)}}+\lambda^2\sum_{k\neq n}\sum_{\ell\neq n}|k^{(0)}\rangle\frac{\langle k^{(0)}|V|\ell^{(0)}\rangle\langle \ell^{(0)}|V|n^{(0)}\rangle}{\left(E_n^{(0)}-E_k^{(0)}\right)\left(E_n^{(0)}-E_\ell^{(0)}\right)}$$
$$-\lambda^2\sum_{k\neq n}|k^{(0)}\rangle\frac{\langle n^{(0)}|V|n^{(0)}\rangle\langle k^{(0)}|V|n^{(0)}\rangle}{\left(E_n^{(0)}-E_k^{(0)}\right)^2}-\frac{1}{2}\lambda^2|n^{(0)}\rangle\sum_{k\neq n}\frac{\langle n^{(0)}|V|k^{(0)}\rangle\langle k^{(0)}|V|n^{(0)}\rangle}{\left(E_n^{(0)}-E_k^{(0)}\right)^2}+O(\lambda^3)$$

Extending the process further, the third-order energy correction can be shown to be

$$E_n^{(3)} = \sum_{k\neq n}\sum_{m\neq n} \frac{\langle n^{(0)}|V|m^{(0)}\rangle\langle m^{(0)}|V|k^{(0)}\rangle\langle k^{(0)}|V|n^{(0)}\rangle}{\left(E_n^{(0)}-E_m^{(0)}\right)\left(E_n^{(0)}-E_k^{(0)}\right)} - \langle n^{(0)}|V|n^{(0)}\rangle \sum_{m\neq n} \frac{|\langle n^{(0)}|V|m^{(0)}\rangle|^2}{\left(E_n^{(0)}-E_m^{(0)}\right)^2}.$$

Effects of Degeneracy

Suppose that two or more energy eigenstates are degenerate. The first-order energy shift is not well defined, since there is no unique way to choose a basis of eigenstates for the unperturbed system. The various eigenstates for a given energy will perturb with different energies, or may well possess no continuous family of perturbations at all.

This is manifested in the calculation of the perturbed eigenstate via the fact that the operator

$$E_n^{(0)} - H_0$$

does not have a well-defined inverse.

Let D denote the subspace spanned by these degenerate eigenstates. No matter how small the perturbation is, in the degenerate subspace D the energy differences between the eigenstates of H are non-zero, so complete mixing of at least some of these states is assured. Typically, the eigenvalues will split, and the eigenspaces will become simple (one-dimensional), or at least of smaller dimension than D.

The successful perturbations will not be "small" relative to a poorly chosen basis of D. Instead, we consider the perturbation "small" if the new eigenstate is close to the subspace D. The new Hamiltonian must be diagonalized in D, or a slight variation of D, so to speak. These perturbed eigenstates in D are now the basis for the perturbation expansion,

$$|n\rangle = \sum_{k\in D} \alpha_{nk}|k^{(0)}\rangle + \lambda|n^{(1)}\rangle.$$

For the first-order perturbation, we need solve the perturbed Hamiltonian restricted to the degenerate subspace D,

$$V|k^{(0)}\rangle = \epsilon_k |k^{(0)}\rangle + \text{small} \qquad \forall |k^{(0)}\rangle \in D,$$

simultaneously for all the degenerate eigenstates, where ϵ_k are first-order corrections to the degenerate energy levels, and "small" is a vector of $O(\lambda)$ orthogonal to D. This amounts to diagonalizing the matrix

$$\langle k^{(0)}|V|l^{(0)}\rangle = V_{kl} \qquad \forall |k^{(0)}\rangle, |l^{(0)}\rangle \in D.$$

This procedure is approximate, since we neglected states outside the D subspace ("small"). The splitting of degenerate energies ϵ_k is generally observed. Although the splitting may be small $O(\lambda)$, compared to the range of energies found in the system, it is crucial in understanding certain details, such as spectral lines in Electron Spin Resonance experiments.

Higher-order corrections due to other eigenstates outside D can be found in the same way as for the non-degenerate case,

$$\left(E_n^{(0)} - H_0\right)|n^{(1)}\rangle = \sum_{k \notin D} \left(\langle k^{(0)} | V | n^{(0)}\rangle\right) | k^{(0)}\rangle.$$

The operator on the left-hand side is not singular when applied to eigenstates outside D, so we can write

$$|n^{(1)}\rangle = \sum_{k \notin D} \frac{\langle k^{(0)} | V | n^{(0)}\rangle}{E_n^{(0)} - E_k^{(0)}} | k^{(0)}\rangle,$$

but the effect on the degenerate states is of $O(\lambda)$.

Near-degenerate states should also be treated similarly, when the original Hamiltonian splits aren't larger than the perturbation in the near-degenerate subspace. An application is found in the nearly free electron model, where near-degeneracy, treated properly, gives rise to an energy gap even for small perturbations. Other eigenstates will only shift the absolute energy of all near-degenerate states simultaneously.

Generalization to Multi-parameter Case

The generalization of the time-independent perturbation theory to the case where there are multiple small parameters $x^\mu = (x^1, x^2, \cdots)$ in place of λ can be formulated more systematically using the language of differential geometry, which basically defines the derivatives of the quantum states and calculates the perturbative corrections by taking derivatives iteratively at the unperturbed point.

Hamiltonian and Force Operator

From the differential geometric point of view, a parameterized Hamiltonian is considered as a function defined on the parameter manifold that maps each particular set of parameters $(x^1, x^2, \quad)$ to an Hermitian operator $H\left(x^\mu\right)$ that acts on the Hilbert space. The parameters here can be external field, interaction strength, or driving parameters in the quantum phase transition. Let $E_n\left(x^\mu\right)$ and $|n(x^\mu)\rangle$ be the n-th eigenenergy and eigenstate of $H\left(x^\mu\right)$ respectively. In the language of differential geometry, the states $|n(x \ )\rangle$ form a vector bundle over the parameter manifold, on which derivatives of these states can be defined. The perturbation theory is to answer the following question: given $E_n(x_0^\mu)$ and $|n(x_0^\mu)\rangle$ at an unperturbed reference point x_0^μ, how to estimate the $E_n\left(x^\mu\right)$ and $|n(x^\mu)\rangle$ at x^μ close to that reference point.

Without loss of generality, the coordinate system can be shifted, such that the reference point $x_0^\mu = 0$ is set to be the origin. The following linearly parameterized Hamiltonian is frequently used

$$H(x^\mu) = H(0) + x^\mu F_\mu.$$

If the parameters x^μ are considered as generalized coordinates, then F_μ should be identified as the generalized force operators related to those coordinates. Different indices μ label the different forces along different directions in the parameter manifold. For example, if x^μ denotes the external magnetic field in the μ-direction, then F_μ should be the magnetization in the same direction.

Perturbation Theory as Power Series Expansion

The validity of the perturbation theory lies on the adiabatic assumption, which assumes the eigenenergies and eigenstates of the Hamiltonian are smooth functions of parameters such that their values in the vicinity region can be calculated in power series (like Taylor expansion) of the parameters:

$$E_n(x^\mu) = E_n + x^\mu \partial_\mu E_n + \frac{1}{2!} x^\mu x^\nu \partial_\mu \partial_\nu E_n + \cdots$$

$$\left|n(x^\mu)\right\rangle = |n\rangle + x^\mu |\partial_\mu n\rangle + \frac{1}{2!} x^\mu x^\nu |\partial_\mu \partial_\nu n\rangle + \cdots$$

Here ∂_μ denotes the derivative with respect to x^μ. When applying to the state $|\partial_\mu n\rangle$, it should be understood as the covariant derivative if the vector bundle is equipped with non-vanishing connection. All the terms on the right-hand-side of the series are evaluated at $x^\mu = 0$, e.g. $E_n \equiv E_n(0)$ and $|n\rangle \equiv |n(0)\rangle$. This convention will be adopted throughout this subsection, that all functions without the parameter dependence explicitly stated are assumed to be evaluated at the origin. The power series may converge slowly or even not converge when the energy levels are close to each other. The adiabatic assumption breaks down when there is energy level degeneracy, and hence the perturbation theory is not applicable in that case.

Hellmann–Feynman Theorems

The above power series expansion can be readily evaluated if there is a systematic approach to calculate the derivates to any order. Using the chain rule, the derivatives can be broken down to the single derivative on either the energy or the state. The Hellmann–Feynman theorems are used to calculate these single derivatives. The first Hellmann–Feynman theorem gives the derivative of the energy,

$$\partial_\mu E_n = \langle n|\partial_\mu H|n\rangle$$

The second Hellmann–Feynman theorem gives the derivative of the state (resolved by the complete basis with $m \neq n$),

$$\langle m|\partial_\mu n\rangle = \frac{\langle m|\partial_\mu H|n\rangle}{E_n - E_m}, \qquad \langle \partial_\mu m|n\rangle = \frac{\langle m|\partial_\mu H|n\rangle}{E_m - E_n}.$$

For the linearly parameterized Hamiltonian, $\partial_\mu H$ simply stands for the generalized force operator F_μ.

The theorems can be simply derived by applying the differential operator ∂_μ to both sides of the Schrödinger equation $H|n\rangle = E_n|n\rangle$, which reads

$$\partial_\mu H|n\rangle + H|\partial_\mu n\rangle = \partial_\mu E_n|n\rangle + E_n|\partial_\mu n\rangle.$$

Then overlap with the state $\langle m|$ from left and make use of the Schrödinger equation $\langle m|H = \langle m|E_m$ again,

$$\langle m|\partial_\mu H|n\rangle + E_m\langle m|\partial_\mu n\rangle = \partial_\mu E_n\langle m|n\rangle + E_n\langle m|\partial_\mu n\rangle.$$

Given that the eigenstates of the Hamiltonian always form an orthonormal basis $\langle m|n\rangle = \delta_{mn}$, the cases of $m = n$ and $m \neq n$ can be discussed separately. The first case will lead to the first theorem and the second case to the second theorem, which can be shown immediately by rearranging the terms. With the differential rules given by the Hellmann–Feynman theorems, the perturbative correction to the energies and states can be calculated systematically.

Correction of Energy and State

To the second order, the energy correction reads

$$E_n(x^\mu) = \langle n|H|n\rangle + \langle n|\partial_\mu H|n\rangle x^\mu + \Re\sum_{m\neq n}\frac{\langle n|\partial_\nu H|m\rangle\,\langle m|\partial_\mu H|n\rangle}{E_n - E_m}x^\mu x^\nu + \cdots,$$

where $\Re$ denotes the real part function. The first order derivative $\partial_\mu E_n$ is given by the first Hellmann–Feynman theorem directly. To obtain the second order derivative $\partial_\mu\partial_\nu E_n$, simply applying the differential operator ∂_μ to the result of the first order derivative $\langle n|\partial_\nu H|n\rangle$, which reads

$$\partial_\mu\partial_\nu E_n = \langle\partial_\mu n|\partial_\nu H|n\rangle + \langle n|\partial_\mu\partial_\nu H|n\rangle + \langle n|\partial_\nu H|\partial_\mu n\rangle.$$

Note that for linearly parameterized Hamiltonian, there is no second derivative $\partial_\mu\partial_\nu H = 0$ on the operator level. Resolve the derivative of state by inserting the complete set of basis,

$$\partial_\mu\partial_\nu E_n = \sum_m\left(\langle\partial_\mu n|m\rangle\,\langle m|\partial_\nu H|n\rangle + \langle n|\partial_\nu H|m\rangle\,\langle m|\partial_\mu n\rangle\right)$$

then all parts can be calculated using the Hellmann–Feynman theorems. In terms of Lie derivatives, $\langle\partial_\mu n|n\rangle = \langle n|\partial_\mu n\rangle = 0$ according to the definition of the connection for the vector bundle. Therefore, the case $m = n$ can be excluded from the summation, which avoids the singularity of the energy denominator. The same procedure can be carried on for higher order derivatives, from which higher order corrections are obtained.

The same computational scheme is applicable for the correction of states. The result to the second order is as follows

$$\begin{aligned}|n(x^\mu)\rangle =& |n\rangle + \sum_{m\neq n}\frac{\langle m|\partial_\mu H|n\rangle}{E_n - E_m}|m\rangle x^\mu \\ &+\left(\sum_{m\neq n}\sum_{l\neq n}\frac{\langle m|\partial_\mu H|l\rangle\,\langle l|\partial_\nu H|n\rangle}{(E_n - E_m)(E_n - E_l)}|m\rangle - \sum_{m\neq n}\frac{\langle m|\partial_\mu H|n\rangle\,\langle n|\partial_\nu H|n\rangle}{(E_n - E_m)^2}|m\rangle - \frac{1}{2}\sum_{m\neq n}\frac{\langle n|\partial_\mu H|m\rangle\,\langle m|\partial_\nu H|n\rangle}{(E_n - E_m)^2}|m\rangle\right)x^\mu x^\nu + \cdots.\end{aligned}$$

Both energy derivatives and state derivatives will be involved in deduction. Whenever a state derivative is encountered, resolve it by inserting the complete set of basis, then the Hellmann-Feynman theorem is applicable. Because differentiation can be calculated systematically, the series expansion approach to the perturbative corrections can be coded on computers with symbolic processing software like Mathematica.

Effective Hamiltonian

Let $H(0)$ be the Hamiltonian completely restricted either in the low-energy subspace $\mathcal{H}_L$ or in the high-energy subspace $\mathcal{H}_H$, such that there is no matrix element in $H(0)$ connecting the low- and the high-energy subspaces, i.e. $\langle m|H(0)|l\rangle = 0$ if $m \in \mathcal{H}_L, l \in \mathcal{H}_H$. Let $F_\mu = \partial_\mu H$ be the coupling terms connecting the subspaces. Then when the high energy degrees of freedoms are integrated out, the effective Hamiltonian in the low energy subspace reads

$$H^{\text{eff}}_{mn}\left(x^\mu\right) = \langle m|H|n\rangle + \delta_{nm}\langle m|\partial_\mu H|n\rangle x^\mu + \frac{1}{2!}\sum_{l \in \mathcal{H}_H}\left(\frac{\langle m|\partial_\mu H|l\rangle\langle l|\partial_\nu H|n\rangle}{E_m - E_l} + \frac{\langle m|\partial_\nu H|l\rangle\langle l|\partial_\mu H|n\rangle}{E_n - E_l}\right)x^\mu x^\nu + \cdots.$$

Here $m, n \in \mathcal{H}_L$ are restricted in the low energy subspace. The above result can be derived by power series expansion of $\langle m|H(x^\mu)|n\rangle$.

In a formal way it is possible to define an effective Hamiltonian that gives exactly the low-lying energy states and wavefunctions. In practice, some kind of approximation (perturbation theory) is generally required.

Time-dependent Perturbation Theory

Method of Variation of Constants

Time-dependent perturbation theory, developed by Paul Dirac, studies the effect of a time-dependent perturbation $V(t)$ applied to a time-independent Hamiltonian H_0.

Since the perturbed Hamiltonian is time-dependent, so are its energy levels and eigenstates. Thus, the goals of time-dependent perturbation theory are slightly different from time-independent perturbation theory. One is interested in the following quantities:

- The time-dependent expectation value of some observable A, for a given initial state.
- The time-dependent amplitudes of those quantum states that are energy eigenkets (eigenvectors) in the unperturbed system.

The first quantity is important because it gives rise to the classical result of an A measurement performed on a macroscopic number of copies of the perturbed system. For example, we could take A to be the displacement in the x-direction of the electron in a hydrogen atom, in which case the expected value, when multiplied by an appropriate coefficient, gives the time-dependent dielectric polarization of a hydrogen gas. With an appropriate choice of perturbation (i.e. an oscillating electric potential), this allows one to calculate the AC permittivity of the gas.

The second quantity looks at the time-dependent probability of occupation for each eigenstate. This is particularly useful in laser physics, where one is interested in the populations of different atomic states in a gas when a time-dependent electric field is applied. These probabilities are also useful for calculating the "quantum broadening" of spectral lines and particle decay in particle physics and nuclear physics.

We will briefly examine the method behind Dirac's formulation of time-dependent perturbation theory. Choose an energy basis $|n\rangle$ for the unperturbed system. (We drop the (0) superscripts for the eigenstates, because it is not useful to speak of energy levels and eigenstates for the perturbed system.)

If the unperturbed system is in eigenstate (of the Hamiltonian) $|j\rangle$ at time t = 0, its state at subsequent times varies only by a phase (in the Schrödinger picture, where state vectors evolve in time and operators are constant),

$$|j(t)\rangle = e^{-iE_j t/\hbar}|j\rangle .$$

Now, introduce a time-dependent perturbing Hamiltonian $V(t)$. The Hamiltonian of the perturbed system is

$$H = H_0 + V(t) .$$

Let $|\psi(t)\rangle$ denote the quantum state of the perturbed system at time t. It obeys the time-dependent Schrödinger equation,

$$H|\psi(t)\rangle = i\hbar\frac{\partial}{\partial t}|\psi(t)\rangle .$$

The quantum state at each instant can be expressed as a linear combination of the complete eigenbasis of $|n\rangle$:

$$|\psi(t)\rangle = \sum_n c_n(t)e^{-iE_n t/\hbar}|n\rangle$$

where the $c_n(t)$ s are to be determined complex functions of t which we will refer to as amplitudes (strictly speaking, they are the amplitudes in the Dirac picture).

We have explicitly extracted the exponential phase factors $\exp(-iE_n t/\hbar)$ on the right hand side. This is only a matter of convention, and may be done without loss of generality. The reason we go to this trouble is that when the system starts in the state $|j\rangle$ and no perturbation is present, the amplitudes have the convenient property that, for all t, $c_j(t) = 1$ and $c_n(t) = 0$ if $n \neq j$.

The square of the absolute amplitude $c_n(t)$ is the probability that the system is in state n at time t, since

$$|c_n(t)|^2 = |\langle n|\psi(t)\rangle|^2 .$$

Plugging into the Schrödinger equation and using the fact that $\partial/\partial t$ acts by a product rule, one obtains

$$\sum_n \left(i\hbar\frac{\partial c_n}{\partial t} - c_n(t)V(t) \right) e^{-iE_n t/\hbar}|n\rangle = 0 .$$

By resolving the identity in front of V and multiplying through by the bra $\langle n|$ on the left, this can be reduced to a set of coupled differential equations for the amplitudes,

$$\frac{\partial c_n}{\partial t} = \frac{-i}{\hbar}\sum_k \langle n|V(t)|k\rangle\, c_k(t)e^{-i(E_k - E_n)t/\hbar} .$$

where we have used equation $|\psi(t)\rangle = \sum_n c_n(t)e^{-iE_n t/\hbar}|n\rangle$ to evaluate the sum on n in the second term, then used the fact that $\langle k|\Psi(t)\rangle = c_k(t)e^{-iE_k t/\hbar}$.

The matrix elements of V play a similar role as in time-independent perturbation theory, being proportional to the rate at which amplitudes are shifted between states. Note, however, that the direction of the shift is modified by the exponential phase factor. Over times much longer than the energy difference $E_k - E_n$, the phase winds around 0 several times. If the time-dependence of V is sufficiently slow, this may cause the state amplitudes to oscillate. (E.g., such oscillations are useful for managing radiative transitions in a laser.)

Up to this point, we have made no approximations, so this set of differential equations is exact. By supplying appropriate initial values $c_n(t)$, we could in principle find an exact (i.e., non-perturbative) solution. This is easily done when there are only two energy levels (n = 1, 2), and this solution is useful for modelling systems like the ammonia molecule.

However, exact solutions are difficult to find when there are many energy levels, and one instead looks for perturbative solutions. These may be obtained by expressing the equations in an integral form,

$$c_n(t) = c_n(0) + \frac{-i}{\hbar}\sum_k \int_0^t dt' \langle n|V(t')|k\rangle \, c_k(t')e^{-i(E_k - E_n)t'/\hbar} .$$

Repeatedly substituting this expression for c_n back into right hand side, yields an iterative solution,

$$c_n(t) = c_n^{(0)} + c_n^{(1)} + c_n^{(2)} + \cdots$$

where, for example, the first-order term is

$$c_n^{(1)}(t) = \frac{-i}{\hbar}\sum_k \int_0^t dt' \langle n|V(t')|k\rangle \, c_k(0)e^{-i(E_k - E_n)t'/\hbar} .$$

Several further results follow from this, such as Fermi's golden rule, which relates the rate of transitions between quantum states to the density of states at particular energies; or the Dyson series, obtained by applying the iterative method to the time evolution operator, which is one of the starting points for the method of Feynman diagrams.

Method of Dyson Series

Time-dependent perturbations can be reorganized through the technique of the Dyson series. The Schrödinger equation

$$H(t)|\psi(t)\rangle = i\hbar \frac{\partial |\psi(t)\rangle}{\partial t}$$

has the formal solution

$$|\psi(t)\rangle = T\exp\left[-\frac{i}{\hbar}\int_{t_0}^t dt' H(t')\right]|\psi(t_0)\rangle ,$$

where T is the time ordering operator,

$$TA(t_1)A(t_2) = \begin{cases} A(t_1)A(t_2) & t_1 > t_2 \\ A(t_2)A(t_1) & t_2 > t_1 \end{cases} .$$

Thus, the exponential represents the following Dyson series,

$$|\psi(t)\rangle = \left[1 - \frac{i}{\hbar}\int_{t_0}^{t} dt_1 H(t_1) - \frac{1}{\hbar^2}\int_{t_0}^{t} dt_1 \int_{t_0}^{t_1} dt_2 H(t_1)H(t_2) + \ldots \right] |\psi(t_0)\rangle .$$

Note that in the second term, the 1/2! factor exactly cancels the double contribution due to the time-ordering operator, etc.

Consider the following perturbation problem

$$[H_0 + \lambda V(t)]|\psi(t)\rangle = i\hbar \frac{\partial |\psi(t)\rangle}{\partial t} ,$$

assuming that the parameter λ is small and that the problem $H_0 |n\rangle = E_n |n\rangle$ has been solved.

Perform the following unitary transformation to the interaction picture (or Dirac picture),

$$|\psi(t)\rangle = e^{-\frac{i}{\hbar}H_0(t-t_0)} |\psi_I(t)\rangle .$$

Consequently, the Schrödinger equation simplifies to

$$\lambda e^{\frac{i}{\hbar}H_0(t-t_0)} V(t) e^{-\frac{i}{\hbar}H_0(t-t_0)} |\psi_I(t)\rangle = i\hbar \frac{\partial |\psi_I(t)\rangle}{\partial t} ,$$

so it is solved through the above Dyson series,

$$|\psi_I(t)\rangle = \left[1 - \frac{i\lambda}{\hbar}\int_{t_0}^{t} dt_1 e^{\frac{i}{\hbar}H_0(t_1-t_0)} V(t_1) e^{-\frac{i}{\hbar}H_0(t_1-t_0)} - \frac{\lambda^2}{\hbar^2}\int_{t_0}^{t} dt_1 \int_{t_0}^{t_1} dt_2 e^{\frac{i}{\hbar}H_0(t_1-t_0)} V(t_1) e^{-\frac{i}{\hbar}H_0(t_1-t_0)} e^{\frac{i}{\hbar}H_0(t_2-t_0)} V(t_2) e^{-\frac{i}{\hbar}H_0(t_2-t_0)} + \ldots \right] |\psi(t_0)\rangle ,$$

as a perturbation series with small λ.

Using the solution of the unperturbed problem $H_0 |n\rangle = E_n |n\rangle$ and $\sum_n |n\rangle\langle n| = 1$ (for the sake of simplicity assume a pure discrete spectrum), yields, to first order,

$$|\psi_I(t)\rangle = \left[1 - \frac{i\lambda}{\hbar}\sum_m \sum_n \int_{t_0}^{t} dt_1 \langle m|V(t_1)|n\rangle e^{-\frac{i}{\hbar}(E_n - E_m)(t_1 - t_0)} |m\rangle\langle n| + \ldots \right] |\psi(t_0)\rangle .$$

Thus, the system, initially in the unperturbed state $|\alpha\rangle = |\psi(t_0)\rangle$,, by dint of the perturbation can go into the state $|\beta\rangle$. The corresponding transition probability amplitude to first order is

$$A_{\alpha\beta} = -\frac{i\lambda}{\hbar}\int_{t_0}^{t} dt_1 \langle\beta|V(t_1)|\alpha\rangle e^{-\frac{i}{\hbar}(E_\alpha - E_\beta)(t_1 - t_0)} ,$$

while the corresponding transition probability to a continuum is furnished by Fermi's golden rule.

As an aside, note that time-independent perturbation theory is also organized inside this time-dependent perturbation theory Dyson series. To see this, write the unitary evolution operator, obtained from the above Dyson series, as

$$U(t)=1-\frac{i\lambda}{\hbar}\int_{t_0}^{t}dt_1 e^{\frac{i}{\hbar}H_0(t_1-t_0)}V(t_1)e^{-\frac{i}{\hbar}H_0(t_1-t_0)}-\frac{\lambda^2}{\hbar^2}\int_{t_0}^{t}dt_1\int_{t_0}^{t_1}dt_2 e^{\frac{i}{\hbar}H_0(t_1-t_0)}V(t_1)e^{-\frac{i}{\hbar}H_0(t_1-t_0)}e^{\frac{i}{\hbar}H_0(t_2-t_0)}V(t_2)e^{-\frac{i}{\hbar}H_0(t_2-t_0)}+\cdots$$

and take the perturbation V to be time-independent.

Using the identity resolution

$$\sum_n |n\rangle\langle n|=1$$

with $H_0|n\rangle = E_n|n\rangle$ for a pure discrete spectrum, write

$$U(t)=1-\left\{\frac{i\lambda}{\hbar}\int_{t_0}^{t}dt_1\sum_m\sum_n\langle m|V|n\rangle e^{-\frac{i}{\hbar}(E_n-E_m)(t_1-t_0)}|m\rangle\langle n|\right\}$$
$$-\left\{\frac{\lambda^2}{\hbar^2}\int_{t_0}^{t}dt_1\int_{t_0}^{t_1}dt_2\sum_m\sum_n\sum_q e^{-\frac{i}{\hbar}(E_n-E_m)(t_1-t_0)}\langle m|V|n\rangle\langle n|V|q\rangle e^{-\frac{i}{\hbar}(E_q-E_n)(t_2-t_0)}|m\rangle\langle q|\right\}+\cdots$$

It is evident that, at second order, one must sum on all the intermediate states. Assume $t_0=0$ and the asymptotic limit of larger times. This means that, at each contribution of the perturbation series, one has to add a multiplicative factor $e^{-\epsilon t}$ in the integrands for ε arbitrarily small. Thus the limit $t\to\infty$ gives back the final state of the system by eliminating all oscillating terms, but keeping the secular ones. The integrals are thus computable, and, separating the diagonal terms from the others yields

$$U(t)=1-\frac{i\lambda}{\hbar}\sum_n\langle n|V|n\rangle t-\frac{i\lambda^2}{\hbar}\sum_{m\neq n}\frac{\langle n|V|m\rangle\langle m|V|n\rangle}{E_n-E_m}t-\frac{1}{2}\frac{\lambda^2}{\hbar^2}\sum_{m,n}\langle n|V|m\rangle\langle m|V|n\rangle t^2+\ldots$$
$$+\lambda\sum_{m\neq n}\frac{\langle m|V|n\rangle}{E_n-E_m}|m\rangle\langle n|+\lambda^2\sum_{m\neq n}\sum_{q\neq n}\sum_n\frac{\langle m|V|n\rangle\langle n|V|q\rangle}{(E_n-E_m)(E_q-E_n)}|m\rangle\langle q|+\ldots$$

where the time secular series yields the eigenvalues of the perturbed problem specified above, recursively; whereas the remaining time-constant part yields the corrections to the stationary eigenfunctions also given above ($|n(\lambda)\rangle=U(0;\lambda)|n\rangle$).)

The unitary evolution operator is applicable to arbitrary eigenstates of the unperturbed problem and, in this case, yields a secular series that holds at small times.

Strong Perturbation Theory

In a similar way as for small perturbations, it is possible to develop a strong perturbation theory. Let us consider as usual the Schrödinger equation

$$H(t)|\psi(t)\rangle=i\hbar\frac{\partial|\psi(t)\rangle}{\partial t}$$

and we consider the question if a dual Dyson series exists that applies in the limit of a perturbation increasingly large. This question can be answered in an affirmative way and the series is the well-known adiabatic series. This approach is quite general and can be shown in the following way. Let us consider the perturbation problem

$$[H_0 + \lambda V(t)]|\psi(t)\rangle = i\hbar\frac{\partial|\psi(t)\rangle}{\partial t}$$

being $\lambda \to \infty$. Our aim is to find a solution in the form

$$|\psi\rangle = |\psi_0\rangle + \frac{1}{\lambda}|\psi_1\rangle + \frac{1}{\lambda^2}|\psi_2\rangle + \ldots$$

but a direct substitution into the above equation fails to produce useful results. This situation can be adjusted making a rescaling of the time variable as $\tau = \lambda t$ producing the following meaningful equations

$$V(t)|\psi_0\rangle = i\hbar\frac{\partial|\psi_0\rangle}{\partial\tau}$$

$$V(t)|\psi_1\rangle + H_0|\psi_0\rangle = i\hbar\frac{\partial|\psi_1\rangle}{\partial\tau}$$

$$\vdots$$

that can be solved once we know the solution of the leading order equation. But we know that in this case we can use the adiabatic approximation. When $V(t)$ does not depend on time one gets the Wigner-Kirkwood series that is often used in statistical mechanics. Indeed, in this case we introduce the unitary transformation

$$|\psi(t)\rangle = e^{-\frac{i}{\hbar}\lambda V(t-t_0)}|\psi_F(t)\rangle$$

that defines a free picture as we are trying to eliminate the interaction term. Now, in dual way with respect to the small perturbations, we have to solve the Schrödinger equation

$$e^{\frac{i}{\hbar}\lambda V(t-t_0)}H_0 e^{-\frac{i}{\hbar}\lambda V(t-t_0)}|\psi_F(t)\rangle = i\hbar\frac{\partial|\psi_F(t)\rangle}{\partial t}$$

and we see that the expansion parameter λ appears only into the exponential and so, the corresponding Dyson series, a dual Dyson series, is meaningful at large λ s and is

$$|\psi_F(t)\rangle = \left[1 - \frac{i}{\hbar}\int_{t_0}^{t} dt_1 e^{\frac{i}{\hbar}\lambda V(t_1-t_0)}H_0 e^{-\frac{i}{\hbar}\lambda V(t_1-t_0)} - \frac{1}{\hbar^2}\int_{t_0}^{t} dt_1\int_{t_0}^{t_1} dt_2 e^{\frac{i}{\hbar}\lambda V(t_1-t_0)}H_0 e^{-\frac{i}{\hbar}\lambda V(t_1-t_0)} e^{\frac{i}{\hbar}\lambda V(t_2-t_0)}H_0 e^{-\frac{i}{\hbar}\lambda V(t_2-t_0)} + \ldots\right]|\psi(t_0)\rangle.$$

After the rescaling in time $\tau = \lambda t$ we can see that this is indeed a series in $1/\lambda$ justifying in this way the name of dual Dyson series. The reason is that we have obtained this series simply interchanging H_0 and V and we can go from one to another applying this exchange. This is called duality principle in perturbation theory. The choice $H_0 = p^2/2m$ yields, as already said, a Wigner-Kirkwood series that is a gradient expansion. The Wigner-Kirkwood series is a semiclassical series with eigenvalues given exactly as for WKB approximation.

Examples

Example of First Order Perturbation Theory – Ground State Energy of the Quartic Oscillator

Let us consider the quantum harmonic oscillator with the quartic potential perturbation and the Hamiltonian

$$H = -\frac{\hbar^2}{2m}\frac{\partial^2}{\partial x^2} + \frac{m\omega^2 x^2}{2} + \lambda x^4$$

The ground state of the harmonic oscillator is

$$\psi_0 = \left(\frac{\alpha}{\pi}\right)^{\frac{1}{4}} e^{-\alpha x^2/2}$$

($\alpha = m\omega/\hbar$) and the energy of unperturbed ground state is

$$E_0^{(0)} = \tfrac{1}{2}\hbar\omega.$$

Using the first order correction formula we get

$$E_0^{(1)} = \lambda\left(\frac{\alpha}{\pi}\right)^{\frac{1}{2}} \int e^{-\alpha x^2/2} x^4 e^{-\alpha x^2/2} dx = \lambda\left(\frac{\alpha}{\pi}\right)^{\frac{1}{2}} \frac{\partial^2}{\partial \alpha^2}\int e^{-\alpha x^2} dx$$

or

$$E_0^{(1)} = \lambda\left(\frac{\alpha}{\pi}\right)^{\frac{1}{2}} \frac{\partial^2}{\partial \alpha^2}\left(\frac{\pi}{\alpha}\right)^{\frac{1}{2}} = \lambda\frac{3}{4}\frac{1}{\alpha^2} = \frac{3}{4}\frac{\hbar^2\lambda}{m^2\omega^2}$$

Example of First and second Order Perturbation Theory – Quantum Pendulum

Consider the quantum mathematical pendulum with the Hamiltonian

$$H = -\frac{\hbar^2}{2ma^2}\frac{\partial^2}{\partial \phi^2} - \lambda\cos\phi$$

with the potential energy $-\lambda\cos\phi$ taken as the perturbation i.e.

$$V = -\cos\phi$$

The unperturbed normalized quantum wave functions are those of the rigid rotor and are given by

$$\psi_n(\phi) = \frac{e^{in\phi}}{\sqrt{2\pi}}$$

and the energies

$$E_n^{(0)} = \frac{\hbar^2 n^2}{2ma^2}$$

The first order energy correction to the rotor due to the potential energy is

$$E_n^{(1)} = -\frac{1}{2\pi}\int e^{-in\phi}\cos\phi e^{in\phi} = -\frac{1}{2\pi}\int \cos\phi = 0$$

Using the formula for the second order correction one gets

$$E_n^{(2)} = \frac{ma^2}{2\pi^2\hbar^2}\sum_k \frac{\left|\int e^{-ik\phi}\cos\phi e^{in\phi}\right|^2}{n^2-k^2}$$

or

$$E_n^{(2)} = \frac{ma^2}{2\hbar^2}\sum_k \frac{\left|\left(\delta_{n,1-k}+\delta_{n,-1-k}\right)\right|^2}{n^2-k^2}$$

or

$$E_n^{(2)} = \frac{ma^2}{2\hbar^2}\left(\frac{1}{2n-1}+\frac{1}{-2n-1}\right) = \frac{ma^2}{\hbar^2}\frac{1}{4n^2-1}$$

Scattering Theory

Scattering theory is essentially timc indcpendent perturbation theory applied to the case of a continuous spectrum.

Typically, the scattering formalism is described in the following way: an incident particle in state $|\psi_0\rangle$ is scattered by the potential V, resulting in a scattered state $|\psi_s\rangle$. The incident state $|\psi_0\rangle$ is assumed to be an eigenstate of the 'background' hamiltonian H_o, with eigenvalue E. This is expressed mathematically as

$$(\mathrm{E} - \mathrm{H}_0)|\psi_0\rangle = 0.$$

Unless otherwise specified, the background Hamiltonian should be taken as that of a free-particle,

$$H_0 = \frac{P^2}{2M},$$

and the incident state taken as a plane wave

$$\langle \vec{r} \mid \psi_0\rangle = \psi_0(\vec{r}) = e^{i\vec{k}\cdot\vec{r}}.$$

As with one-dimensional scattering, we do not need to worry about the normalization of the incident state. Furthermore, the potential $V\left(\vec{R}\right)$ is assumed to be 'localized', so that

$$\lim_{r\to\infty} V\left(\vec{r}\right)=0.$$

The goal of scattering theory is then to solve the full energy-eigenstate problem

$$\left(E\ -H\ \ -V\ \right)\left|\psi\right\rangle=\ \ ,$$

where $E > 0$ (unless otherwise specified), and $\left|\psi\right\rangle$ is the eigenstate of the full Hamiltonian $H = H_0 + V$ with energy E. It should be clear that there is a different $\left|\psi_0\right\rangle$ and correspondingly, a different $\left|\psi\right\rangle$ for each energy E, even though our notation does not indicate this explicitly.

Time Independent Scattering Theory

Consider time-independent, energy conserving scattering in which the Hamiltonian of the system is written

$$H = H_0 + V(\mathrm{r}),$$

Where

$$H_0 = \frac{p^2}{2m} \equiv -\frac{\hbar^2}{2m}\nabla^2$$

is the Hamiltonian of a free particle of mass m, and $V(r)$ the scattering potential. This potential is assumed to only be non-zero in a fairly localized region close to the origin. Let

$$\psi_0(\mathrm{r}) = \sqrt{n}\mathrm{e}^{\mathrm{ik}\cdot\mathrm{r}}$$

represent an incident beam of particles, of number density n, and velocity $\mathrm{v} = \hbar\mathrm{k}/m$. Of course,

$$H_0\psi_0 = E\psi_0,$$

where $E = \hbar^2k^2/2m$ is the particle energy. Schrödinger's equation for the scattering problem is

$$(H_0 + V)\psi = E\psi,$$

subject to the boundary condition $\psi \to \psi_0$ as $V \to 0$.

The above equation can be rearranged to give

$$(\nabla^2 + k^2)\psi = \frac{2m}{\hbar^2}V\psi.$$

Now,

$$(\nabla^2 + k^2)u(\mathrm{r}) = \rho(\mathrm{r})$$

is known as the Helmholtz equation. The solution to this equation is well-known:

$$u(\mathrm{r}) = u_0(\mathrm{r}) - \int \frac{\mathrm{e}^{ik\pi|\mathrm{r}-\mathrm{r}'|}}{4\pi|r-r'|} \rho(\mathrm{r}')d^3\mathrm{r}'.$$

Here, $u_0(\mathrm{r})$ is any solution of $(\nabla^2 + k^2)u_0 = 0$. Hence, equation $(\nabla^2 + k^2)\psi = \frac{2m}{\hbar^2}V\psi$ can be inverted, subject to the boundary condition $\psi \to \psi_0$ as $V \to 0$, to give

$$\psi(\mathrm{r}) = \psi_0(\mathrm{r}) - \frac{2m}{\hbar^2}\int \frac{\mathrm{e}^{ik\pi|\mathrm{r}-\mathrm{r}'|}}{4\pi|r-r'|} V(\mathrm{r}')\psi(\mathrm{r}')d^3\mathrm{r}'.$$

Let us calculate the value of the wave function $\psi(\mathrm{r})$ well outside the scattering region. Now, if $r \gg r'$ then

$$|\mathrm{r}-\mathrm{r}'| \simeq r - \hat{\mathrm{r}}\cdot\mathrm{r}'$$

to first-order in r'/r, where $\hat{\mathrm{r}}/r$ is a unit vector which points from the scattering region to the observation point. It is helpful to define $\mathrm{k}' = k\hat{\mathrm{r}}.$. This is the wave vector for particles with the same energy as the incoming particles (*i.e.*, $k' = k$) which propagate from the scattering region to the observation point.

Equation $\psi(\mathrm{r}) = \psi_0(\mathrm{r}) - \frac{2m}{\hbar^2}\int \frac{\mathrm{e}^{ik\pi|\mathrm{r}-\mathrm{r}'|}}{4\pi|r-r'|} V(\mathrm{r}')\psi(\mathrm{r}')d^3\mathrm{r}'$ reduces to

$$\psi(\mathrm{r}) \simeq \sqrt{n}\left[\mathrm{e}^{i\mathrm{k}\cdot\mathrm{r}} + \frac{\mathrm{e}^{ikr}}{\mathrm{r}} \mathrm{f}(\mathrm{k}, \mathrm{k}')\right],$$

Where

$$f(\mathrm{k}, \mathrm{k}') = -\frac{m}{2\pi\sqrt{n}\hbar^2}\int \mathrm{e}^{-i\mathrm{k}'\cdot\mathrm{r}'} V(\mathrm{r}')\psi(\mathrm{r}')d^3\mathrm{r}'.$$

The first term on the right-hand side of equation $\psi(\mathrm{r}) \simeq \sqrt{n}\left[\mathrm{e}^{i\mathrm{k}\cdot\mathrm{r}} + \frac{\mathrm{e}^{ikr}}{\mathrm{r}} \mathrm{f}(\mathrm{k}, \mathrm{k}')\right]$ represents the incident particle beam, whereas the second term represents an outgoing spherical wave of scattered particles.

The differential scattering cross-section $d\sigma/d\Omega$ is defined as the number of particles per unit time scattered into an element of solid angle $d\Omega$, divided by the incident particle flux. From the probability flux (*i.e.*, the particle flux) associated with a wave function ψ is

$$\mathrm{j} = \frac{\hbar}{m}\mathrm{Im}(\psi^*\nabla\psi).$$

Thus, the particle flux associated with the incident wave function ψ_0 is

$$\mathrm{j} = n\mathrm{v}$$

where $\mathrm{v} = \hbar \mathrm{k} / m$ is the velocity of the incident particles. Likewise, the particle flux associated with the scattered wave function $\psi - \psi_0$ is

$$\mathrm{j}' = n \frac{|f(\mathrm{k}, \mathrm{k}')|^2}{r^2} \mathrm{v}',$$

where $\mathrm{v}' = \hbar \mathrm{k}' / m$ is the velocity of the scattered particles. Now,

$$\frac{d\sigma}{d\Omega} d\Omega = \frac{r^2 d\Omega |\mathrm{j}'|}{|\mathrm{j}|},$$

which yields

$$\frac{d\sigma}{d\Omega} = |f(\mathrm{k}, \mathrm{k}')|^2 .$$

Thus, $|f(\mathrm{k}, \mathrm{k}')|^2$ gives the differential cross-section for particles with incident velocity $\mathrm{v} = \hbar \mathrm{k} / m$ to be scattered such that their final velocities are directed into a range of solid angles $d\Omega$ about $\mathrm{v}' = \hbar \mathrm{k}' / m$. Note that the scattering conserves energy, so that $|\mathrm{v}'| = |\mathrm{v}|$ and $|\mathrm{k}'| = |\mathrm{k}|$.

Types of Scattering Experiments

The following is an incomplete list of the most important families of scattering experiments.

- Elastic scattering off immobile targets: Here, the target is effectively so heavy that its mass can be considered infinite. In such cases, scattering is dominantly elastic, $\Delta E = -\Delta_\epsilon = 0$. The scattering target can be modelled by some potential distribution, fixed in real space, and arbitrary momentum exchange Δk is possible. Scattering processes of this type form the basis of, e.g., crystallography. This is the simplest type of scattering, and we will explore its theory first.
- Elastic scattering off general targets: Still no energy exchange, $\Delta E = -\Delta_\epsilon = 0$. but we abandon the condition of infinite target mass. Rutherford scattering, i.e. the elastic scattering of α -particles off atoms and molecules is an important example.
- Inelastic scattering: When the energies carried by the incident particles become comparable to the energies of excitations of the target, energy exchange becomes an option. The information carried by inelastic scattering data is particularly rich. For example, the existence of quarks in hadronic particles was detected 'indirectly' through the unconventional energy absorption characteristics of the latter in 'deep inelastic' scattering experiments.
- Rearrangement scattering: Scattering processes involving an actual 'reaction' of the participating targets. For example, in the collision of two beams of nucleons N_1 and N_2 (figure right) may be elastic, $N_1 + N_2 \to N_1 + N_2$, or lead to particle production such as $N_1 + N_2 \to N_1 + N_2 + \pi$ (pion production), or $N_1 + N_2 \to N_1 + N_2 + K + \bar{K}_a$ on production.
- Resonance scattering: Scattering leads to the formation and subsequent decay of an unstable yet long lived particle. Resonance scattering plays an important role in, e.g., nuclear physics where the scattering of nucleons off heavy nuclei may generate long lived transient states.

Elastic and Inelastic Scattering

Elastic scattering occurs when two or more particles collide without any loss of energy. This means that while the directions of the particles may change, the total kinetic energy of the system, or movement energy, is always conserved. The term elastic scattering is typically used in particle physics, which is the study of microscopic particles, but an elastic collision can also take place between macroscopic objects. An inelastic collision occurs when energy is lost during the collision.

The term elastic scattering comes from scattering theory, which is a set of rules and equations which describe how particles and waves interact. In the macroscopic world, when two objects collide it is usually through a physical collision. In particle physics, however, the objects may collide through other forces, including electromagnetic collisions. An elastic collision can occur between any objects and in any type of collision.

Elastic scattering is very important in particle physics. When electrons collide with other particles, for example, the collision is elastic as no energy is lost. This is known as Rutherford scattering and is a phenomenon that led to the discovery of the structure of the atom.

In the macroscopic or physical world, it is exceptionally unlikely for a true elastic collision to occur between two large objects. This is due to surrounding forces as well as the vibrations that occur inside large objects. There are some situations, however, where a collision can be approximated as elastic. This is useful as it allows for the predicted speed and direction of two objects after a collision to be estimated using simpler methods.

A common example of elastic scattering in the physical world is the collision of two billiard balls. Although a small amount of energy will be lost in this collision due to friction, this is small enough to be negligible. When two billiard balls collide, the second ball gains almost exactly the amount of energy that the first ball loses, so the total kinetic energy of the system is conserved.

Inelastic scattering occurs when the total energy of the two or more colliding objects is not conserved. In collisions between real world objects, inelastic collisions are much more common. This is mainly due to the extra vibrations which will occur when a large object collides with another, as these vibrations take up energy that would have been used for movement. Momentum, however, is always conserved in both elastic and inelastic collisions.

References

- Bir, Gennadiĭ Levikovich; Pikus, Grigoriĭ Ezekielevich (1974). "Chapter 15: Perturbation theory for the degenerate case". Symmetry and Strain-induced Effects in Semiconductors. ISBN 978-0-470-07321-6
- Quantum-field-theory, science: britannica.com, Retrieved 16 April 2018
- Frasca, M. (1998). "Duality in Perturbation Theory and the Quantum Adiabatic Approximation". Physical Review A. 58 (5): 3439. arXiv:hep-th/9801069 . Bibcode:1998PhRvA..58.3439F. doi:10.1103/PhysRevA.58.3439.
- Sakurai, J.J., and Napolitano, J. (1964,2011). Modern quantum mechanics (2nd ed.), Addison Wesley ISBN 978-0-8053-8291-4
- What-is-elastic-scattering: wisegeek.com, Retrieved 25 June 2018
- Mostafazadeh, A. (1997). "Quantum adiabatic approximation and the geometric phase,". Physical Review A. 55 (3): 1653. arXiv:hep-th/9606053 . Bibcode:1997PhRvA..55.1653M. doi:10.1103/PhysRevA.55.1653

Quantum Mechanics Effects

Quantum effects are generally witnessed in small energy scales of the atom. This chapter discusses some of the significant quantum mechanical effects such as Casimir effect, Zeeman effect, Stark effect, quantum Hall effect, photoelectric effect, Aharonov-Bohm effect, etc. for a holistic understanding of quantum mechanics.

Casimir Effect

The Casimir effect is a small attractive force caused by quantum fluctuations of the electromagnetic field in vacuum. According to Quantum field theory, a vacuum contains particles (photons), the numbers of which are in a continuous state of fluctuation and can be thought of as popping in and out of existence. These particles can cause a force of attraction. Most generally, the quantum Casimir effect is thought about in regards to two closely parallel plates. As the plates are brought together, Casimir realized that between them, only those virtual photons whose wavelengths fit a whole number of times should be counted whilst calculating the vacuum energy. This leads to a decrease in energy density between the plates as they are moved closer which implies that a small force is drawing them together. Similarly you can say that due to the smaller space between the plates only smaller exotic particles can exist between them. From this difference in particles outside the plates and those between the plates, a small pressure change can be calculated which creates a force pushing the plates towards one another. This force is the Casimir effect.

Casimir force derivation comes about from a situation in which the differences in infinities cancel out which arises from very interesting mathematics. There are inconsistencies and puzzles that arise from the existence of this effect, especially when applying it to the theories of quantum gravity. The solutions to these inconsistencies are however expected to be found within the solution to a theory of quantum gravity.

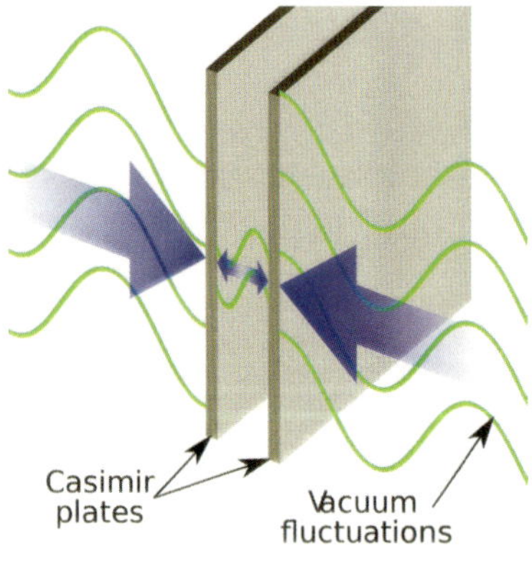

Figure: A simple diagram of two parallel plates and a representation of vacuum fluctuations.

In 2005, Jaffe made it clear that the zero-point fluctuations formulated in quantum field theory was not observable in any laboratory experiments though the vacuum value of the stress tensor (energy density of the vacuum) $\langle T_{\mu\upsilon} \rangle \equiv -\varepsilon g_{\mu\upsilon}$ even appears in the right hand side of Einstein's equation for gravity in general theory of relativity

$$\frac{1}{2} g_{\mu\upsilon} R - R_{\mu\upsilon} = 8\pi G\left(\tilde{T}_{\mu\upsilon} - \varepsilon g_{\mu\upsilon}\right).$$

Jaffe also demonstrates in his paper that you can calculate the Casimir force without reference to the zero-point energy, which suggests that the zero-point energy may simply just be a nice mathematical construct in this situation to arrive at a measurable result.

Let k_x, k_y, and k_z represent the wave numbers in the x, y and z directions respectively. If we allow two plates to be parallel in the x – y plane at a distance a apart, then we can define the cavity between the plates by

$$\begin{aligned} &\le x \le \sqrt{A} \\ &\le y \le \sqrt{A} \\ 0 &\le z \le a, \end{aligned}$$

where the plates are a square of area A. If we adopt a periodic boundary condition, then we can show

$$\begin{aligned} k_x &= \frac{2\pi n_x}{\sqrt{A}} \Rightarrow dn_x = \frac{\sqrt{A}}{2\pi} dk_x \\ k_y &= \frac{2\pi n_y}{\sqrt{A}} \Rightarrow dn_y = \frac{\sqrt{A}}{2\pi} dk_y \\ k_z &= \frac{n_z \pi}{a}, \end{aligned}$$

With $(n_x, n_y, n_z) \in Z$. The frequency of this wave is $\omega_n = \upsilon|\vec{k}| = \upsilon\sqrt{k_x^2 + k_y^2 + k_z^2}$. If we assume we are in a vacuum, then the speed of any electromagnetic wave is just c and thus $\omega_{n_z} = c\sqrt{k_x^2 + k_y^2 + k_z^2}$. The vacuum energy is the sum over all possible modes. The zero-point (ground state) energy associated with the n_z^{th} mode is given by $E_{n_z} = \frac{\hbar\omega_{n_z}}{2}$. The energy of all combined modes is then the sum over all n_z or $E = \sum_{n_z=1}^{\infty} \frac{\hbar\omega_{n_z}}{2}$. For simplicity we can allow $n \equiv n$. Taking the expectation value of the energy over the entire area of the plates can be done by integrating over all possible values of n, n and all possible expectation modes which yields

$$\begin{aligned} \langle E \rangle &= \frac{\hbar}{2} \int\int \sum_{n=1}^{\infty} w_n dn_x dn_y \\ &= \frac{A\hbar}{8\pi^2} \int\int \sum_{n=1}^{\infty} w_n dk_x dk_y. \end{aligned}$$

This expression is clearly infinite due to the diverging sum. If we use zeta-regulation, we can find a finite energy per unit area by defining a quantity $\langle E(s) \rangle$ which goes to equation $\frac{A\hbar}{8\pi^2}\iint \sum_{n=1}^{\infty} w_n dk_x dk_y$ when s = 0.

$$\frac{\langle E(s) \rangle}{A} = \frac{\hbar}{8\pi^2}\iint \sum_{n=1}^{\infty} w_n |w_n|^{-s} dk_x dk_y$$

$$= \frac{\hbar}{8\pi^2}\sum_{n=1}^{\infty}\iint w_n |w_n|^{-s} dk_x dk_y.$$

Simplifying the above expression (Using Mathematica to take the integral over dk_x and dk_y) gives us

$$\frac{\langle E(s) \rangle}{A} = \frac{\hbar c^{1-s}\pi^{2-s}}{2a^{3-s}(3-s)}\sum_{n=1}^{\infty}|n|^{3-s}.$$

This may then be analytically continued to s = 0 where it becomes finite when using equation,

$$\frac{\langle E \rangle}{A} = \lim_{s\to 0}\frac{\langle E(s) \rangle}{A} = \frac{\hbar c \pi^2}{6a^3}\zeta(-3).$$

Now, plugging in equation $\zeta(-3) = \sum_{n-1}^{\infty}\frac{1}{n^{-3}} = 1 + 2^3 + \cdots \to \frac{1}{120}$ in the above expression gives us

$$\frac{\langle\ \rangle}{A} \quad \frac{}{720a}$$

The Casimir force per unit area between two parallel plates within a vacuum is therefore given by $F = -\nabla\langle E \rangle$ which is

$$\frac{F_c}{A} = -\frac{d}{da}\frac{\langle E \rangle}{A} = \frac{-\hbar c \pi^2}{240a^4}.$$

As we can clearly see, this result would not have come about without the use of analytical continuation. In a sense, this is due to nature not containing apparent infinities. Rather, the continuation allowed us to arrive at a finite solution which is experimentally confirmed. The fact that our expression came out negative suggests that the force is an attractive force and due to the presence of h, we can see that the force is of a quantum origin. In the original derivation, Casimir computed non-convergent sums using Euler-Maclaurin summation with a regularizing function.

Zeeman Effect

Zeeman effect in physics and astronomy is the splitting of a spectral line into two or more components of slightly different frequency when the light source is placed in a magnetic field.

It was first observed in 1896 by the Dutch physicist Pieter Zeeman as a broadening of the yellow D-lines of sodium in a flame held between strong magnetic poles. Later the broadening was found to be a distinct splitting of spectral lines into as many as 15 components.

Zeeman's discovery earned him the 1902 Nobel Prize for Physics, which he shared with a former teacher, Hendrik Antoon Lorentz, another Dutch physicist. Lorentz, who had earlier developed a theory concerning the effect of magnetism on light, hypothesized that the oscillations of electrons insidc an atom produce light and that a magnetic field would affect the oscillations and thereby the frequency of the light emitted. This theory was confirmed by Zeeman's research and later modified by quantum mechanics, according to which spectral lines of light are emitted when electrons change from one discrete energy level to another. Each of the levels, characterized by an angular momentum (quantity related to mass and spin), is split in a magnetic field into substates of equal energy. These substates of energy are revealed by the resulting patterns of spectral line components.

The Zeeman effect has helped physicists determine the energy levels in atoms and identify them in terms of angular momenta. It also provides an effective means of studying atomic nuclei and such phenomena as electron paramagnetic resonance. In astronomy, the Zeeman effect is used in measuring the magnetic field of the Sun and of other stars.

Stark Effect

Stark effect is the splitting of spectral lines observed when the radiating atoms, ions, or molecules are subjected to a strong electric field.

The electric analogue of the Zeeman effect (*i.e.*, the magnetic splitting of spectral lines), it was discovered by a German physicist, Johannes Stark (1913). Earlier experimenters had failed to maintain a strong electric field in conventional spectroscopic light sources because of the high electrical conductivity of luminous gases or vapors. Stark observed the hydrogen spectrum emitted just behind the perforated cathode in a positive-ray tube. With a second charged electrode parallel and close to this cathode, he was able to produce a strong electric field in a space of a few millimeters. At electric field intensities of 100,000 volts per centimeter, Stark observed with a spectroscope that the characteristic spectral lines, called Balmer lines, of hydrogen were split into a number of symmetrically spaced components, some of which were linearly polarized (vibrating in one plane) with the electric vector parallel to the lines of force, the remainder being polarized perpendicular to the direction of the field except when viewed along the field. This transverse Stark effect resembles in some respects the transverse Zeeman effect, but, because of its complexity, the Stark effect has relatively less value in the analysis of complicated spectra or of atomic structure. Historically, the satisfactory explanation of the Stark effect (1916) was one of the great triumphs of early quantum mechanics.

Aharonov–Bohm effect

The Aharonov-Bohm effect is a configuration of the electromagnetic field which has vanishing electric/magnetic field strength (vanishing Faraday tensor F=0) and but is nevertheless non-trivial, in that the vector potential A is non-trivial. Since the vector potential affects the quantum mechanical phase on the wavefunction of electrons moving in an electromagnetic field, in such a configuration classical physics sees no effect, but the phase of quantum particles, which may be observed as an interference pattern on some screen, does.

There are two kinds of AB effect, electric and magnetic. In the magnetic AB effect, an electron beam having passed through the field-free regions on both sides of an infinitely long solenoid is physically influenced as a displacement of interference fringers. The fringe displacement was attributed to the vector potentials, which are directed in the opposite directions on both sides of the solenoid for the Coulomb gauge, and shift the phases of the electron beams passing through on both sides differently. The vector potentials, and also the phase S, are not uniquely determined for a given physical situation, but have an arbitrariness of gauge transformation. However, it does not mean that vector potentials and electron phases have no physical significance. It is true that the phase S can have arbitrary values. However, what can be observed experimentally is a relative phase shift ΔS between two beams starting from a point source, passing through electromagnetic potentials, and ending at another point. This ΔS, which can be observed from the interference pattern, is given by,

$$\Delta S = \frac{1}{\hbar}\oint (mv - eA)\, ds.$$

This time the integral is carried out along a closed path determined by connecting the two electron paths.

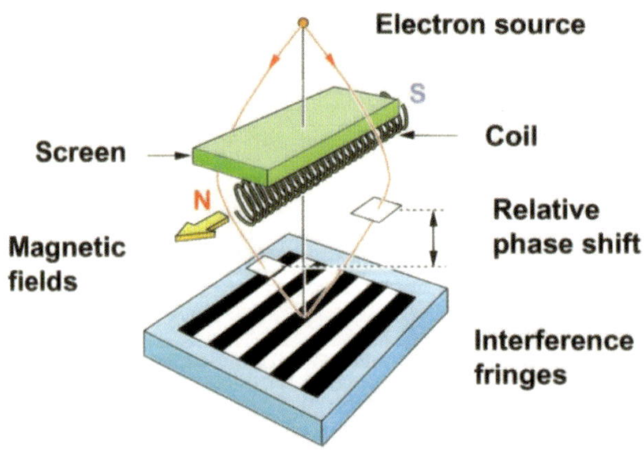

In a pure magnetic case,

$$\Delta S = \frac{e}{\hbar}\oint A\, ds = -\frac{e}{\hbar}\int B\, dS,$$

where the latter surface integral is performed over the surface bordered by the closed path. Since ΔS is thus proportional to the magnetic flux passing through the closed path, it is uniquely determined.

The AB effect became regarded as a fundamental in quantum mechanics and furthermore increased in significance in relation to the unified theories of gauge fields.

Quantum Hall Effect

Quantum Hall is the quantum-level version of the hall effect: a voltage produced as electrons flow through a magnetic field. In two-dimensional materials and at low temperatures the quantum mechanical nature of this phenomenon is evident as a series of steps in the Hall voltage with increasing applied magnetic field. It's useful to distinguish between two different quantum Hall effects which are associated to two related phenomena. These are called the integer and fractional quantum hall effects.

Integer Quantum Hall Effect

The first experiments exploring the quantum regime of the Hall effect were performed in 1980 by von Klitzing, using samples prepared by Dorda and Pepper. The resistivities look like this:

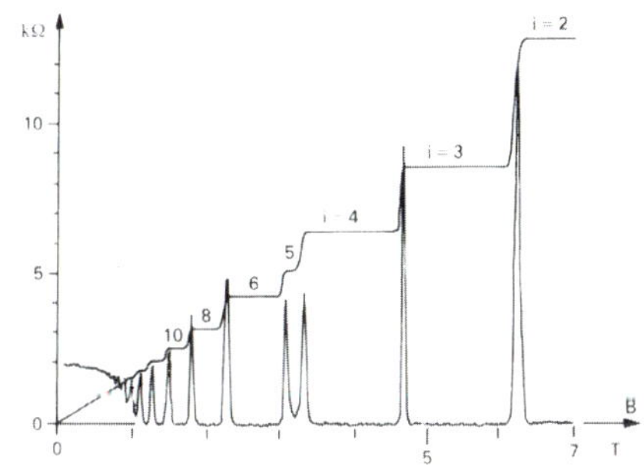

This is the integer quantum Hall effect.

Both the Hall resistivity ρ_{xy} and the longitudinal resistivity ρ_{xx} exhibit interesting behavior. Perhaps the most striking feature in the data is that Hall resistivity ρ_{xy} sits on a plateau for a range of magnetic field, before jumping suddenly to the next plateau. On these plateau, the resistivity takes the value

$$\rho_{xy} = \frac{2\pi\hbar}{e^2}\frac{1}{v} \quad v \in Z$$

The value of v is measured to be an integer to an extraordinary accuracy — something like one part in 10. The quantity $2\pi\hbar / e^2$ is called the quantum of resistivity (with −e, the electron charge). It is now used as the standard for measuring of resistivity. Moreover, the integer quantum Hall effect is now used as the basis for measuring the ratio of fundamental constants $2\pi\hbar / e^2$ sometimes referred to as the von Klitzing constant. This means that, by definition, the v = 1 state in $\rho_{xy} = \frac{2\pi\hbar}{e^2}\frac{1}{v} \quad v \in z$ is exactly integer.

The centre of each of these plateaux occurs when the magnetic field takes the value

$$B = \frac{2\pi\hbar n}{ve} = \frac{n}{v}\Phi_0$$

where n is the electron density and $\Phi_0 = 2\pi\hbar / e$ is known as the flux quantum. These are the values of the magnetic field at which the first v ∈ Z Landau levels are filled. In fact, as we will see, it is very easy to argue that the Hall resistivity should take value $\rho_{xy} = \frac{2\pi\hbar}{e^2}\frac{1}{\nu}$ $\nu \in Z$ when ν Landau levels are filled. The surprise is that the plateau exists, with the quantisation persisting over a range of magnetic fields.

There is a clue in the experimental data about the origin of the plateaux. Experimental systems are typically dirty, filled with impurities. The technical name for this is disorder. Usually one wants to remove this dirt to get at the underlying physics. Yet, in the quantum Hall effect, as you increase the amount of disorder (within reason) the plateaux become more prominent, not less. In fact, in the absence of disorder, the plateaux are expected to vanish completely.

The longitudinal resistivity ρ_{xx} also exhibits a surprise. When ρ_{xy} sits on a plateau, the longitudinal resistivity vanishes: $\rho_{xx} = 0$. It spikes only when ρ_{xy} jumps to the next plateau.

Usually we would think of a system with $\rho_{xx} = 0$ as a perfect conductor. But there's something a little counter-intuitive about vanishing resistivity in the presence of a magnetic field.

$$\sigma_{xx} = \frac{\rho_{xx}}{\rho_{xx}^2 + \rho_{xy}^2} \text{ and } \sigma_{xy} = \frac{-\rho_{xy}}{\rho_{xx}^2 + \rho_{xy}^2}$$

If $\rho_{xy} = 0$ then we get the familiar relation between conductivity and resistivity: $\sigma_{xx} = 1/\rho_{xx}$. But if $\rho_{xy} \neq 0$, then we have the more interesting relation above. In particular, we see

$$\rho_{xx} = 0 \Rightarrow \sigma_{xx} = 0 \quad \left(\text{if } \rho_{xy} \neq 0\right)$$

While we would usually call a system with $\rho_{xx} = 0$ a perfect conductor, we would usually call a system with $\sigma_{xx} = 0$ a perfect insulator.

This particular surprise has more to do with the words we use to describe the phenomena than the underlying physics. In particular, it has nothing to do with quantum mechanics: this behavior occurs in the Drude model in the limit $\tau \to \infty$ where there is no scattering. In this situation, the current is flowing perpendicular to the applied electric field, so E · J = 0. But recall that E · J has the interpretation as the work done in accelerating charges. The fact that this vanishes means that we have a steady current flowing without doing any work and, correspondingly, without any dissipation. The fact that $\sigma_{xx} = 0$ is telling us that no current is flowing in the longitudinal direction (like an insulator) while the fact that $\rho_{xx} = 0$ is telling us that there is no dissipation of energy (like in a perfect conductor).

Fractional Quantum Hall Effect

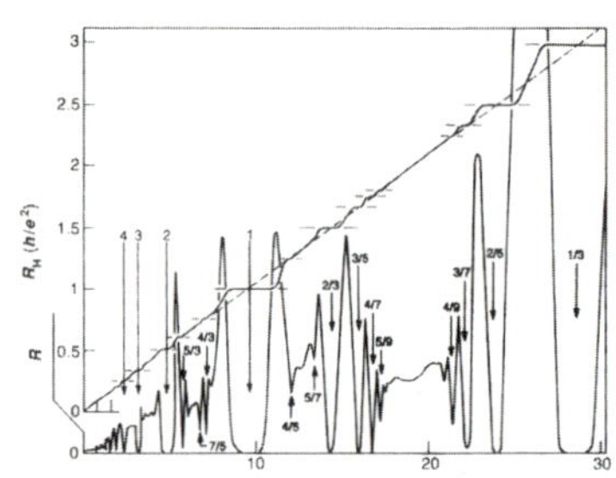

As the disorder is decreased, the integer Hall plateaux become less prominent. But other plateaux emerge at fractional values. The resistivities look like this:

The is the fractional quantum Hall effect. On the plateaux, the Hall resistivity again takes the simple form $\rho_{xy} = \frac{2\pi\hbar}{e^2}\frac{1}{\nu}$ $\nu \in z$, but now with ν a rational number

$\nu \in Q$

Not all fractions appear. The most prominent plateaux sit at ν = 1/3, 1/5 (not shown above) and 2/5 but there are many more. The vast majority of these have denominators which are odd. But there are exceptions: in particular a clear plateaux has been observed at ν = 5/2. As the disorder is decreased, more and more plateaux emerge. It seems plausible that in the limit of a perfectly clean sample, we would get an infinite number of plateaux which brings us back to the classical picture of a straight line for ρ_{xy}.

The integer quantum Hall effect can be understood using free electrons. In contrast, to explain the fractional quantum Hall effect we need to take interactions between electrons into account. This makes the problem much harder and much richer. The basics of the theory were first laid down by Laughlin, but the subject has since expanded in a myriad of different directions.

Photoelectric Effect

When light strikes materials, it can eject electrons from them. This is called the *photoelectric effect*, meaning that light produces electricity. One common use of the photoelectric effect is in light meters, such as those that adjust the automatic iris on various types of cameras. In a similar way, another use is in solar cells, as you probably have in your calculator or have seen on a roof top or a roadside sign. These make use of the photoelectric effect to convert light into electricity for running different devices.

Figure: The photoelectric effect can be observed by allowing light to fall on the metal plate in this evacuated tube. Electrons ejected by the light are collected on the collector wire and measured as a current. A retarding voltage between the collector wire and plate can then be adjusted so as to determine the energy of the ejected electrons.

This effect has been known for more than a century and can be studied using a device such as that shown in figure. This figure shows an evacuated tube with a metal plate and a collector wire that are connected by a variable voltage source, with the collector more negative than the plate. When light (or other EM radiation) strikes the plate in the evacuated tube, it may eject electrons. If the electrons have energy in electron volts (eV) greater than the potential difference between the plate and the wire in volts, some electrons will be collected on the wire. Since the electron energy in eV is eV, where q is the electron charge and V is the potential difference, the electron energy can be measured by adjusting the retarding voltage between the wire and the plate. The voltage that stops the electrons from reaching the wire equals the energy in eV. For example, if –3.00 V barely stops the electrons, their energy is 3.00 eV. The number of electrons ejected can be determined by measuring the current between the wire and plate. The more light, the more electrons; a little circuitry allows this device to be used as a light meter.

What is really important about the photoelectric effect is what Albert Einstein deduced from it. Einstein realized that there were several characteristics of the photoelectric effect that could be explained only if EM radiation is itself quantized: the apparently continuous stream of energy in an EM wave is actually composed of energy quanta called photons. In his explanation of the photoelectric effect, Einstein defined a quantized unit or quantum of EM energy, which we now call a *photon*, with an energy proportional to the frequency of EM radiation. In equation form, the photon energy is $E = hf$, where E is the energy of a photon of frequency f and h is Planck's constant. This revolutionary idea looks similar to Planck's quantization of energy states in blackbody oscillators, but it is quite different. It is the quantization of EM radiation itself. EM waves are composed of photons and are not continuous smooth waves as described in previous chapters on optics. Their energy is absorbed and emitted in lumps, not continuously. This is exactly consistent with Planck's quantization of energy levels in blackbody oscillators, since these oscillators increase and decrease their energy in steps of hf by absorbing and emitting photons having $E = hf$. We do not observe this with our eyes, because there are so many photons in common light sources that individual photons go unnoticed.

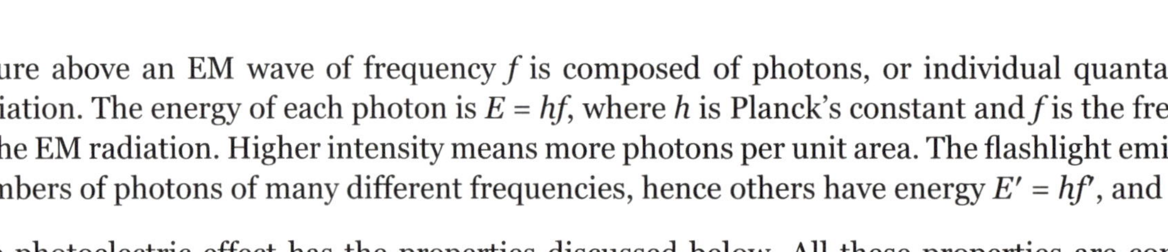

Figure above an EM wave of frequency f is composed of photons, or individual quanta of EM radiation. The energy of each photon is $E = hf$, where h is Planck's constant and f is the frequency of the EM radiation. Higher intensity means more photons per unit area. The flashlight emits large numbers of photons of many different frequencies, hence others have energy $E' = hf'$, and so on.

The photoelectric effect has the properties discussed below. All these properties are consistent with the idea that individual photons of EM radiation are absorbed by individual electrons in a material, with the electron gaining the photon's energy. Some of these properties are inconsistent with the idea that EM radiation is a simple wave. For simplicity, let us consider what happens with monochromatic EM radiation in which all photons have the same energy hf.

1. If we vary the frequency of the EM radiation falling on a material, we find the following: For a given material, there is a threshold frequency f_0 for the EM radiation below which no

electrons are ejected, regardless of intensity. Individual photons interact with individual electrons. Thus if the photon energy is too small to break an electron away, no electrons will be ejected. If EM radiation was a simple wave, sufficient energy could be obtained by increasing the intensity.

2. Once EM radiation falls on a material, electrons are ejected without delay. As soon as an individual photon of a sufficiently high frequency is absorbed by an individual electron, the electron is ejected. If the EM radiation were a simple wave, several minutes would be required for sufficient energy to be deposited to the metal surface to eject an electron.

3. The number of electrons ejected per unit time is proportional to the intensity of the EM radiation and to no other characteristic. High-intensity EM radiation consists of large numbers of photons per unit area, with all photons having the same characteristic energy *hf*.

4. If we vary the intensity of the EM radiation and measure the energy of ejected electrons, we find the following: The maximum kinetic energy of ejected electrons is independent of the intensity of the EM radiation. Since there are so many electrons in a material, it is extremely unlikely that two photons will interact with the same electron at the same time, thereby increasing the energy given it. Instead, increased intensity results in more electrons of the same energy being ejected. If EM radiation were a simple wave, a higher intensity could give more energy, and higher-energy electrons would be ejected.

5. The kinetic energy of an ejected electron equals the photon energy minus the binding energy of the electron in the specific material. An individual photon can give all of its energy to an electron. The photon's energy is partly used to break the electron away from the material. The remainder goes into the ejected electron's kinetic energy. In equation form, this is given by $KE_e = hf - BE$, where KE_e is the maximum kinetic energy of the ejected electron, hf is the photon's energy, and BE is the binding energy of the electron to the particular material. (BE is sometimes called the work function of the material.) This equation, due to Einstein in 1905, explains the properties of the photoelectric effect quantitatively. An individual photon of EM radiation (it does not come any other way) interacts with an individual electron, supplying enough energy, BE, to break it away, with the remainder going to kinetic energy. The binding energy is $BE = hf_0$, where f_0 is the threshold frequency for the particular material. Figure shows a graph of maximum KE_e versus the frequency of incident EM radiation falling on a particular material.

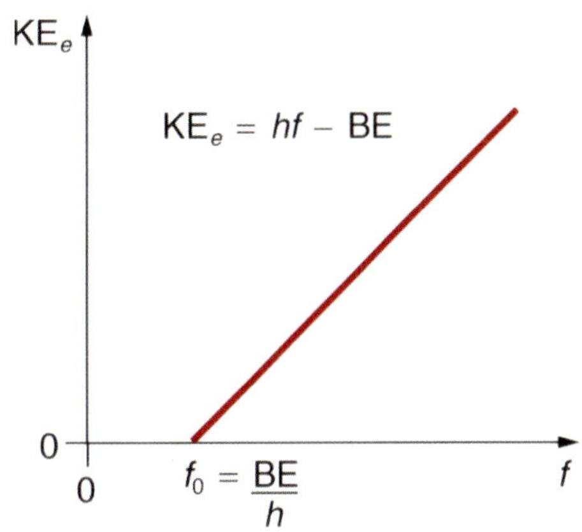

Figure: Photoelectric effect.

This graph of the kinetic energy of an ejected electron, KEe, versus the frequency of EM radiation impinging on a certain material. There is a threshold frequency below which no electrons are ejected, because the individual photon interacting with an individual electron has insufficient energy to break it away. Above the threshold energy, KEe increases linearly with f, consistent with KEe = hf – BE. The slope of this line is h—the data can be used to determine Planck's constant experimentally. Einstein gave the first successful explanation of such data by proposing the idea of photons—quanta of EM radiation.

Einstein's idea that EM radiation is quantized was crucial to the beginnings of quantum mechanics. It is a far more general concept than its explanation of the photoelectric effect might imply. All EM radiation can also be modeled in the form of photons, and the characteristics of EM radiation are entirely consistent with this fact. (Many aspects of EM radiation, such as the hazards of ultraviolet (UV) radiation, can be explained *only* by photon properties.) More famous for modern relativity, Einstein planted an important seed for quantum mechanics in 1905, the same year he published his first paper on special relativity. His explanation of the photoelectric effect was the basis for the Nobel Prize awarded to him in 1921. Although his other contributions to theoretical physics were also noted in that award, special and general relativity were not fully recognized in spite of having been partially verified by experiment by 1921. Although hero-worshipped, this great man never received Nobel recognition for his most famous work—relativity.

Example

Calculating Photon Energy and the Photoelectric Effect: A Violet Light

1. What is the energy in joules and electron volts of a photon of 420-nm violet light?

Solution

Photon energy is given by $E = hf$.

Since we are given the wavelength rather than the frequency, we solve the familiar relationship $c = f\lambda$ for the frequency, yielding

$$f = \frac{c}{\lambda}$$

Combining these two equations gives the useful relationship

$$E = \frac{hc}{\lambda}$$

Now substituting known values yields

$$E = \frac{(6.63 \times 10^{-34}\ \mathrm{J.s})(3.00 \times 10^{8}\ \mathrm{m/s})}{420 \times 10^{-9}\ m} = 4.74 \times 10^{-19}\ \mathrm{J\ a}$$

Converting to eV, the energy of the photon is

$$E = (4.47 \times 10^{-19}\ \text{J})\frac{1\ \text{eV}}{1.6 \times 10^{-19}\ \text{J}} = 2.96\ \text{eV}.$$

2. What is the maximum kinetic energy of electrons ejected from calcium by 420-nm violet light, given that the binding energy (or work function) of electrons for calcium metal is 2.71 eV?

Solution

Finding the kinetic energy of the ejected electron is now a simple application of the equation $\text{KE}e = hf - \text{BE}$. Substituting the photon energy and binding energy yields $\text{KE}e = hf - \text{BE} = 2.96\ \text{eV} - 2.71\ \text{eV} = 0.246\ \text{eV}$.

In conclusion, the energy of this 420-nm photon of violet light is a tiny fraction of a joule, and so it is no wonder that a single photon would be difficult for us to sense directly—humans are more attuned to energies on the order of joules. But looking at the energy in electron volts, we can see that this photon has enough energy to affect atoms and molecules. A DNA molecule can be broken with about 1 eV of energy, for example, and typical atomic and molecular energies are on the order of eV, so that the UV photon in this example could have biological effects. The ejected electron (called a *photoelectron*) has a rather low energy, and it would not travel far, except in a vacuum. The electron would be stopped by a retarding potential of but 0.26 eV. In fact, if the photon wavelength were longer and its energy less than 2.71 eV, then the formula would give a negative kinetic energy, an impossibility. This simply means that the 420-nm photons with their 2.96-eV energy are not much above the frequency threshold. You can show for yourself that the threshold wavelength is 459 nm (blue light). This means that if calcium metal is used in a light meter, the meter will be insensitive to wavelengths longer than those of blue light. Such a light meter would be completely insensitive to red light, for example.

8

Mathematical Formulation and Structure of Quantum Mechanics

The mathematical formulation of quantum mechanics allows a rigorous of quantum phenomena and quantum effects. Abstract mathematical structures like quantum operators and the infinite-dimensional Hilbert spaces are used. This chapter has been carefully designed to provide an extensive understanding of bra-ket notation, Hilbert space and quantum operators such as position operator, momentum operator, Hamiltonian, angular momentum operator, etc.

Mathematical Formulation of Quantum Mechanics

The mathematical formulation of quantum mechanics refers to the mathematical formalism that allows the construction of a comprehensive description of quantum mechanics. Such formalism is a mathematical construct of some abstract mathematical structures, like the Hilbert operator in an infinite dimensional Hilbert space. Observable and measurable quantities such as energy or momentum are considered as eigenvalues and are obtained as a result of the action of a linear operator operating on a wave function in Hilbert space.

At the core of the mathematical approach to quantum states and quantum observables is the concept of the Heisenberg uncertainty principle, which establishes a fundamental limit to the accuracy of determining the values of a pair of complementary variables of a particle, such as its position and momentum. This principle is reflected in terms of the non-commutativity of linear operators representing these quantum variables.

Before the emergence of quantum mechanics, the mathematics for the description of quantum phenomena involved formal mathematical analysis, complex differential geometry and differential equations. Quantum theory was treated within the confines of classical physics and involved the same set of mathematical structures, as used in classical domains. However, with time, the mathematical formalism advanced and a new description emerged.

The Postulates

One of the experimental observations leading to the development of quantum mechanics was the fact that certain physical quantities, which should take values from a continuum according to classical physics, take only discrete (or quantized) values in certain experiments. This is the case, for example, for the energy emitted by an electron in an atom. Free electrons, on the other hand, have a continuous range of energies. This should be a motivation for the first two postulates:

Postulate 1: To every quantum mechanical system there corresponds a separable Hilbert space $\mathcal{H}$.

Postulate 2: Every physical quantity (observable) is represented by a self adjoint operator A in $\mathcal{H}$. The physical quantity can only take values in the spectrum of A.

Another experimental observation shows that the outcome of experiments of quantum mechanical systems cannot be predicted deterministically, but only in a probabilistic sense. The same could be said within the framework of classical mechanics as a result of measurement errors and/or the complexity of a system. In quantum mechanics, however, the probabilistic nature is seen as a basic principle and, thereforc, has to bc part of every quantum mechanical theory.

The state of a system is therefore always understood as the description of an ensemble of systems with identical experimental conditions. To motivate the representation of states, let us consider a situation with a finite dimensional $\mathcal{H}$ and a self adjoint operator A with spectral decomposition $A = \sum_{k=1}^{n} \lambda_k P_k$. We shall compute the expected value of the observable corresponding to A. By postulate 2, the observable can only take one of the values $\lambda_1, \ldots, \lambda_n$. If the system is in a fixed state, these values are taken with probabilities $p_1, \ldots, p_n \geq 0$ $(p_1 + \cdots + p_n = 1)$. The expected value of A is, thus,

$$\langle A \rangle = \sum_{k=1}^{n} p_k \lambda_k = \sum_{k=1}^{n} p_k \langle Au_k, u_k \rangle = \mathrm{tr}(\varrho A)$$

$$\text{with } \varrho ij = \sum_{k=1}^{n} p_k u_k, \overline{i^u k, j}.$$

The operator ϱ % is nonnegative and satisfies tr(%) = 1.

Postulatc 3: Thc statc of a quantum mechanical system is represented by a nonnegative trace operator % with tr(%) = 1, the state operator (or density matrix). The expectation value of an observable A is given by

$$\langle A \rangle = \mathrm{tr}(\varrho A).$$

The definition of the probabilistic setting is completed by using the spectral family E(λ) corresponding to the observable A. We define the distribution function for A by

$$\mathrm{prob}(A \subset (-\infty, \lambda]) = \mathrm{tr}(\varrho E(\lambda)).$$

The spectral theorem shows that this is consistent with the formula for the expectation value in postulate 3.

The properties of the state operator immediately imply $\sigma(\varrho) \subset [0, 1]$. The projection onto a normalized vector $\psi \in H$, $\|\psi\| = 1$, defines a pure state

$$\varrho u = \langle u, \psi \rangle \psi.$$

For a pure state, the formula for the distribution function and for the expectation value is simple:

$$\mathrm{prob}(A \subset (-\infty, \lambda)) = \langle E(\lambda)\psi, \psi \rangle, \quad \langle A \rangle = \langle A\psi, \psi \rangle.$$

Non-pure states are called mixed states. The spectral theorem applied to state operators has the interpretation that mixed states can be written as convex combinations of pure states (at least in the case of a discrete $\sigma(\varrho)=\sigma_p(\varrho)$)

Note that for arbitrary real w, ψ and $e^{iw}\psi$ define the same state. For a pure state, the vector ψ is only unique up to a phase factor $e^{iw}\psi$.

Bra–ket Notation

Paul Dirac invented a powerful and intuitive mathematical notation to describe quantum states, known as bra-ket notation. For instance, one can refer to an $|\text{excited atom}>$ or to $|\uparrow\rangle$ for a spin-up particle, hiding the underlying complexity of the mathematical description, which is revealed when the state is projected onto a coordinate basis. For instance, the simple notation $|1s>$ describes the first hydrogen atom bound state, but becomes a complicated function in terms of Laguerre polynomials and spherical harmonics when projected onto the basis of position vectors |r>. The resulting expression $\Psi(\mathrm{r})=<\mathrm{r}|1s>$, which is known as the wave function, is a special representation of the quantum state, namely, its projection into position space. Other representations, such as projection into momentum space, are possible. The various representations are simply different expressions of a single physical quantum state.

Bra-ket Formalism

The bra-ket formalism was introduced by Paul Dirac in order to describe in a uniform manner vectors and linear operators both in the abstract Hilbert space of the state vectors, H, and in the extensions of H which are used to solve the eigenvector problem for observables with continuous spectrum.

From Riesz's representation theorem we know that a Hilbert space H and its dual space are isometrically conjugate isomorphic. Thus we can associate to any vector $\psi \in H$ the linear functional $\psi\dagger$, continuous in the norm topology, defined on H as follows:

$$\psi\dagger[\phi]=\langle\,\psi,\phi.\rangle\,.$$

The space dual to the dual of H can be identified with H as it is isomorphic to H. Thus we can associate to any vector $\phi \in \mathrm{H}$ the linear functional, indicated with the same symbol ϕ, continuous in the norm topology, defined on $\mathrm{H}\dagger$ as follows:

The vectors of H are called kets and are denoted by $|\,\rangle$; the vectors of $\mathrm{H}\dagger$ are called bras and are denoted by $\langle\,|$.

Therefore bras, respectively kets, are continuous linear functionals defined on the space of kets, respectively bras; the value that the bra $\langle\psi|$, respectively ket $|\phi\rangle$, has on the ket $|\phi\rangle$, respectively bra $\langle\psi|$ is denoted

$$\langle\psi|\phi\rangle \in \mathrm{C}.$$

This also denotes the scalar product of the ordered pair of kets $|\psi\rangle$, $|\phi\rangle$ and the scalar product of the ordered pair of bras $\langle\phi|$, $\langle\psi|$.

It follows that the scalar product between kets and that between bras is Hermitian, i.e.

$$\langle\psi|\phi\rangle = \langle\phi|\psi\rangle^*.$$

Moreover, all the operations that determine the structure of a complex vectorial space (multiplication by a complex number, sum of elements) can be defined for bras and kets.

So, the isomorphism due to Riesz's theorem defines a one-to-one mapping between bras and kets that we can indicate as †:

$$|\psi\rangle\dagger = \langle\psi|, \quad \langle\psi|\dagger = |\psi\rangle,$$

$$(\alpha 1|\psi 1\rangle + \alpha 2|\psi 2\rangle)\dagger = \alpha 1^*\langle\psi 1| + \alpha 2^*\langle\psi 2|,$$

$$(\alpha 1\langle\psi 1| + \alpha 2\langle\psi 2|)\dagger = \alpha 1^*|\psi 1\rangle + \alpha 2^*|\psi 2\rangle.$$

With $A|\psi\rangle$ we indicate the action of a bounded linear operator A on a ket $|\psi\rangle$ belonging to its domain, whereas with $\langle\psi|A$, which is the conjugate of $A\dagger|\psi\rangle$, we indicate the action of a bounded linear operator A on a bra $\langle\psi|$ of its domain. Thus, from the definition of the adjoint of an operator, we have:

$(\langle\psi|A)|\phi\rangle = \langle\psi|(A|\phi\rangle) \quad |\phi\rangle, |\psi\rangle \in \mathrm{H}$ hence, the parentheses are superfluous and we currently write:

$$(\langle\psi|A)|\phi\rangle = \langle\psi|(A|\phi\rangle) = (\langle\psi|A)|\phi\rangle.$$

The adjoint $A\dagger$ of the operator A is defined in the Dirac formalism as follows:

$\langle\psi|(A\dagger|\phi\rangle) = (\langle\phi|A)|\psi\rangle^*$, which can simply be rewritten

$$\langle\psi|A\dagger|\phi\rangle = \langle\phi|A|\psi\rangle^*.$$

External Product and Projectors

It turns out useful to define the external product in the Dirac formalism, as follows.

The external product $|\phi\rangle\langle\psi|$ of a ket times a bra, interpreted as the symbol of a linear operator on the ket space, is defined as

$(|\phi\rangle\langle\psi|)|\phi\rangle = |\phi\rangle\langle\psi|\phi\rangle$, while as the symbol of a linear operator on the bra space is defined as

$$\langle\phi|(|\phi\rangle\langle\psi|) = \langle\phi|\phi\rangle\langle\psi|.$$

So the conjugation operation defined for bras and kets can be extended to linear operators and we have the following general rule: The conjugate of any composite symbol is obtained by simply exchanging the order of the elementary symbols and taking the Hermitian conjugate for operators and the complex conjugate for numbers.

The external product turns out to be useful in the definition of projectors.

If $|\phi\rangle$ represents a vector of H normalized to unity, then $|\phi\rangle\langle\phi|$ is a projector in the direction of $|\phi\rangle$; indeed we have:

$(|\phi\rangle\langle\phi|)|\psi\rangle = |\phi\rangle\langle\phi|\psi\rangle$, and the r.h.s. term is the vectorial component of ψ in the subspace of H in the direction of $|\phi\rangle$.

As any projector, the linear operator $|\phi\rangle\langle\phi|$ has the following evident properties:

1. It is Hermitian, i.e. $(|\phi\rangle\langle\phi|)\dagger = |\phi\rangle\langle\phi|$;
2. It is idempotent, i.e. $(|\phi\rangle\langle\phi|)2 = |\phi\rangle\langle\phi|\phi\rangle\langle\phi| = |\phi\rangle\langle\phi|$.

The definition of projection operators in the Dirac formalism can be generalized to any closed finite or infinite-dimensional subspace Ho of H. Let $\{|n\rangle, n=1,\ldots,N\}$, with N possibly infinite, be an orthonormal basis in $H0: |\langle n|m\rangle = \text{dnm, n, m} = 1, 2, \ldots, N$. Then it is straightforward to verify that the operator $Q0 = \sum n = 1N|n\rangle\langle n|$ is a projector on Ho. Notice that the defintion of the projector Qo is independent from the particular basis chosen to construct it.

If, in particular, Ho = H, then Qo = 1, where 1 is the identity operator, and $\sum n = 1N|n\rangle\langle n| = 1$ gives us the completeness relation.

Eigenvalue Equation

The bra-ket notation turns out useful even for the eigenvalue equation for observables.

Consider an observable A represented by the self-adjoint operator A; the eigenvalue problem for A admits solutions both in the space of bras and in that of kets. Therefore the eigenvalue equation for A can be formulated in the bra-ket formalism, as follows:

$A|a;r\rangle = a|a;r\rangle \quad \langle a;r|A = \langle a;r|a$, where $a \in Sp(A)$ and *r* indicates the possible degeneration of the eigenvalues. The eigenvectors of A satisfy the orthonormal relation.

$\langle a|b\rangle = \delta(a,b)$ where $\delta(a,b)$ indicates the Kroenecker, respectively Dirac, delta if both *a* and *b* belong to the discrete, respectively continuous, spectrum of *A*, and is zero if a and b belong one to the discrete part and one to the continuous part of the spectrum of *A*.

Hilbert Space

A Hilbert space is a vector space V equipped with an *inner product*, which can be thought of as a generalization of the dot product in Euclidean space, with the additional property that the metric coming from the inner product makes V into a complete metric space. The basic example of a Hilbert space is $\mathbb{R}^n$ (or $\mathbb{C}^n$) with the standard dot product, but many other problems and structures in mathematics and physics turn out to be best described by other types of Hilbert spaces, most notably spaces of certain types of functions.

Definition of Inner Product

An inner product on a vector space V over a field $F = \mathbb{R}$ or $\mathbb{C}$ is a function $(.\,,.): V \times V \to F$ satisfying the following properties:

(1) $\langle x, y\rangle = \overline{\langle y, x\rangle}$ for all $x, y \in V$.

(2) It is linear in the first argument: $\langle ax_i + bx_2, y\rangle = a\langle x_1, y\rangle + b\langle x_2, y\rangle$ *for all* $a, b \in F, x_1, x_2, y \in V$.

(3) For $x \in V$, the inner product of x with itself is positive definite: $\langle x, x\rangle \geq 0$, and equality holds if and only if $x = 0$.

Notes:

- (1) implies that $\langle x, x\rangle = \overline{\langle x, x\rangle}$ so $\langle x,x\rangle$ is a real number, so the inequality in (3) makes sense.
- (1) and (2) imply that the inner product is antilinear in the second argument:

  $\langle x, ay_1 + by_2\rangle = \bar{a}\langle x, y_1\rangle + \bar{b}\langle x, y_2\rangle$.

A vector space with an inner product is called an inner product space.

The norm of a vector in an inner product space is $\|x\| = \sqrt{\langle x, x\rangle}$. The distance between two elements x, y in an inner product space is defined to be $\|x, y\|$.

Metric Space Properties of Inner Product Space

The goal is to show that any inner product space is a metric space. The proof will use the following fundamental theorem:

Cauchy-Schwarz inequality:

$$|\langle x, y\rangle| \leq \|x\|\|y\|,$$

and equality holds if and only if x, y are linearly dependent.

The proof follows the same path as the proof in $\mathbb{R}^n$ First note that the theorem is trivial for x or $y = 0$. So assume $x,y \neq 0$. Now let $t = \dfrac{|\langle x, y\rangle|}{\langle x, y\rangle}$. Let $u = \dfrac{x}{\|x\|}$, $v = \dfrac{y}{\|y\|}$. Note that u and v are unit vectors (their norms are both 1), and $|t| = 1$. Then

$$\begin{aligned} 0 \leq \|tu - v\|^2 &= \langle tu - v, tu - v\rangle = \langle tu, u\rangle - \langle v, tu\rangle - \langle tv, v\rangle + \langle v, v\rangle \\ &= |t|^2 \|u\|^2 - 2\,\mathrm{Re}(\langle tu, v\rangle) + \|v\|^2 \\ &= 2 - 2\,\mathrm{Re}(t\langle u, v\rangle) \end{aligned}$$

but $\langle u, v\rangle = \frac{\langle x, y\rangle}{\|x\|\,\|y\|}$, so this becomes

$$0 \le 2 - 2\,\mathrm{Re}\left(\frac{|\langle x, y\rangle|}{\|x\|\,\|y\|}\right)$$

and since the quantity in parentheses is real, we can drop the Re :

$$0 \le 2 - 2\frac{|\langle x, y\rangle|}{\|x\|\,\|y\|}$$

$$\frac{|\langle x, y\rangle|}{\|x\|\,\|y\|} \le 1$$

$$|\langle x, y\rangle| \le \|x\|\,\|y\|.$$

Equality holds if and only if $tu = v$, which certainly implies that x, y are linearly dependent; but it is easy to check that equality holds if x, y are dependent.

Theorem: The distance function $d(x, y) = \|x - y\|$ is a metric.

First, $d(x, y) = \|x - y\|$ is nonnegative by definition of the norm (which makes sense because $\langle x - y, x - y\rangle$ is nonnegative) And it is 0 if and only if $\langle x - y, x - y\rangle = 0$, which happens if and only if $x - y = 0$, or $x = y$, by property (1) of the inner product.

Second, $d(x, y) = d(y, x)$ is immediately clear; in general

$$\|tu\| = |t|\,\|u\| \text{ for } t \in F \text{ and } u \in V, \text{ so}$$
$$\|x - y\| = \|(-1)(y - x)\| = |-1|\,\|y - x\| = \|y - x\|.$$

The triangle inequality is where Cauchy-Schwarz comes in: let $u = x - y, v = y - z$. Then $u + v = x - z$. Now

$$\begin{aligned}(\|u\| + \|v\|)^2 - \|u + v\|^2 &= \|u\|^2 + \|v\|^2 + 2\|u\|\,\|v\| - \langle u + v,\ u + v\rangle \\ &= \|u\|^2 + \|v\|^2 + 2\|u\|\,\|v\| - \|u\|^2 - \|v\|^2 - 2\,\mathrm{Re}(\langle u, v\rangle) \\ &= 2(\|u\|\,\|v\| - \mathrm{Re}(\langle u, v\rangle)) \\ &\ge 0\end{aligned}$$

by Cauchy-Schwarz. Note that equality holds for nonzero u, v if and only if u is a positive real multiple of v

So $\|u\| + \|v\| \ge \|u + v\|$; and substituting back gives $d(x, y) + d(y, z) \ge d(x, z)$.

Example of Hilbert Space - Lebesgue Spaces

Let $L^2(\mathbb{R})$ be the set of functions $f : \mathbb{R} \to \mathbb{C}$ such that

$$\int_{-\infty}^{\infty} \left| f(x)^2 \right| dx$$

exists and is finite. Then there is an inner product defined on $L^2(\mathbb{R})$ by

$$\langle f, g \rangle = \int_{-\infty}^{\infty} f(x) \overline{g(x)}\, dx.$$

(So the L^2 requirement is precisely that $\|f\|$ is finite.) It turns out that if the integral is the Lebesgue integral, then "enough" functions are integrable so that Cauchy sequences of functions will converge. So this makes $L^2(\mathbb{R})$ into a Hilbert space.

Similarly, $L^2([0, 1])$ is a Hilbert space, where the definition is the same except that functions and integrals are taken over the interval $[0, 1]$.

The alert reader will have noted an apparent inaccuracy in the above discussion; it is possible for a nonzero function f to have L^2-norm equal to 0, i.e. if f is zero "almost everywhere," except on a set of measure zero. The solution is to let $L^2(X)$ consist of equivalence classes of functions, rather than functions, where two functions f and g are considered equivalent if $f - g$ is zero almost everywhere. This is an important detail to keep in mind, especially since it is (unfortunately) often suppressed in discussions of Hilbert spaces for convenience's sake.

Example of Hilbert Space - Sequence Spaces

The discrete version of the previous example is as follows: define ℓ^2 to be the set of sequences $(z_1, z_2, \ldots)$ of complex numbers such that

$$\sum_{n=1}^{\infty} |z_n|^2$$

exists and is finite. Then there is an inner product on ℓ^2 given by

$$\langle (z_n), (w_n) \rangle = \sum_{n=1}^{\infty} z_n \overline{w_n}.$$

It is a standard exercise to show that ℓ^2 is in fact a Hilbert space (i.e. that it is complete with respect to the metric induced by this inner product).

Orthonormal Bases in Hilbert Spaces

In the Hilbert space ℓ^2, let e_k be the sequence with all 0's except for a 1 in the kth term. Then any sequence $z = z_n$ can be written as an infinite sum

$$z = \sum_{n=1}^{\infty} z_n e_n.$$

That is, the infinite sum on the right converges to z, using the metric defined by the ℓ^2. Note that this representation is unique.

There is an obvious analogy with a basis of a vector space, which is a set of elements of the vector space such that every element can be written uniquely as a linear combination of *finitely many* members of the basis. For instance, the same construction gives a basis $e_1, ..., e_n$ of $\mathbb{C}^n$.

Since Hilbert spaces are vector spaces, they have regular vector space bases (by the axiom of choice). To avoid ambiguity, these are often referred to as *Hamel bases*. Hamel bases of Hilbert spaces are generally useless for computation and difficult to construct, unless they are finite $\left(\text{e.g.}\, V = \mathbb{C}^n\right)$. For "larger" Hilbert spaces, the more natural notion is that of an *orthonormal basis*, which generalizes the example given above.

A set (y_n) of vectors in a Hilbert space is orthonormal if

$$\langle y_i, y_j \rangle = \begin{cases} 1 & \text{if } i = j \\ 0 & \text{if } i \neq j \end{cases}$$

It is an *orthonormal basis* if, in addition, the only vector x satisfying $\langle x, y_n \rangle = 0$ for all n is the zero vector. (Equivalently, the span of the y_i is dense.)

Note that an orthonormal basis is not necessarily a Hamel basis. For instance, the orthonormal basis (e_n) of ℓ^2 is not a Hamel basis, since expressing an arbitrary element of as a linear combination of basis vectors requires an infinite (convergent) sum.

As in the motivating example, countable orthonormal bases have various nice properties.

Theorem: Let $(x_n)_{n\in\mathbb{N}}$ be an orthonormal basis for a Hilbert space X. Then

(1) $x = \sum_{n=1}^{\infty} \langle x, x_n \rangle x_n$ for all $x \in X$; in particular, every x has a unique representation as a (possibly infinite) linear combination of the x_n, with coefficients given by this formula.

(2) $\|x\|^2 = \sum_{n=1}^{\infty} |\langle x, x_n \rangle|^2$ (Parseval's identity)

(3) $\langle x, y \rangle = \sum_{n=1}^{\infty} \langle x, x_n \rangle \overline{\langle y, x_n \rangle}$.

Example: The space $L^2([0, 1])$ has an orthonormal basis consisting of the functions $e^{2\pi i n\theta}$, for all integers n. The coefficients in the sum $f(\theta) = \sum_{n=-\infty}^{\infty} a_n e^{2\pi i n\theta}$ are called the Fourier coefficients of f.

The axiom of choice implies that every Hilbert space has an orthonormal basis.

Operators of Quantum Mechanics

The numerical quantities that the old Newtonian physics uses, (position, momentum, energy,) are just shadows of what really describes nature: operators. The operators described are the key to quantum mechanics.

As the first example, while a mathematically precise value of the position x of a particle never exists, instead there is an x-position *operator* $\hat{x}$. It turns the wave function Ψ into $x\Psi$:

$$\Psi(x, y, z, t) \xrightarrow{\hat{x}} x\Psi(x, y, z, t)$$

The operators $\hat{y}$ and $\hat{z}$ are defined similarly as $\hat{x}$.

Instead of a linear momentum $p_x = mu$, there is an x-momentum operator.

$$\hat{p}_x = \frac{\hbar}{\mathrm{i}}\frac{\partial}{\partial x}$$

that turns Ψ into its x-derivative:

$$\Psi(x, y, z, t) \xrightarrow{\hat{p}_x = \frac{\hbar}{\mathrm{i}}\frac{\partial}{\partial x}} \frac{\hbar}{\mathrm{i}}\Psi_x(x, y, z, t)$$

The constant $\hbar$ is called "Planck's constant." (Or rather, it is Planck's original constant h divided by 2π.) If it would have been zero, all these troubles with quantum mechanics would not occur. The blobs would become points. Unfortunately, $\hbar$ is very small, but nonzero. It is about 10^{-34} kg m^2/s.

The factor i in $\hat{p}_x$ makes it a Hermitian operator. All operators reflecting macroscopic physical quantities are Hermitian.

The operators $\hat{p}_y$ and $\hat{p}_z$ are defined similarly as $\hat{p}_x$:

$$\hat{p}_y = \frac{\hbar}{\mathrm{i}}\frac{\partial}{\partial y} \qquad \hat{p}_z = \frac{\hbar}{\mathrm{i}}\frac{\partial}{\partial z}$$

The *kinetic energy operator* $\hat{T}$ is:

$$\hat{T} = \frac{\hat{p}_x^2 + \hat{p}_y^2 + \hat{p}_z^2}{2m}$$

Its shadow is the Newtonian notion that the kinetic energy equals:

$$T = \frac{1}{2}m\left(u^2 + v^2 + w^2\right) = \frac{(mu)^2 + (mv)^2 + (mw)^2}{2m}$$

This is an example of the "Newtonian analogy": the relationships between the different operators in quantum mechanics are in general the same as those between the corresponding numerical values in Newtonian physics. But since the momentum *operators* are gradients, the actual kinetic energy operator is, from the momentum operators above:

$$\hat{T} = -\frac{\hbar^2}{2m}\left(\frac{\partial^2}{\partial x^2} + \frac{\partial^2}{\partial y^2} + \frac{\partial^2}{\partial z^2}\right).$$

Mathematicians call the set of second order derivative operators in the kinetic energy operator the Laplacian, and indicate it by ∇^2:

$$\nabla^2 \equiv \left(\frac{\partial^2}{\partial x^2} + \frac{\partial^2}{\partial y^2} + \frac{\partial^2}{\partial z^2} \right)$$

In those terms, the kinetic energy operator can be written more concisely as:

$$\hat{T} = -\frac{\hbar^2}{2m} \nabla^2$$

Following the Newtonian analogy once more, the total energy operator, indicated by H, is the the sum of the kinetic energy operator above and the potential energy operator $V(x, y, z, t)$:

$$H = -\frac{\hbar^2}{2m} \nabla^2 + V$$

This total energy operator H is called the *Hamiltonian* and it is very important. Its eigenvalues are indicated by E (for energy), for example E_1, E_2, E_3 ... with:

$$H\psi_n = E_n \psi_n \quad \text{for } n = 1, 2, 3, \ldots$$

where ψ_n is eigenfunction number n of the Hamiltonian.

It is seen later that in many cases a more elaborate numbering of the eigenvalues and eigenvectors of the Hamiltonian is desirable instead of using a single counter n. For example, for the electron of the hydrogen atom, there is more than one eigenfunction for each different eigenvalue E_n and additional counters l and m are used to distinguish them.

Momentum Operator

The momentum operator can be introduced into quantum mechanics by a general approach based on the space displacement operator.

We determine the momentum operator by requiring that, when we operate with $p_x^{(op)}$ on $u_{p^o}(x,t)$ we get p_o times the same wave function.

$$p^{(op)} u_{p^o}(x,t) = p_o \, u_{p^o}(x,t)$$

This means that for these definite momentum states, multiplying by $p_x^{(op)}$ is the same as multiplying by the variable p. We find that this is true for the following momentum operator.

$$p^{(op)} = \frac{\hbar}{i} \frac{\partial}{\partial x}$$

If we take our momentum operator and act on a arbitrary state,

$$p^{(op)} \psi(x,t) = p^{(op)} \int_{-\infty}^{\infty} \phi(p) \; u_p(x,t) dp = \int_{-\infty}^{\infty} \phi(p) \, p \; u_p(x,t) dp$$

it gives us the right p for each term in the integral. This will allow us to compute expectation values for any variable we can represent by an operator.

Hamiltonian

Associated with each measurable parameter in a physical system is a quantum mechanical operator, and the operator associated with the system energy is called the Hamiltonian. The Hamiltonian contains the operations associated with the kinetic and potential energies and for a particle in one dimension can be written:

$$H_{operator} = \frac{-\hbar^2}{2m}\frac{\partial^2}{\partial x^2} + V(X)$$

Operator associated with kinetic energy — Potential energy

Operating on the wave function with the Hamiltonian produces the Schrodinger equation. In the time independent Schrodinger equation, the operation may produce specific values for the energy called energy eigenvalues. This situation can be shown in the form

$$H_{op}\psi_i = E_i\psi_i$$

where the specific values of energy are called energy eigenvalues and the functions ψ_i are called eigenfunctions.

In addition to its role in determining system energies, the Hamiltonian operator generates the time evolution of the wave function in the form

$$H\psi = i\hbar\frac{\partial}{\partial t}\psi$$

The full role of the Hamiltonian is shown in the time dependent Shrodinger equation where both its spatial and time operations manifest themselves.

Rotation Operator

Consider a particle whose position is described by the spherical polar coordinates (r, θ, φ). The classical momentum conjugate to the azimuthal angle φ is the z -component of angular momentum L_z. In quantum mechanics we can always adopt the Schrödinger representation, for which ket space is spanned by the simultaneous eigenkets of the position operators r, θ, and φ, and L_z takes the form

$$L_z = -i\hbar\frac{\partial}{\partial\varphi}.$$

We can do this because there is nothing, which specifies that we have to use Cartesian coordinates--the representation $p_i = -i\hbar\frac{\partial}{\partial q_i}$ works for any well-defined set of coordinates.

Consider an operator $R(\Delta\varphi)$ that rotates the system through an angle $\Delta\varphi$ about the z -axis. This operator is very similar to the operator $D(\Delta x)$, which translates the system a distance Δx along the x -axis. We were able to demonstrate that

$$p_x = \mathrm{i}\hbar \lim_{\delta x \to 0} \frac{D(\delta x) - 1}{\delta x}$$

where p_x is the linear momentum conjugate to x. There is nothing in our derivation of this result which specifies that x has to be a Cartesian coordinate. Thus, the result should apply just as well to an angular coordinate. We conclude that

$$L_z = \mathrm{i}\hbar \lim_{\delta\varphi \to 0} \frac{R(\delta\varphi) - 1}{\delta\varphi}.$$

According to equation above we can write

$$R(\delta\varphi) = 1 - \mathrm{i}\, L_z \delta\varphi / \hbar$$

in the limit $\delta\varphi \to 0$. In other words, the angular momentum operator L_z can be used to rotate the system about the z -axis by an infinitesimal amount. We say that L_z is the *generator* of rotations about the z -axis. The above equation implies that

$$R(\Delta\varphi) = \lim_{N\to\infty} \left(1 - i \frac{\Delta\varphi}{N} \frac{L_z}{\hbar} \right)^N ,$$

which reduces to

$$R(\Delta\varphi) = \exp(-iL_z \Delta\varphi / \hbar).$$

Note that $R(\Delta\varphi)$ has all of the properties we would expect of a rotation operator: i.e.,

$$R(0) = 1,$$
$$R(\Delta\varphi) R(-\Delta\varphi) = 1,$$
$$R(\Delta\varphi_1) R(\Delta\varphi_2) = R(\Delta\varphi_1 + \Delta\varphi_2).$$

Suppose that the system is in a simultaneous eigenstate of L^2 and L_z. As before, this state is represented by the eigenket $|l, m\rangle$, where the eigenvalue of L^2 is $l(l+1)\hbar^2$ and the eigenvalue of L_z is $m\hbar$. We expect the wave function to remain unaltered if we rotate the system 2π degrees about the z -axis. Thus,

$$R(2\pi)|l, m\rangle = \exp(-\mathrm{i}\, L_z 2\pi / \hbar)|l, m\rangle = \exp(-\mathrm{i}\, 2\pi m)|l, m\rangle = |l, m\rangle$$

We conclude that m must be an integer. This implies, that l must also be an integer. Thus, an *orbital* angular momentum can only take *integer* values of the quantum numbers l and m.

Consider the action of the rotation operator $R(\Delta\varphi)$ on an eigenstate possessing zero angular momentum about the z -axis (i.e., an $m = 0$ state). We have

$$R(\Delta\varphi)|l, 0\rangle = \exp(0)|l, 0\rangle = |l, 0\rangle.$$

Thus, the eigenstate is invariant to rotations about the z -axis. Clearly, its wave function must be symmetric about the z -axis.

There is nothing special about the z -axis, so we can write

$$R_x(\Delta\varphi_x) = \exp(-\mathrm{i}\, L_x \Delta\varphi_x / \hbar),$$
$$R_y(\Delta\varphi_y) = \exp(-\mathrm{i}\, L_y \Delta\varphi_y / \hbar),$$
$$R_z(\Delta\varphi_y) = \exp(-\mathrm{i}\, L_z \Delta\varphi_z / \hbar),$$

by analogy with equation $R(\Delta\varphi) = \exp(-iL_z\Delta\varphi / \hbar)$. Here, $R_x(\Delta\varphi_x)$ denotes an operator that rotates the system by an angle $\Delta\varphi_x$ about the x -axis, etc. Suppose that the system is in an eigenstate of zero overall orbital angular momentum (i.e., an $l = 0$ state). We know that the system is also in an eigenstate of zero orbital angular momentum about any particular axis. This follows because $l = 0$ implies $m = 0$, we can choose the z -axis to point in any direction. Thus,

$$R_x(\Delta\varphi_x)|0, 0\rangle = \exp(0)|0, 0\rangle = |0, 0\rangle,$$
$$R_y(\Delta\varphi_y)|0, 0\rangle = \exp(0)|0, 0\rangle = |0, 0\rangle,$$
$$R_z(\Delta\varphi_z)|0, 0\rangle = \exp(0)|0, 0\rangle = |0, 0\rangle.$$

Clearly, a zero angular momentum state is invariant to rotations about *any* axis. Such a state must possess a spherically symmetric wave function.

Note that a rotation about the x -axis does not commute with a rotation about the y -axis. In other words, if the system is rotated an angle $\Delta\varphi_x$ about the x -axis, and then $\Delta\varphi_y$ about the y -axis, it ends up in a different state to that obtained by rotating an angle $\Delta\varphi_y$ about the y -axis, and then $\Delta\varphi_x$ about the x -axis. In quantum mechanics, this implies that $R_y(\Delta\varphi_y)R_x(\Delta\varphi_x) \neq R_x(\Delta\varphi_x)R_y(\Delta\varphi_y)$, or $L_y L_x \neq L_x L_y$. Thus, the non-commuting nature of the angular momentum operators is a direct consequence of the fact that rotations do not commute.

Angular Momentum Operator

Classical angular momentum is a vector quantity denoted $\vec{L} = \vec{r}\ X\ \vec{p}$. A common mnemonic to calculate the components is

$$\vec{L} = \begin{vmatrix} \hat{i} & \hat{j} & \hat{k} \\ x & y & z \\ p_x & p_y & p_z \end{vmatrix} = (yp_z - zp_y)i + (zp_x - xp_z)\hat{j} + (xp_y - yp_x)\hat{j}$$
$$= L_x i + L_y \hat{j} + L_z \hat{j}.$$

Let's focus on one component of angular momentum, say $L_x = yp_z - zp_y$. On the right side of the equation are two components of position and two components of linear momentum. Quantum mechanically, all four quantities are operators. Since the product of two operators is an operator, and the difference of operators is another operator, we expect the components of angular momentum to be operators. In other words, quantum mechanically

$$\mathcal{L}_x = \mathcal{Y}\mathcal{P}_z - \mathcal{Z}\mathcal{P}_y, \quad \mathcal{L}_y = \mathcal{Z}\mathcal{P}_x - \mathcal{X}\mathcal{P}_z, \quad \mathcal{L}_z = \mathcal{X}\mathcal{P}_y - \mathcal{Y}\mathcal{P}_x.$$

These are the components. Angular momentum is the vector sum of the components. The sum of operators is another operator, so angular momentum is an operator. We have not encountered an operator like this one, however, this operator is comparable to a vector sum of operators; it is essentially a ket with operator components. We might write

$$|\mathcal{L}\rangle = \begin{pmatrix} \mathcal{L}_x \\ \mathcal{L}_y \\ \mathcal{L}_z \end{pmatrix} = \begin{pmatrix} \mathcal{Y}\mathcal{P}_z - \mathcal{Z}\mathcal{P}_y \\ \mathcal{Z}\mathcal{P}_x - \mathcal{X}\mathcal{P}_z \\ \mathcal{X}\mathcal{P}_y - \mathcal{Y}\mathcal{P}_x \end{pmatrix}. \tag{1}$$

A word of caution concerning common notation—this is usually written just $\mathcal{L}$, and the ket/vector nature of quantum mechanical angular momentum is not explicitly written but implied. Equation above shows that abstract Hilbert space and is completely devoid of a representation. We will want to pick a basis to perform a calculation. In position space, for instance

$$\mathcal{X} \to x, \quad \mathcal{Y} \to y, \quad \text{and} \quad \mathcal{Z} \to z,$$

And

$$\mathcal{P}_x \to -i\hbar\frac{\partial}{\partial x}, \qquad \mathcal{P}_y \to -i\hbar\frac{\partial}{\partial y}, \quad \text{and} \quad \mathcal{P}_z \to -i\hbar\frac{\partial}{\partial z}.$$

Equation above in position space would then be written

$$|\mathcal{L}\rangle = \begin{pmatrix} -i\hbar y\frac{\partial}{\partial z} & + & i\hbar z\frac{\partial}{\partial y} \\ -i\hbar z\frac{\partial}{\partial x} & + & i\hbar x\frac{\partial}{\partial z} \\ -i\hbar x\frac{\partial}{\partial y} & + & i\hbar y\frac{\partial}{\partial x} \end{pmatrix}.$$

The operator nature of the components promise difficulty, because unlike their classical analogs which are scalars, the angular momentum operators do not commute.

Example: Show the components of angular momentum in position space do not commute. Let the commutator of any two components, say $\left[\mathcal{L}_x, \mathcal{L}_y\right]$, act on the function x. This means

$$\left[\mathcal{L}_x, \mathcal{L}_y\right]x = \left(\mathcal{L}_x\mathcal{L}_y - \mathcal{L}_y\mathcal{L}_x\right)x$$

$$\to \left(-i\hbar y\frac{\partial}{\partial z} + i\hbar z\frac{\partial}{\partial y}\right)\left(-i\hbar z\frac{\partial}{\partial x} + i\hbar x\frac{\partial}{\partial z}\right)x - \left(-i\hbar z\frac{\partial}{\partial x} + i\hbar x\frac{\partial}{\partial z}\right)\left(-i\hbar y\frac{\partial}{\partial y} + i\hbar z\frac{\partial}{\partial x}\right)x$$

$$= \left(-i\hbar y \frac{\partial}{\partial z} + i\hbar z \frac{\partial}{\partial y} \right)(-i\hbar z) - \left(-i\hbar z \frac{\partial}{\partial x} + i\hbar x \frac{\partial}{\partial z} \right)(0)$$
$$= \left((-i\hbar z)^2 y \right) = -\hbar^2 y \neq 0,$$

therefore $\mathcal{L}_x$ and $\mathcal{L}_y$ do not commute. Using functions which are simply appropriate position space components, other components of angular momentum can be shown not to commute similarly.

Example: What is equation in the momentum basis?

In momentum space, the operators are

$$\mathcal{X} \to i\hbar \frac{\partial}{\partial p_x}, \quad \mathcal{Y} \to i\hbar \frac{\partial}{\partial p_y}, \quad \text{and } \mathcal{Z} \to i\hbar \frac{\partial}{\partial p_z},$$

And

$$\mathcal{P}_x \to p_x, \quad \mathcal{P}_y \to p_y, \quad \mathcal{P}_z \to p_z$$

Equation (1) in momentum space would be written

$$|\mathcal{L}\rangle = \begin{pmatrix} i\hbar \frac{\partial}{\partial p_y} p_z & - & i\hbar \frac{\partial}{\partial p_z} p_y \\ i\hbar \frac{\partial}{\partial p_z} p_x & - & i\hbar \frac{\partial}{\partial p_x} p_z \\ i\hbar \frac{\partial}{\partial p_x} p_y & - & i\hbar \frac{\partial}{\partial p_y} p_x \end{pmatrix}.$$

Canonical Commutation Relations in Three Dimensions

We indicated in equation $[\mathcal{X}_i, \mathcal{X}_j] = 0$, the fundamental canonical commutator is

$$[\mathcal{X}, \mathcal{P}] = i\hbar.$$

This is fine when working in one dimension, however, descriptions of angular momentum are generally three dimensional. The generalization to three dimensions is

$$[\mathcal{X}_i, \mathcal{X}_j] = 0,$$

which means any position component commutes with any other position component including itself,

$$[\mathcal{P}_i, \mathcal{P}_j] = 0,$$

which means any linear momentum component commutes with any other linear momentum component including itself,

$$[\mathcal{X}_i, \mathcal{P}_j] = i\hbar\delta_{i,j},$$

and the meaning of this equation requires some discussion. This means a position component will commute with an unlike component of linear momentum,

$$\left[\mathcal{X},\mathcal{P}_y\right]=\left[\mathcal{X},\mathcal{P}_z\right]=\left[\mathcal{Y},\mathcal{P}_z\right]=\left[\mathcal{Z},\mathcal{P}_x\right]=\left[\mathcal{Z},\mathcal{P}_y\right]=0,$$

but a position component and a like component of linear momentum are canonical commutators, i.e.,

$$\left[\mathcal{X}_x,\mathcal{P}_x\right]=\left[\mathcal{Y},\mathcal{P}_y\right]=\left[\mathcal{Z},\mathcal{P}_z\right]=i\hbar.$$

Angular Momentum Commutation Relation

Given the relations of equations $\left[\mathcal{X}_i,\mathcal{X}_j\right]=0$ through $\left[\mathcal{X}_i,\mathcal{P}_j\right]=i\hbar\delta_{i,j}$, it follows that

$$\left[\mathcal{L}_x,\mathcal{L}_y\right]=i\hbar\,\mathcal{L}_z,\qquad \left[\mathcal{L}_y,\mathcal{L}_z\right]=i\hbar\,\mathcal{L}_x,\qquad \text{and}\ \left[\mathcal{L}_z,\mathcal{L}_x\right]=i\hbar\,\mathcal{L}_y.$$

Example: Show $\left[\mathcal{L}_x,\mathcal{L}_y\right]=i\hbar\,\mathcal{L}_z$.

$$\begin{aligned}
\left[\mathcal{L}_x,\mathcal{L}_y\right]&=\left[\mathcal{Y}\mathcal{P}_z-\mathcal{Z}\mathcal{P}_y,\ \mathcal{Z}\mathcal{P}_x-\mathcal{X}\mathcal{P}_z\right]\\
&=\left(\mathcal{Y}\mathcal{P}_z-\mathcal{Z}\mathcal{P}_y\right)\left(\mathcal{Z}\mathcal{P}_x-\mathcal{X}\mathcal{P}_z\right)-\left(\mathcal{Z}\mathcal{P}_x-\mathcal{X}\mathcal{P}_z\right)\left(\mathcal{Y}\mathcal{P}_z-\mathcal{Z}\mathcal{P}_y\right)\\
&=\mathcal{Y}\mathcal{P}_z\mathcal{Z}\mathcal{P}_x-\mathcal{Y}\mathcal{P}_z\mathcal{X}\mathcal{P}_z-\mathcal{Z}\mathcal{P}_y\mathcal{Z}\mathcal{P}_x+\mathcal{Z}\mathcal{P}_y\mathcal{X}\mathcal{P}_z-\mathcal{Z}\mathcal{P}_x\mathcal{Y}\mathcal{P}_z+\mathcal{Z}\mathcal{P}_x\mathcal{Z}\mathcal{P}_y+\mathcal{X}\mathcal{P}_z\mathcal{Y}\mathcal{P}_z-\mathcal{X}\mathcal{P}_z\mathcal{Z}\mathcal{P}_y\\
&=\left(\mathcal{Y}\mathcal{P}_z\mathcal{Z}\mathcal{P}_x-\mathcal{Z}\mathcal{P}_x\mathcal{Y}\mathcal{P}_z\right)+\left(\mathcal{Z}\mathcal{P}_y\mathcal{X}\mathcal{P}_z-\mathcal{X}\mathcal{P}_z\mathcal{Z}\mathcal{P}_y\right)\\
&\qquad+\left(\mathcal{Z}\mathcal{P}_x\mathcal{Z}\mathcal{P}_y-\mathcal{Z}\mathcal{P}_y\mathcal{Z}\mathcal{P}_x\right)+\left(\mathcal{X}\mathcal{P}_z\mathcal{Y}\mathcal{P}_z-\mathcal{Y}\mathcal{P}_z\mathcal{X}\mathcal{P}_z\right)\\
&=\left[\mathcal{Y}\mathcal{P}_z,\ \mathcal{Z}\mathcal{P}_y\right]+\left[\mathcal{Z}\mathcal{P}_y,\ \mathcal{X}\mathcal{P}_z\right]+\left[\mathcal{Z}\mathcal{P}_x,\mathcal{Z}\mathcal{P}_y\right]+\left[\mathcal{X}\mathcal{P}_z,\mathcal{Y}\mathcal{P}_z\right].
\end{aligned}$$

Using the result of example above, the plan is to express these commutators in terms of individual operators, and then evaluate those using the commutation relations of equations $\left[\mathcal{X}_i,\mathcal{X}_j\right]=0$, $\left[\mathcal{X}_i,\mathcal{P}_j\right]=i\hbar\delta_{i,j}$, through. In example, one commutator of the products of two operators turns into four commutators. Since we start with four commutators of the products of two operators, we are going to get 16 commutators in terms of individual operators. The good news is 14 of them are zero from equations $\left[\mathcal{X}_i,\mathcal{X}_j\right]=0$, $\left[\mathcal{X}_i,\mathcal{P}_j\right]=i\hbar\delta_{i,j}$, $\left[\mathcal{P}_i,\mathcal{P}_j\right]=0$, so will be struck.

$$\begin{aligned}
\left[\mathcal{L}_x,\mathcal{L}_y\right]&=\mathcal{Y}\left[\mathcal{P}_z,\mathcal{Z}\right]\mathcal{P}_x+\left[\mathcal{Y},/\mathcal{Z}\right]\mathcal{P}_z\mathcal{P}_x+\mathcal{Z}\mathcal{Y}\left[\mathcal{P}_z/\mathcal{P}_x\right]+\mathcal{Z}\left[\mathcal{Y}/\mathcal{P}_x\right]\mathcal{P}_z\\
&\quad+\mathcal{Z}\left[\mathcal{P}_y,/\mathcal{X}\right]\mathcal{P}_z+\left[\mathcal{Z}/\mathcal{X}\right]\mathcal{P}_y\mathcal{P}_z+\mathcal{X}\mathcal{Z}\left[\mathcal{P}_y,/\mathcal{P}_z\right]+\mathcal{X}\left[\mathcal{Z},\mathcal{P}_z\right]\mathcal{P}_y\\
&\quad+\mathcal{Z}\left[\mathcal{P}_x,/\mathcal{Z}\right]\mathcal{P}_y+\left[\mathcal{Z},/\mathcal{Z}\right]\mathcal{P}_x\mathcal{P}_y+\mathcal{Z}\mathcal{Z}\left[\mathcal{P}_x,/\mathcal{P}_y\right]+\mathcal{Z}\left[\mathcal{Z},/\mathcal{P}_y\right]\mathcal{P}_x\\
&\quad+\mathcal{X}\left[\mathcal{P}_z,/\mathcal{Y}\right]\mathcal{P}_z+\left[\mathcal{X},/\mathcal{Y}\right]\mathcal{P}_z\mathcal{P}_z+\mathcal{Y}\mathcal{X}\left[\mathcal{P}_z,/\mathcal{P}_z\right]+\mathcal{Y}\left[\mathcal{X},/\mathcal{P}_z\right]\mathcal{P}_z\\
&=\mathcal{Y}\left[\mathcal{P}_z,\mathcal{Z}\right]\mathcal{P}_x+\mathcal{X}\left[\mathcal{Z},\mathcal{P}_z\right]\mathcal{P}_y\\
&=\mathcal{Y}\left(-i\hbar\right)\mathcal{P}_x+\mathcal{X}\left(i\hbar\right)\mathcal{P}_y\\
&=i\hbar\left(\mathcal{X}\mathcal{P}_y-\mathcal{Y}\mathcal{P}_x\right)\\
&=i\hbar\mathcal{L}_z.
\end{aligned}$$

The other two relations, $\left[\mathcal{L}_y, \mathcal{L}_z\right] = i\hbar\mathcal{L}_x$ and $\left[\mathcal{L}_z, \mathcal{L}_x\right] = i\hbar\mathcal{L}_y$ can be calculated using similar procedures.

A Representation of Angular Momentum Operators

We would like to have matrix operators for the angular momentum operators $\mathcal{L}_x$, $\mathcal{L}_y$, and $\mathcal{L}_z$ .In the form $\mathcal{L}_x$, $\mathcal{L}_y$, and $\mathcal{L}_z$ these are abstract operators in an infinite dimensional Hilbert space. Remember from chapter 2 that a subspace is a specific subset of a general complex linear vector space. In this case, we are going to find relations in a subspace C 3 of an infinite dimensional Hilbert space. The idea is to find three 3 X 3 matrix operators that satisfy relations, which are

$$\left[\mathcal{L}_x, \mathcal{L}_y\right] = i\hbar\mathcal{L}_z, \quad \left[\mathcal{L}_y, \mathcal{L}_z\right] = i\hbar\mathcal{L}_x, \quad \text{and} \quad \left[\mathcal{L}_z, \mathcal{L}_x\right] = i\hbar\mathcal{L}_y.$$

One such group of objects is

$$\mathcal{L}_x = \frac{1}{\sqrt{2}}\begin{pmatrix} 0 & 1 & 0 \\ 1 & 0 & 1 \\ 0 & 1 & 0 \end{pmatrix}\hbar, \quad \mathcal{L}_y = \frac{1}{\sqrt{2}}\begin{pmatrix} 0 & -i & 0 \\ i & 0 & -i \\ 0 & i & 0 \end{pmatrix}\hbar, \quad \mathcal{L}_z = \begin{pmatrix} 1 & 0 & 0 \\ 0 & 0 & 0 \\ 0 & 0 & -1 \end{pmatrix}\hbar. \tag{2}$$

There are other ways to express these matrices in C ^{3}. Relations (2) are dominantly the most popular. Since the three operators do not commute, we arbitrarily have selected a basis for one of them, and then expressed the other two in that basis. Notice $\mathcal{L}_z$ is diagonal. That means the basis selected is natural for $\mathcal{L}_z$ .The terminology usually used is the operators in equations (2) are in the $\underline{\mathcal{L}_z}$ basis.

We could have selected a basis which makes $\mathcal{L}_x$ or $\mathcal{L}_y$ and expressed the other two in terms of the natural basis for $\mathcal{L}_x$ or $\mathcal{L}_y$ If we had done that, the operators are different than those seen in relations (2). The mathematics of this is not important at the moment, but it is important that you understand there are other self consistent ways to express these operators as 3 X 3 matrices.

Example: Show $\left[\mathcal{L}_x \; \mathcal{L}_y\right] = i\hbar\mathcal{L}_z$ using relations (2).

$$\left[\mathcal{L}_x, \mathcal{L}_y\right] = \frac{1}{\sqrt{2}}\begin{pmatrix} 0 & 1 & 0 \\ 1 & 0 & 1 \\ 0 & 1 & 0 \end{pmatrix}\hbar \frac{1}{\sqrt{2}}\begin{pmatrix} 0 & -i & 0 \\ i & 0 & -i \\ 0 & i & 0 \end{pmatrix}\hbar - \frac{1}{\sqrt{2}}\begin{pmatrix} 0 & -i & 0 \\ i & 0 & -i \\ 0 & i & 0 \end{pmatrix}\hbar \frac{1}{\sqrt{2}}\begin{pmatrix} 1 & 0 & 0 \\ 0 & 0 & 0 \\ 0 & 0 & -1 \end{pmatrix}\hbar.$$

$$= \frac{\hbar^2}{2}\begin{pmatrix} i & 0 & -i \\ 0 & -i+i & 0 \\ i & 0 & -i \end{pmatrix} - \frac{\hbar^2}{2}\begin{pmatrix} -i & 0 & -i \\ 0 & i-i & 0 \\ i & 0 & i \end{pmatrix} = \frac{\hbar^2}{2}\begin{pmatrix} i+i & 0 & -i+i \\ 0 & 0 & 0 \\ i-i & 0 & -i-i \end{pmatrix}$$

$$= \frac{\hbar^2}{2}\begin{pmatrix} 2i & 0 & 0 \\ 0 & 0 & 0 \\ 0 & 0 & -2i \end{pmatrix} = i\hbar\begin{pmatrix} 1 & 0 & 0 \\ 0 & 0 & 0 \\ 0 & 0 & -1 \end{pmatrix}\hbar$$

$$= i\hbar\mathcal{L}_z.$$

Again, the other two relations can be calculated using similar procedures. In fact, the arithmetic for the other two relations is simpler. Why would this be so? Because $\mathcal{L}_z$ is a diagonal operator.

Remember $\mathcal{L}$ is comparable to a vector sum of the three component operators, so in vector/matrix notation would look like

$$|\mathcal{L}\rangle = \begin{pmatrix} \mathcal{L}_x \\ \mathcal{L}_y \\ \mathcal{L}_z \end{pmatrix} = \begin{pmatrix} \frac{1}{\sqrt{2}} \begin{pmatrix} 0 & 1 & 0 \\ 1 & 0 & 1 \\ 0 & 1 & 0 \end{pmatrix} \hbar \\ \frac{1}{\sqrt{2}} \begin{pmatrix} 0 & -i & 0 \\ i & 0 & -i \\ 0 & i & 0 \end{pmatrix} \hbar \\ \begin{pmatrix} 1 & 0 & 0 \\ 0 & 0 & 0 \\ 0 & 0 & -1 \end{pmatrix} \hbar \end{pmatrix}$$

Again, this operator will normally be denoted just $\mathcal{L}$. The $\mathcal{L}$ operator is a different sort of object than the component operators. It is a different object in a different space. Yet, we would like a way to address angular momentum with a 3 X 3 matrix which is in the same subspace as the components. We can do this if we use $\mathcal{L}^2$. This operator is

$$\mathcal{L}^2 = 2\hbar^2 \mathcal{I} = 2\hbar^2 \begin{pmatrix} 1 & 0 & 0 \\ 0 & 1 & 0 \\ 0 & 0 & 1 \end{pmatrix}.$$

Example: Show $\mathcal{L}^2 = 2\hbar^2 \mathcal{I}$

$$\mathcal{L}^2 =< \mathcal{L}|\mathcal{L} >$$

$$\rightarrow < \frac{1}{\sqrt{2}} \begin{pmatrix} 0 & 1 & 0 \\ 1 & 0 & 1 \\ 0 & 1 & 0 \end{pmatrix} \hbar, \frac{1}{\sqrt{2}} \begin{pmatrix} 0 & -i & 0 \\ i & 0 & -i \\ 0 & i & 0 \end{pmatrix} \hbar, \begin{pmatrix} 1 & 0 & 0 \\ 0 & 0 & 0 \\ 0 & 0 & -1 \end{pmatrix} \hbar \Bigg| \begin{pmatrix} \frac{1}{\sqrt{2}} \begin{pmatrix} 0 & 1 & 0 \\ 1 & 0 & 1 \\ 0 & 1 & 0 \end{pmatrix} \hbar \\ \frac{1}{\sqrt{2}} \begin{pmatrix} 0 & -i & 0 \\ i & 0 & -i \\ 0 & i & 0 \end{pmatrix} \hbar \\ \begin{pmatrix} 1 & 0 & 0 \\ 0 & 0 & 0 \\ 0 & 0 & -1 \end{pmatrix} \hbar \end{pmatrix} >$$

$$= \frac{1}{\sqrt{2}} \begin{pmatrix} 0 & 1 & 0 \\ 1 & 0 & 1 \\ 0 & 1 & 0 \end{pmatrix} \hbar \frac{1}{\sqrt{2}} \begin{pmatrix} 0 & 1 & 0 \\ 1 & 0 & 1 \\ 0 & 1 & 0 \end{pmatrix} \hbar + \frac{1}{\sqrt{2}} \begin{pmatrix} 0 & -i & 0 \\ i & 0 & -i \\ 0 & i & 0 \end{pmatrix} \hbar \frac{1}{\sqrt{2}} \begin{pmatrix} 0 & -i & 0 \\ i & 0 & -i \\ 0 & i & 0 \end{pmatrix} \hbar$$

$$+ \begin{pmatrix} 1 & 0 & 0 \\ 0 & 0 & 0 \\ 0 & 0 & -1 \end{pmatrix} \hbar \begin{pmatrix} 1 & 0 & 0 \\ 0 & 0 & 0 \\ 0 & 0 & -1 \end{pmatrix} \hbar$$

$$= \frac{1}{2}\begin{pmatrix} 1 & 0 & 1 \\ 0 & 1+1 & 0 \\ 1 & 0 & 1 \end{pmatrix}\hbar^2 + \frac{1}{2}\begin{pmatrix} 1 & 0 & -1 \\ 0 & 1+1 & 0 \\ -1 & 0 & 1 \end{pmatrix}\hbar^2 + \begin{pmatrix} 1 & 0 & 0 \\ 0 & 0 & 0 \\ 0 & 0 & 1 \end{pmatrix}\hbar^2$$

$$= \begin{pmatrix} 1/2 & 0 & 1/2 \\ 0 & 1 & 0 \\ 1/2 & 0 & 1/2 \end{pmatrix}\hbar^2 + \begin{pmatrix} 1/2 & 0 & -1/2 \\ 0 & 1 & 0 \\ -1/2 & 0 & 1/2 \end{pmatrix}\hbar^2 = \begin{pmatrix} 1 & 0 & 0 \\ 0 & 0 & 0 \\ 0 & 0 & 1 \end{pmatrix}\hbar^2$$

$$= \begin{pmatrix} 1 & 0 & 0 \\ 0 & 2 & 0 \\ 0 & 0 & 1 \end{pmatrix}\hbar^2 + \begin{pmatrix} 1 & 0 & 0 \\ 0 & 0 & 0 \\ 0 & 0 & 1 \end{pmatrix}\hbar^2 = \begin{pmatrix} 2 & 0 & 0 \\ 0 & 2 & 0 \\ 0 & 0 & 2 \end{pmatrix}\hbar^2$$

$$= 2\hbar^2 \mathcal{I}$$

Permissions

All chapters in this book are published with permission under the Creative Commons Attribution Share Alike License or equivalent. Every chapter published in this book has been scrutinized by our experts. Their significance has been extensively debated. The topics covered herein carry significant information for a comprehensive understanding. They may even be implemented as practical applications or may be referred to as a beginning point for further studies.

We would like to thank the editorial team for lending their expertise to make the book truly unique. They have played a crucial role in the development of this book. Without their invaluable contributions this book wouldn't have been possible. They have made vital efforts to compile up to date information on the varied aspects of this subject to make this book a valuable addition to the collection of many professionals and students.

This book was conceptualized with the vision of imparting up-to-date and integrated information in this field. To ensure the same, a matchless editorial board was set up. Every individual on the board went through rigorous rounds of assessment to prove their worth. After which they invested a large part of their time researching and compiling the most relevant data for our readers.

The editorial board has been involved in producing this book since its inception. They have spent rigorous hours researching and exploring the diverse topics which have resulted in the successful publishing of this book. They have passed on their knowledge of decades through this book. To expedite this challenging task, the publisher supported the team at every step. A small team of assistant editors was also appointed to further simplify the editing procedure and attain best results for the readers.

Apart from the editorial board, the designing team has also invested a significant amount of their time in understanding the subject and creating the most relevant covers. They scrutinized every image to scout for the most suitable representation of the subject and create an appropriate cover for the book.

The publishing team has been an ardent support to the editorial, designing and production team. Their endless efforts to recruit the best for this project, has resulted in the accomplishment of this book. They are a veteran in the field of academics and their pool of knowledge is as vast as their experience in printing. Their expertise and guidance has proved useful at every step. Their uncompromising quality standards have made this book an exceptional effort. Their encouragement from time to time has been an inspiration for everyone.

The publisher and the editorial board hope that this book will prove to be a valuable piece of knowledge for students, practitioners and scholars across the globe.

Index